# Neuroimaging I

## Basic Science

# HUMAN BRAIN FUNCTION
## Assessment and Rehabilitation

SERIES EDITORS:

Antonio Puente, *University of North Carolina at Wilmington North Carolina*

Gerald Goldstein, *Veterans Administration Medical Center Pittsburgh, Pennsylvania*

Erin D. Bigler, *Brigham Young University, Provo, Utah*

---

NEUROIMAGING I: Basic Science
Edited by Erin D. Bigler

NEUROIMAGING II: Clinical Applications
Edited by Erin D. Bigler

---

A Continuation Order Plan is available for this series. A continuation order will bring delivery of each new volume immediately upon publication. Volumes are billed only upon actual shipment. For further information please contact the publisher.

# Neuroimaging I

## Basic Science

Edited by

Erin D. Bigler

*Brigham Young University*
*Provo, Utah*

PLENUM PRESS • NEW YORK AND LONDON

Library of Congress Cataloging-in-Publication Data

Neuroimaging / edited by Erin D. Bigler.
p. cm. -- (Human brain function)
Includes bibliographical references and index.
Contents: v. 1. Basic science.
ISBN 0-306-45228-6
1. Brain--Imaging. I. Bigler, Erin D., 1949- . II. Series.
[DNLM: 1. Brain. 2. Diagnostic Imaging--methods. 3. Brain Mapping--methods. 4. Cognition--physiology. 5. Brain--atlases.
WL 141 N4829 1996]
RC386.6.D52N48 1996
616.8'04754--dc20
DNLM/DLC
for Library of Congress 96-29400
CIP



ISBN 0-306-45228-6


A Division of Plenum Publishing Corporation
233 Spring Street, New York, N. Y. 10013

10 9 8 7 6 5 4 3 2 1



Printed in the United States of America

# Contributors

CAROL ANDERSON, *Department of Psychology, Brigham Young University, Provo, Utah 84602-5543*

ERIN D. BIGLER, *Department of Psychology, Brigham Young University, Provo, Utah 84602-5543*

DUANE D. BLATTER, *Division of Neuroradiology, Department of Radiology, LDS Hospital, Salt Lake City, Utah 84143*

M. BOMANS, *Institute of Mathematics and Computer Science in Medicine (IMDM), Martinistrasse 52, 20246 Hamburg, Germany*

SUSAN Y. BOOKHEIMER, *Division of Brain Mapping, UCLA School of Medicine, Los Angeles, California 90024*

AMIT CHAKRABORTY, *Department of Electrical Engineering, Yale University, New Haven, Connecticut 06520*

SHAWN D. GALE, *Department of Psychology, Brigham Young University, Provo, Utah 84602-5543*

ALAN GEVINS, *EEG Systems Laboratory and SAM Technology, San Francisco, California 94105*

RUBEN C. GUR, *Brain Behavior Laboratory, Neuropsychiatry Section, Department of Psychiatry, University of Pennsylvania, Philadelphia, Pennsylvania 19104*

JUAN MANUEL GUTIERREZ, *Brain Behavior Laboratory, Neuropsychiatry Section, Department of Psychiatry, University of Pennsylvania, Philadelphia, Pennsylvania 19104*

K. H. HÖHNE, *Institute of Mathematics and Computer Science in Medicine (IMDM), Martinistrasse 52, 20246 Hamburg, Germany*

JAMES A. HOLDNACK, *Brain Behavior Laboratory, Neuropsychiatry Section, Department of Psychiatry, University of Pennsylvania, Philadelphia, Pennsylvania 19104*

TERRY L. JERNIGAN, *Veterans Affairs Medical Center, San Diego, and Departments of Psychiatry and Radiology, University of California, San Diego, La Jolla, California 92093*

C. Sterling Johnson, *Department of Psychology, Brigham Young University, Provo, Utah 84602-5543*

Andrew Kertesz, *St. Joseph's Health Centre, Department of Clinical Neurological Sciences, University of Western Ontario, London, Ontario N6A 4V2, Canada*

W. Lierse,† *Department of Neuroanatomy, University of Hamburg, and University Hospital Eppendorf, Martinistrasse 52, 20246 Hamburg, Germany*

Richard N. Mahr, *Brain Behavior Laboratory, Neuropsychiatry Section, Department of Psychiatry, University of Pennsylvania, Philadelphia, Pennsylvania 19104*

Andrew C. Papanicolaou, *Department of Neurosurgery, University of Texas Health Science Center–Houston, Houston, Texas 77030*

A. Pommert, *Institute of Mathematics and Computer Science in Medicine (IMDM), Martinistrasse 52, 20246 Hamburg, Germany*

M. Riemer, *Institute of Mathematics and Computer Science in Medicine (IMDM), Martinistrasse 52, 20246 Hamburg, Germany*

Th. Schiemann, *Institute of Mathematics and Computer Science in Medicine (IMDM), Martinistrasse 52, 20246 Hamburg, Germany*

R. Schubert, *Institute of Mathematics and Computer Science in Medicine (IMDM), Martinistrasse 52, 20246 Hamburg, Germany*

Robert T. Schultz, *Child Study Center, Yale University, New Haven, Connecticut 06520-7900*

Elizabeth R. Sowell, *San Diego State University/University of California, San Diego Joint Doctoral Program in Clinical Psychology, University of California, San Diego, School of Medicine, La Jolla, California 92093*

Marc Steed, *Department of Psychology, Brigham Young University, Provo, Utah 84602-5543*

Ina M. Tarkka, *Department of Neurosurgery, University of Texas Health Science Center–Houston, Houston, Texas 77030*

U. Tiede, *Institute of Mathematics and Computer Science in Medicine (IMDM), Martinistrasse 52, 20246 Hamburg, Germany*

†*Deceased.*

# Preface

Until recent advents in neuroimaging, the brain had been inaccessible to *in vivo* visualization, short of neurosurgical procedures or some unfortunate traumatic exposure. It is a tribute to the early contributors to clinical neuroscience that through what, by today's standards, would be deemed extremely crude measurements, advancements in understanding brain function were made. For example, the theories of higher cortical functions of the brain by Aleksandr Luria or Hans-Lukas Teuber in the 1950s were essentially based on military subjects who sustained traumatic head wounds during World War II. These researchers could inspect the patient and determine where penetrating entrance and exit wounds were on the head; sometimes they had skull films to identify entrance and exit fracture wounds, sometimes neurosurgical reports were available, and Luria even had the opportunity to acutely examine some patients with exposed wounds. Thus, one would take whatever information might be available and *infer* what regions of the brain were involved but could never actually visualize the brain. Of course, this changed dramatically with the introduction of brain imaging in the 1970s, but it really was not until the 1990s that analysis and image display technologies finally caught up with the basic brain-imaging methods of computerized tomography (CT) and magnetic resonance imaging (MRI). In contemporary neuroimaging, in addition to the standard arsenal of CT and MRI, methods of functional imaging now include positron emission tomography (PET), single photon emission computed tomography (SPECT), and functional magnetic resonance imaging (fMRI), along with more sophisticated methods for electrophysiological and magnetoencephalographic assessment of brain function. All of these imaging methods are addressed in this text by leaders in their respective fields. With these technologies, we are no longer relegated to the role of our predecessors—wondering about what the brain may actually look like in the living individual and what regions are pathologically involved.

Despite these marvelous developments in brain imaging, one must not lose sight of the fact that brain imaging only represents a molar view of brain structure and function. Thus, although the science of brain imaging and function advances, current technology still has some major limitations. Hopefully, both the clinical utility and scientific advancement of contemporary brain imaging, along with its limitations, will be evident as one reads this text. As significant as the advancements have been in the last two decades, we are only at the threshold of some of the most important and exciting research in imaging and understanding human brain function. Accordingly, this volume attempts to capture some of the current progress in the area of human brain imaging as it relates to function.

Knowing full well how rapidly this field changes, I have made an effort to keep this two-volume series as contemporary as possible. Part of the rapidity of change in neuroimaging has to do with improvements in technology. Brain imaging is technology driven. Even as I write this Preface, new methods of image acquisition and display are being published, superseding older technologies. Understandably, researchers and writers in this field always have a sense of being "behind" and "outdated." It is analogous to purchasing the ultimate personal computer only to find that 6 months after purchase it is outdated. Thus, being outdated represents an unfortunate risk of researching and writing in this area. As such, the various chapters that constitute this volume have been written from the perspective of an overview of the field and content area, not as the most up-to-date treatise on the subject.

This volume contains three sections, which to a certain extent are all interrelated: Overview, Basic Methods and Techniques, and Appendix. In Chapter 1, I review some of the history of neuroimaging, with an emphasis on contemporary displays of brain imaging. Since the standard and most exquisite method for gross brain imaging is magnetic resonance (MR), the second chapter, by Schultz and Chakraborty, overviews MR image analysis. Chapter 3, by Sowell and Jernigan, examines the developing brain as assessed via neuroimaging. Over the lifespan, there are developmental changes that can be detected and demonstrated by brain imaging methods. This also is addressed in Chapter 4, in which using the MR method, a normative database for brain structure, is discussed. Further analysis techniques using MRI to assess brain structure are provided by Kertesz in Chapter 6. The technologies of PET imaging, electrophysiological techniques to display neurocognitive networks, and magnetoencephalography are reviewed by Bookheimer (Chapter 5), Gevins (Chapter 7), and Papanicolaou and Tarkka (Chapter 8). Chapter 9 by Tiede and colleagues demonstrates the magnificent power of contemporary image display, with which essentially any major neural structure can now be presented with three-dimensional computer display. Chapter 10 by Gur and colleagues provides a brief foray into the next level of image analysis, wherein the precision of MRI is used to not only examine structure–function relationships but also to assess a more dynamic assessment of structure–function.

The text concludes with an Appendix that features an MRI brain atlas. The Appendix provides the reader with a general, but not exhaustive, atlas of major brain structures discussed in the two volumes in this series. With the exception of Chapter 9, all of the chapters presented in Parts I and II of this volume assume that the reader has some familiarity with brain imaging and anatomy, but none take the time to explicitly point out where given structures may be located. To provide the reader with structural localization, the MRI atlas is presented in three planes—axial, sagittal, and coronal. This is not to be considered a detailed atlas; rather, it should be used for general location of the various structures discussed in this volume and in Volume II in this series. For a more thorough atlas of MRI brain anatomy, the reader is referred to Duvernoy (*The Human Brain Surface, Three-Dimensional Anatomy and MRI,* New York: Springer-Verlag, 1991) or Truwitt and Lempert (*High Resolution Atlas of Cranial Neuroanatomy,* Baltimore, MD: Williams & Wilkins, 1994), which served as the basis for the current atlas.

*Erin D. Bigler*

# Contents

## PART II BASIC METHODS AND TECHNIQUES

1

# Introduction

ERIN D. BIGLER

Oldendorf (1980), in his book titled *The Quest for an Image of Brain,* cites William Morton's 1896 medical text, *The X-ray,* as the source of the first published skull radiograph. Although over the next several decades skull radiography improved with advances in x-ray technology (see review by Eisenberg, 1992), none of these techniques led to any direct image of the brain. Only indirect inferences based on asymmetries of the skull, calcified lesions, or some pathology at the brain–skull interface that resulted in opacities (e.g., neoplasms of the skull, skull erosion secondary to chronic increase of intracranial pressure, etc.) could be made about the brain.

Although the first case of radiographically described pneumocephalus was reported in 1913 (Luckett, 1913), it would be another five years before the neurosurgeon Walter Dandy introduced "pneumoventriculography" (Dandy, 1918). Pneumoencephalography (PEG), as it became known, provided a direct view of the ventricular system, but any "view" of brain paraenchyma was inferred indirectly. Pneumoencephalography would continue as the dominant neuroimaging method of the century along with cerebral angiography, introduced by Moniz (1927), until the advent of computerized tomography in the 1970s. Cerebral angiography would permit a view of the architecture of the cerebral vasculature, and from that, inferences could be made about brain surface anatomy. Thus, PEG would give some indication of internal brain structures via imaging of the ventricle, and cerebral angiography would give an indication of surface anatomy. In the late 1940s, radioactive isotope imaging of the brain was developed, which would identify certain brain tissue abnormalities (i.e., tumor) but would not permit specific imaging of brain tissue in any detail be-

ERIN D. BIGLER Department of Psychology, Brigham Young University, Provo, Utah 84602-5543.

*Neuroimaging I: Basic Science*, edited by Erin D. Bigler. Plenum Press, New York, 1996.

cause of limitations in resolution. Radioactive isotope imaging, along with PEG and cerebral angiography constituted the bulk of neuroimaging until the development of computerized tomography (CT), as discussed below.

Houndsfield began his work on CT in 1967, performed the first clinical scan in 1972, and published his major paper fully introducing CT methods in 1973 (Houndsfield; 1973). The rest, as they say, is history.

The other major advance at the end of this century, which is the main focus of this text, is nuclear magnetic resonance (NMR) imaging. Although the basic physic was reported as early as 1946 (with F. Bloch and E. M. Purcell being awarded the 1952 Nobel Prize in physics for NMR), it would be nearly three decades before medical applications followed. The acronym NMR was replaced by "magnetic resonance imaging (MRI)" because of the lay misunderstanding of the term "nuclear," with the misperception that patients were being exposed in some way to radiation. Physicists have continued to discuss NMR with that acronym, whereas radiologists and other physicians, as well as neuroscientists involved in imaging, tend to use the term MR. Magnetic resonance is now the premier imaging method, with many derivatives that permit not only anatomic precision both for brain and vessels but also spectorscopic analysis of tissue and functional imaging of the brain, either through functional MRI or by combining MR with other methods such as positron emission tomography (PET), advanced EEG spatial image analysis, or single photon emission computed tomography (SPECT) (Damasio, 1994; Greenberg, 1995; Kucharczyk, Mosely, & Barkovich, 1994; Osborn, 1994; Thatcher, Hallett, Zeffiro, John, & Huerta, 1994; Truwit & Lempert, 1994).

To show how far the field has come, I would like to review two general statements. The first is from Oldendorf (1980):

> There are currently (1979) several major medical imaging firms actively pursuing NMR. It is difficult to predict, based on the crude images now available, how far this new and radically different approach to imaging will go and how quickly, but its development has only begun. (p. 143)

Second, in the conclusions of the first text devoted to integrating neuroimaging with behavior, we commented on the "explosion" of research and clinical application of these areas just a decade since the statement by Oldendorf (Bigler, Yeo, & Turkheimer, 1989). At that time, we envisioned the melding of neuroimaging with other aspects of diagnostic medicine and behavioral assessment.

> Such an evaluation could document anatomic integrity via specific measures of major brain structures and nuclei and integrate this with neuropsychological function and associated metabolic and physiological activity that correspond with certain cognitive states. This is the future. Such clinical advances will be paralleled with advances in the basic neurosciences, allowing simultaneous assessment of differing facets of brain activity and the development of a more integrated science of the brain. (p. 344)

Obviously, the area or neuroimaging has progressed rapidly in a very short period of time.

Based on the above quotes, we are truly witnessing amazing developments in this, the decade of the brain. What has occurred since 1980 has been exponential and was not even comprehended by the best visionary minds of the late 1970s. Many of us feel that we are like scientists of prior generations, knowing that

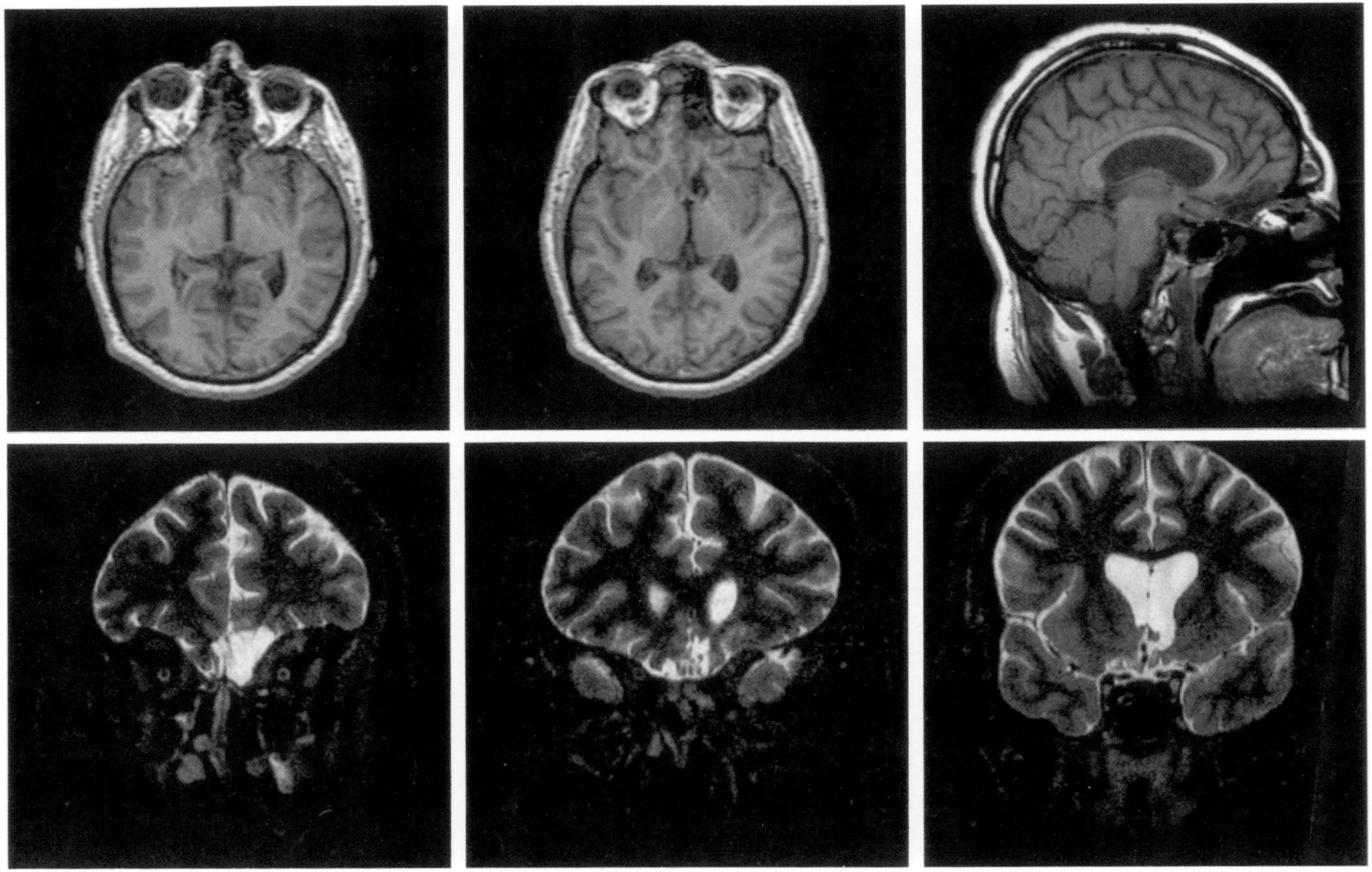

Figure 1. MR depiction of inferior frontal lobe damage (images given in radiologic perspective where left is on the viewer's right). Top left and center: T1 weighted MR images depicting inferior right frontal lesion. Top right: T1 weighted midsagittal MR depicting inferior frontal damage. Bottom: T2 weighted coronal images in sequence from front to back. Note the communication of the lesion site, which is now cystic, with the anterior horn.

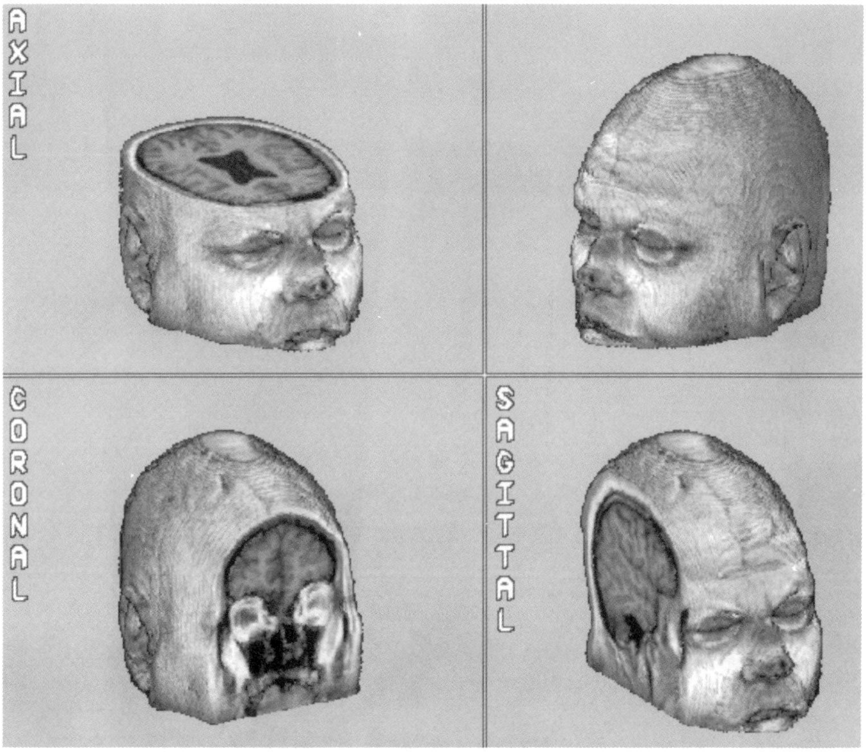


Figure 2. Three-dimensional MR image acquisition in the patient depicted in Figure 1. Using this type of image acquisition, the brain can be sliced/sectioned in any desired plane.

extremely important discoveries are being made that will literally change the way we think about the brain and its function, but being limited by our own lack of understanding of the true implications of this scientific revolution that is occurring. Many (Corsi, 1991) have described the brain as the "enchanted loom"—mysteriously complex, formidable to study. Early behaviorists refer to the brain as the "black box," giving acquiescence to its complexity and the inability to study this most vital organ. For the behaviorist during prior eras, the brain was "unmeasurable" and was of no concern because their focus was on observable behavior and not on biological control or regulatory factors. However, because of current neuroimaging methods the black box inference is no longer appropriate and the intricacies of the "enchanted loom" are being discovered and described.

This brief overview and introduction can be best summarized with the illustrations given in Figures 1–3. Current MR imaging provides a means of acquiring a three-dimensional representation of the brain so that any perspective, plane, or level can be examined with excellent specification of gross anatomy. In Figure 3, the brain is sectioned in multiple planes to identify the area of damage in the inferior frontal region of this patient's brain. Having this type of localization precision can now permit the detailing of behavioral and cognitive changes that may accompany such lesions in the brain. Thus, from a gross anatomic perspective, we can now study, *in vivo,* the general anatomy of the brain and relate anatomic findings to behavior (see Kertesz, 1994). Likewise, using various

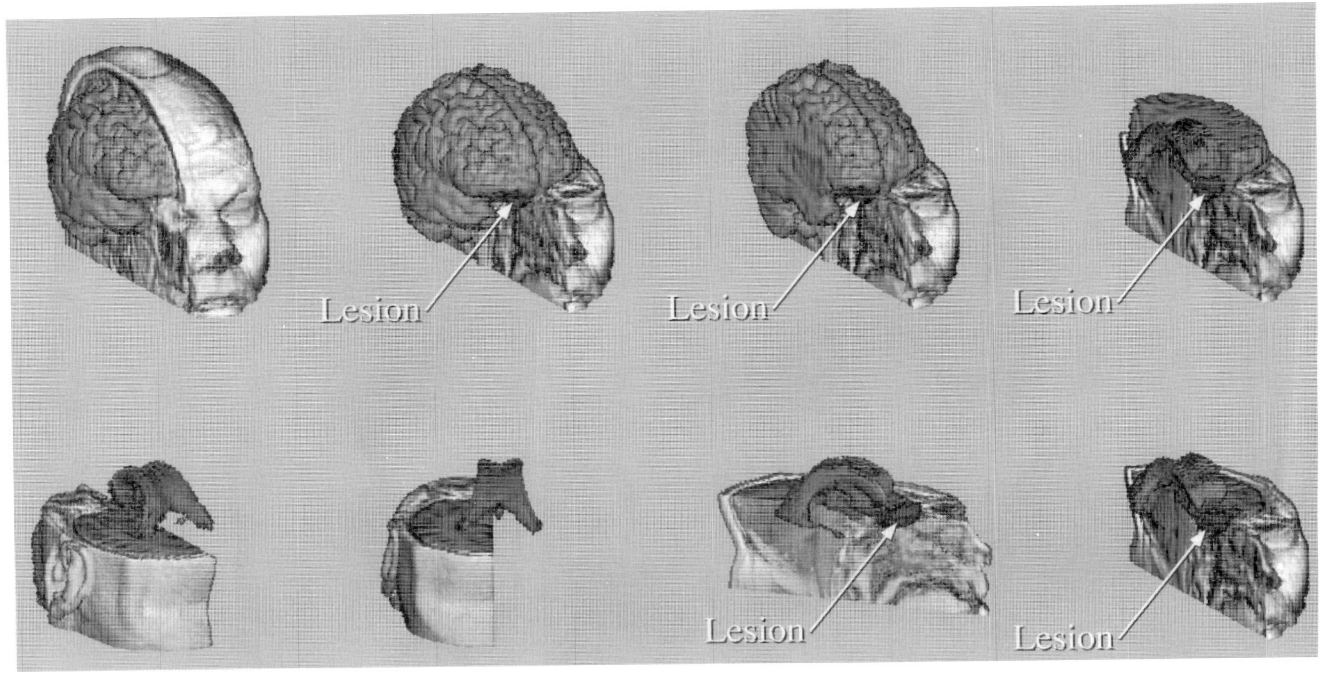


Figure 3. Using the three-dimensional image acquisition sequence depicted in Figure 2, the skull and meninges can be stripped away (upper left) and the brain surface architecture exposed (upper left middle) and then the brain parenchyma sectioned and the ventricle exposed to show the site of the lesion in combination with the ventricle. This series of illustrations demonstrates the elegant techniques for lesion localization that can now be utilized with MR imaging.

cerebral blood flow or metabolic imaging studies, we can look at various physiologic parameters and behavior.

Although this book attempts to address the numerous areas of brain imaging, the focus is clearly on magnetic resonance imaging. As exciting and contemporary as the current chapters are in their review of the progress of neuroimaging and behavior, this book probably represents just a preview of what is in store. The future is going to be most exciting from the perspective of neuroimaging capabilities. Nonetheless, from a contemporary perspective, the current handbook examines in detail the current status of brain imaging in understanding and assessing human function.

## References

Bigler, E. D., Yeo, R., & Turkheimer, E. (1989). *Neuropsychological function and brain imaging.* New York: Plenum Press.

Corsi, P. (1991). *The enchanted loom: Chapters in the history of neuroscience.* New York: Oxford University Press.

Damasio, H. (1994). *Human brain anatomy and computerized images.* New York: Oxford University Press.

Dandy, W. E. (1918). Ventriculography following the injection of air into the cerebral ventricles. *Annals of Surgery, 68,* 5–11.

Eisenberg, R. L. (1992). *Radiology: An illustrative history.* St. Louis, MO: Mosby.

Greenberg, J. O. (1995). *Neuroimaging.* New York: McGraw-Hill.

Houndsfield, G. N. (1973). Computerized transverse axial scanning (tomography). I. Description of system. *British Journal of Radiology, 46,* 1016–1022.

Kertesz, A. (1994). *Localization and neuroimaging in neuropsychology.* San Diego: Academic Press.

Kucharczyk, J., Moseley, M., & Barkovich, A. J. (1994). *Magnetic resonance neuroimaging.* Boca Raton, FL: CRC Press.

Luckett, W. H. (1913). Air in the ventricles of brain, following a fracture of the skull. *Surgery, Gynecology, and Obstetrics, 17,* 237–240.

Moniz, E. (1927). L'encephalographie arterielle. Son importance dans la localisation des tumeurs cérébrales. *Review in Neurology* (Paris), 34, 72–90.

Oldendorf, W. H. (1980). *The quest for an image of brain.* New York: Raven Press.

Osborn, A. G. (1994). *Diagnostic neuroradiology.* St. Louis, MO: Mosby.

Thatcher, R. W., Hallett, M., Zeffiro, T., John, E. R., & Huerta, M. (1994). *Functional neuroimaging: Technical foundations.* San Diego: Academic Press.

Truwit, C. L., & Lempert, T. E. (1994). *High resolution atlas of cranial neuroanatomy.* Baltimore, MD: Williams & Wilkins.

# I

# Overview

# 2

# Magnetic Resonance Image Analysis

Robert T. Schultz and Amit Chakraborty

## Introduction

The great advances made in neuroimaging technology during the last decade provide an unparalleled opportunity for neuroscientists to measure gross brain structure with a high degree of precision. However, this opportunity has yet to be fully realized, largely because the development of reliable, valid, and efficient means of quantifying all the information contained within the digital display has lagged far behind the ability to acquire these images. Bigler's review (1994) demonstrated that even though many studies have employed neuroimaging methods to investigate brain–behavior relationships, the vast majority have not employed quantitative techniques (Figure 1) but rather have relied on binary, clinical readings of pathology (present or absent) and hemisphere or quadrant analyses. While such approaches are appropriate for detecting gross structural abnormalities (e.g., infarct) and active pathological processes (e.g., tumor) responsible for frank nondevelopmental neurologic disorders, the literature in neuropsychology suggests that this approach is sorely inadequate for illuminating the neuropathology of disorders where there is likely congenital or developmental disturbance (e.g., schizophrenia, dyslexia). The neuropathologic pro-

Robert T. Schultz Child Study Center, Yale University, New Haven, Connecticut 06520-7900. Amit Chakraborty Department of Electrical Engineering, Yale University, New Haven, Connecticut 06520.

*Neuroimaging I: Basic Science*, edited by Erin D. Bigler. Plenum Press, New York, 1996.

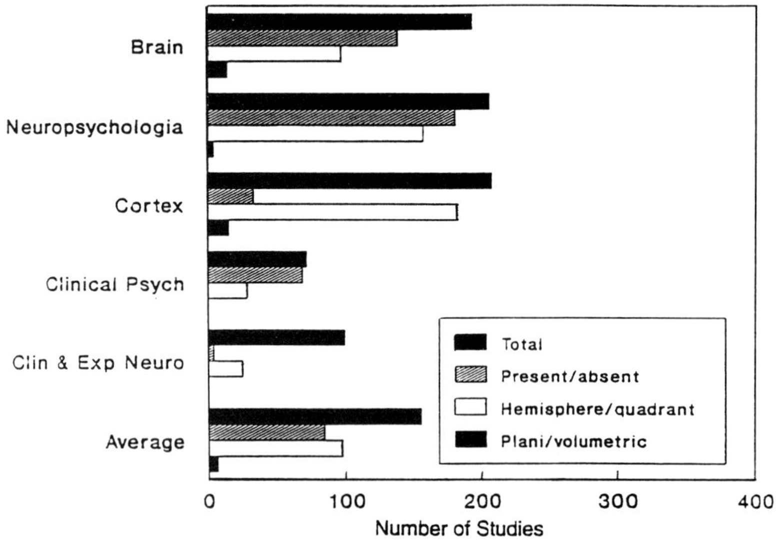


Figure 1. A comparison of brain image analysis methodology from the publications of five leading journals. *Brain, Cortex,* and *Neuropsychologia* are from 1971 and 1990, while *Journal of Clinical Psychology* is from 1975 to 1990, and *Journal of Clinical and Experimental Neuropsychology* is from 1979 to 1990. Across all five journals, the majority of published studies prior to 1991 failed to use quantitative techniques of image analysis. Reprinted from Bigler (1994) with permission.

cesses underlying most psychiatric diseases are apparently more subtle, and if these diseases are observable at all on a gross anatomical level, researchers will need to take a much more sophisticated approach toward brain morphometry. For many disorders it seems likely that the underlying neuropathology is not localizable to a single brain region, but rather is widespread, affecting many brain systems, thus requiring comprehensive measurement of the entire brain and statistical analysis of multiple variables with large samples of patients and controls.

A user-friendly set of techniques for quantifying brain anatomy with wide accessibility is needed. Until recently, however, most scientists were forced to develop their own software for quantifying brain MRI data, resulting in inconsistencies in approach across laboratories and some inherent difficulty in resolving discrepant research findings in the literature. Fortunately, in recent years several commercial software packages with wide appeal have been developed (e.g., ANALYZE, Robb & Hansen, 1990; Scilimage, Van Balen, Kate, Koelma, Smeulders, & Groen, 1991; VoxBlast, 1993; Interactive Data Language, 1987), including some free/public domain programs (e.g., NIH IMAGE). However, none provides a fully automated approach to the problem of brain morphometry. The computer is at an extreme disadvantage compared to man, for the human observer can mentally incorporate models of what the local anatomical boundaries *should* look like. This is especially important when the boundaries are blurred by noise or by the averaging of tissues across the volume element. Despite the use of *a priori* models in computer vision, such tools have not yet been applied to complex, convoluted boundaries, such as those found at the intersection of cortical gray and white matter. At this stage of development, image processing still often requires human judgment, and the process of quantifying the enormous amount of image data typically acquired becomes a very time-consuming endeavor. Software limitations, cost, and time constraints are undoubtedly factors contributing to the paucity of quantitative approaches used in the literature to date (see Figure 1).

This chapter will be confined to discussion of image processing techniques for quantitative assessment of structural brain features from MRI. Other neuroimaging modalities, such as functional MRI or positron emission tomography (PET), present unique measurement problems that are outside the scope of this current presentation. The chapter begins with an overview of magnetic resonance image construction and the physics of MRI, since a basic understanding of how the images are constructed is a useful context for beginning to devise an appropriate approach to MR image processing problems. For example, with MRI one can derive quantitative measures of intrinsic tissue characteristics and use these indices as part of the overall image segmentation strategy (Gore, O'Donnell, Pope, Majumdar, & Adams, 1988). This is followed by a brief discussion of measurement reliability. Next we review elementary image processing techniques, including image thresholding and the use of calculated T2 images, synthetic images, and multispectral approaches. This is followed by a discussion of edge detection and region-based techniques for gray matter–white matter segmentation. The chapter concludes with a discussion of methods used to segment the cerebral cortex into anatomically valid subregions.

The chapter does not include discussion of the measurement of specific, noncortical structures such as the hippocampus. Measurement of these structures inevitably requires some type of gray–white matter segmentation, along with manually guided tracing, following a set of operational rules regarding structure boundary. While many techniques discussed here will be useful in approaching these measurement problems, the accuracy of these measurements ultimately depends on the reliability of the operator's tracings and the validity of the operational definitions. For a discussion of rules specific to boundary definition of specific structures, the reader should consult current publications on the topic of interest. For example, Watson *et al.* (1992); Press, Amarad, and Squire (1989); and Jack, Bentley, Twomey, and Zinmeister (1990) provide good discussions of morphometry of the hippocampus and related mesial temporal lobe structures. Courschesne, Yeung-Courschesne, and Egaas (1994) and Raz (Chapter 6, Volume II of this series; 1992) have developed approaches specific to the cerebellum. Allen, Richey, Chai, and Gorski (1991) and Denenberg, Kertesz, and Cowell (1991) describe methods for assessing the morphology of the corpus callosum. Steinmetz *et al.* (1990); Kulynych, Vladar, Jones, and Weinberger (1993); Leonard *et al.* (1993); and Schultz *et al.* (1994) discuss planum temporale measurement, while Jernigan *et al.* (1991), Peterson *et al.* (1993), and Singer *et al.* (1993) discuss the morphometry of basal ganglia. This list is not exhaustive and is provided here only to serve as an initial source of reference materials.

## Magnetic Resonance Image Construction

Image quantification is really a problem of pixel (picture element) classification—which pixels represent brain tissue and which represent nonbrain, or what pixels define the boundary of one structure and differentiate it from a neighboring structure? An image is merely a matrix of quantitative values or pixels. Each pixel can take on a range of gray scale values. In the case of an allocation of one byte per pixel, pixels are commonly normalized within the operating range of 0 to 255 with higher intensities representing brighter values (for a two-byte pixel the upper end is not $2^8$ but $2^{16}$ or 65,535). All image processing strategies use this digital gray scale information, and although under normal magnification the human observer cannot resolve the individual pixels, the relationships between the pixel intensities form the basis for the contrast between different classes of tissue.

Typically, a single MRI brain slice is composed of 256 × 256 pixels[1], and the dimensions of this matrix, along with the field of view used to collect the image data, determine the in-plane image resolution. The field of view (FOV) is the size of the space in which the data are collected inside the MR scanner. The greater the matrix size or the smaller the FOV, the greater the image resolution and definition between structures. The in-plane pixel dimensions are a function of the FOV divided by the matrix size. For example, with a 240-mm FOV and a 256 × 256 matrix, each pixel in the image will encompass 0.9375 by 0.9375 mm of tissue. Structural differences smaller than this will be obscured, since the signal arising from such small structures will be averaged into a single pixel. Likewise, structural boundaries usually blend within certain pixels so that there are pixels at the border whose intensity is jointly determined by two different classes of tissue. This represents in-plane volume averaging (sometimes termed a partial volume effect), resulting in blurred boundaries. In many studies, however, the most significant problem is tissue averaging across the depth of the pixel. For example, clinically, it is unusual to collect images with a slice thickness of less than 3 mm. Thus, across this 3-mm depth different tissue types might be averaged into a given pixel. This is not a problem if the structure of interest is more than 3 mm deep and completely perpendicular to the plane of imaging, but this is rarely the case. Usually, the geometry of a structural boundary is oblique to the imaging plane, and thus different tissue types are averaged at the boundary, producing intermediate intensities and fuzzy edges.

Naturally the question arises as to why one should not collect brain image data of minimal thickness and minimal in-plane pixel dimensions. The problem is that the smaller the voxel (volume element), the less the signal generated from each volume of tissue. Image quality is also dependent on the signal-to-noise ratio (SNR). Reducing the voxel size reduces the strength of the signal, and relative to a constant background of electronic noise (random fluctuations of pixel intensity), this degrades the image quality, giving rise to "grainy" images. Thus, there is a trade-off between image resolution and the SNR.

There are, however, means of increasing the SNR without lowering the image resolution. One can increase the number of excitations (NEX) that each voxel receives, thus increasing the signal strength. The SNR increases as a function of the square root of the NEX. This has practical limitations, however. Going from an NEX of 1 to an NEX of 2 will increase the SNR by 1.41, resulting in less grainy images, but at the cost of doubling the imaging time. For some subjects the experience of a brain MRI is unpleasant. This is especially true of individuals who are claustrophobic, since the bore of the magnet is typically about 70 cm in diameter. While normal adult control subjects almost always tolerate the procedure quite well (and in our experience generally fall asleep during longer experiments), children, the elderly, and various patient popula-

[1] Data collected with a 192 × 256 matrix is "zero filled" evenly on either side of 192 to achieve a square matrix (256 × 256) and square in-plane pixel dimensions.

tions sometimes have more difficulty remaining still. While one can improve the image quality by increasing the NEX, one does so at the risk of patient movement during the longer scan, which will seriously distort the image, or worse still, patient noncontinuation with the procedure.

In order to assist the research subject in remaining still during data collection, the individual's head is braced on either side by styrofoam or other padding. During imaging, the subjects lie flat on their backs and the motorized table is moved inside the bore of the scanner. A further measure to remind the patient to remain motionless is the use of paper tape across the subject's forehead, anchored on either side of the head cradle; sometimes a piece of tape is also placed across the chin. While this paper tape is easily broken by head movement, thus providing little real restraint, it does deny small, involuntary movements and actually allows the subject to relax and devote less conscious effort to keeping still. One might also employ real restraints, such as a cloth or nylon strap across the head, though the inability to move may frighten many subjects.

A second method of increasing the SNR is to increase the strength of the magnet. While most commercial scanners now utilize a 1.5-tesla magnet, older magnets on the order of 1.0 tesla or less exist and are sometimes used in research. They do not provide as high an SNR and these grainy images are more difficult to process later. This problem may be alleviated by employing a smaller matrix (e.g., 128 × 128), but this sometimes results in an artifact arising from the Fourier transform used in the image construction—the Gibbs artifact (Wood & Henkleman, 1985; Runge, 1994), which appears as dark bands of concentric rings within the brain, emanating from high contrast interfaces such as that of the skin and skull. On the other hand, several sites now have 4.0-tesla magnets, providing a greatly improved SNR. While software on 1.5-tesla machines usually places an upper limit on the matrix size of 512, these more powerful magnets are equipped to collect 1024 × 1024 images, thus greatly increasing image resolution while still maintaining a good SNR.

## The Physics of Magnetic Resonance Imaging

Pixel brightness is determined by a procedural scaling constant ($K$) and three characteristics of the voxel of tissue—the amount of water (N(H) or hydrogen proton density) and the two proton relaxation parameters, T1 and T2. This section will review the physics of MR necessary to understand these three critical tissue parameters.[2] A general understanding of the physical processes underlying MRI is important for two reasons. First, we present a segmentation strategy (below) which utilizes T2 relaxation estimates, and the reader needs a basic understanding of this process to employ the technique. Second, a fundamental understanding of MRI is necessary for choosing pulse sequence parameters for image data collection that will optimize image characteristics needed to adequately address the study's hypotheses. There are many parameters which can be manipulated during data collection to influence image appearance and ultimately the sensitivity of the measurement and the power of the experimental design. Thus, decisions about how the data are collected can exert an enormous

[2] While pixel intensity can also be influenced by magnetic susceptibility and the microscopic (diffusion) and macroscopic (flow) motional states of the protons, these effects are limited to select regions in the image and will not be discussed here.

influence on the later success or failure of quantitative MRI analysis. While scientific, pragmatic, and ethical concerns dictate that a neuroradiologist (and often a nuclear magnetic resonance [NMR] physicist) will always be a consultant to other scientists interested in using MRI to address brain–behavior questions, most clinical neuroradiologists are inexperienced in quantifying image data, since their use of MRI involves qualitative assessments. They may suggest methods of data collection which are not optimal for quantitative image processing, thus compromising the researcher's ability to adequately address the study's hypotheses. Psychologists, psychiatrists, neurologists, and other neuroscientists who have knowledge of the imaging parameters used to create the image data *and* of the image analysis techniques that will be used to quantify the data can participate with the neuroradiologist in this initial stage of the study's design, when decisions about pulse sequence parameters are made.

Structural brain MRI relies exclusively on the diamagnetic proton of the hydrogen atom ($^1H$). While many molecules in biologic tissue contain $^1H$, because of electron shielding effects, only the hydrogen protons of water generate the signal in structural brain MRI under normal imaging pulse sequences (Bottomly *et al.*, 1984; Kamman *et al.*, 1984). The pixel gray scale value is determined by both the proportion of hydrogen within a voxel of tissue and the interactions of water molecules with each other and other molecules of the voxel's chemical matrix (e.g., lipids, proteins). Thus, pixel intensities code information about the proportion (density) and molecular behavior of water in brain tissue. Differences in proton density and proton motion/behavior between adjacent tissue voxels produce differences in relative brightness between different tissue types. Absolute image brightness, however, is also a function of a scaling constant which is determined by scanning parameters used to construct the image (e.g., the receiver gain). While the mean pixel intensity may rise or fall with the changing of scanning parameters the relative brightness of adjacent tissues, or tissue *contrast,* is wholly dependent on tissue biochemistry. Image contrast is ultimately the primary source of information used by human operators during hand-guided tracings of structure boundaries and by computer algorithms employed for image segmentation.

Contrast in MRI is extremely sensitive to changes in the density and distribution of water in the brain during neonatal and child development. In the newborn, gray–white image contrast is absent or reversed compared to adult images (Holland, Hass, Norman, Brant-Zawadzki, & Newton, 1986; Dietrich *et al.*, 1988; Bird *et al.*, 1989). Gray and white matter become differentiable at about 6 months to one year of age, but adult intensity levels are not achieved until about age 10 or early adolescence (Holland *et al.*, 1986; Dietrich *et al.*, 1988; Bird *et al.*, 1989). These MRI developments mirror biochemical changes found during direct anatomic investigation of brain tissue. Brain water content decreases rapidly during the first two years of life (from 88% to 82% in the first six months alone) and then more gradually until adolescence, whereas cholesterol concentration, a marker of myelinization, doubles during the first six months and then increases steadily until adolescence (Dobbing & Sands, 1973). Thus, in addition to morphometric data (i.e., structure size and shape), investigators using MRI with an understanding of the parameters which give rise to pixels of different intensities might glean information about the brain's biochemistry (Schultz, 1991).

Magnetic resonance imaging is based on the phenomenon of nuclear magnetic resonance (NMR). All components of atoms with an odd number of nucleons (protons + neutrons) "spin." The nucleus of the hydrogen atom is composed of just one proton and no neutrons ($^1$H), hence it is one of many atoms which spin. A spinning electrical charge produces a magnetic field, termed a nuclear magnetic dipole moment. The spinning $^1$H acts like a bar magnet, with a positive and negative pole. In addition to spinning about an axis, $^1$H (hereafter referred to simply as a proton) also "precesses" when within the scanner's magnetic field. Protons are positively charged, but this charge is not evenly distributed within it. When in the presence of an external magnetic field, such as the magnet of the MR scanner, the proton wobbles or precesses a few degrees off center and about its axis of orientation as it spins. The rate of precession is defined as the gyromagnetic ratio, a constant unique to each proton species. Fortunately, the magnetization of a voxel of tissue containing hydrogen protons can be manipulated to learn about the relative amount of hydrogen, its interactions with other molecules in the voxel's chemical environment, and its spatial location within a tissue sample so that an image can be constructed.

Ordinarily, protons are randomly oriented and a sample of tissue possesses no net magnetization. However, when an individual's head is placed inside the static external magnetic field ($B_0$) of the scanner, the protons align along the field's axis (referred to as the $z$-axis or longitudinal axis). A very small majority of protons align parallel to the $z$-axis and create a net magnetic force in the longitudinal direction. The size of the plurality of aligned protons depends on the strength of the magnetic field (among other things), with stronger fields producing more disproportionate alignment. It is the excess of protons aligned with the main field that later produces the signal in MRI. The vector quantity $M_z$ ("longitudinal magnetization") is used to signify the strength of the net magnetization along the $z$-axis.

The protons aligned along the $z$-axis in an antiparallel manner are in a higher quantum energy state than those in the parallel orientation. However, because the energy difference between the two states is relatively small at field strengths used in MRI (about $10^{-7}$ eV), thermal motion is sufficient to cause the protons to flip back and forth between the two energy states. This fluctuation is a principal source of noise in MRI. As field strength increases, more protons align with the main field, and the difference in the two energy state increases. Increasing the field strength will increase the signal strength (number of protons aligned with $B_0$ and will decrease the noise (rate of random proton exchange between the two energy states).

In addition to the strength of the external magnetic field ($B_0$), the magnitude of $M_z$ is determined by the gyromagnetic ratio of the hydrogen proton and the density and mobility of hydrogen in the tissue voxel. For hydrogen protons, the gyromagnetic ratio is 42.58 MHz/tesla, where tesla is a measure of the scanner's magnetic field strength. The frequency of precession is a function of the strength of $B_0$ times the gyromagnetic ratio. This product is known as the resonant or Larmor precession frequency. The Larmor frequency signifies how quickly the proton precesses under the influence of the static external magnetic field. Most current MR scanners utilize a 1.5-tesla magnet. In comparison, the earth's magnetic field varies from $3 \times 10^{-5}$ tesla at the equator to $7 \times 10^{-5}$ tesla at the poles. At 1.5 tesla hydrogen protons precess at the rate of 63.87 MHz (about 64 million times per second).

Unfortunately, $M_z$ cannot be measured at equilibrium. By tipping the net magnetization vector 90°, however, a signal can be received.[3] The essence of MR imaging is the application of electromagnetic radiation at the Larmor frequency, which is absorbed by the protons, followed by measurement of a signal as the system returns to equilibrium. The magnetization of the protons in a subject is perturbed by applying a radio frequency (RF) pulse by means of an RF antenna within the bore of the magnet. With a 1.5-tesla main field, a 63.87 MHz RF pulse of the proper phase, duration, and amplitude will tip the net magnetization vector from its longitudinal alignment with the static external magnetic field 90° into the transverse or $xy$-plane. An RF pulse of a frequency more or less than 63.87 MHz may affect other protons but not hydrogen protons in a 1.5-tesla static external field. Protons which had been precessing about the $z$-axis continue to do so, but now within the $xy$-plane, orthogonal to $z$. The magnitude of the transverse magnetization vector, $M_{xy}$, is measured as it passes by the NMR RF receiver coil. The magnitude of the transverse magnetization within a given voxel of tissue is directly proportional to the brightness of the corresponding pixel. Gray matter–white matter tissue contrast in the brain is therefore a function of the difference between the $M_{xy}$ (gray matter) and $M_{xy}$ (white matter) vector quantities.

T1 AND T2 RELAXATION. The RF pulse is quite brief (3.2 msec on General Electric Signa MR scanners), and once terminated, the individual protons begin to realign with the longitudinal axis. The rate at which protons realign with the longitudinal axis drops off exponentially over time, and this process is described by the phenomenological equations of Block (described below). The time it takes for $M_z$ to regain 63.2% of its equilibrium magnetization is defined as T1 (i.e., the time for $1 - e^{-1}$ of the protons to realign longitudinally). It is also referred to as the longitudinal relaxation time and as the spin lattice relaxation, because it is the interaction with other molecules in the "lattice" which induces relaxation. The quantity T1 is defined by the Block equation:

$$M_z = (M_z)^0[1 - e^{-TR/T1}] \qquad (1)$$

where $M_z$ is the net magnetization in the $z$ direction at time TR; $(M_z)^0$ is the net magnetization of the $M_z$ vector at equilibrium, prior to rotation into the transverse plane (it is a function of N(H) and $B_0$); and TR is the time between repetitions of the RF pulse. Thus, when T1 = TR, $M_z$ equals 63.2% of $(M_z)^0$. This is the defining condition for T1.

Another important effect of the RF pulse is that it puts all the individual proton spins into precessional phase with one another. Thus, immediately after the 90° RF pulse the protons are phase-locked in a 360° precession (still at 63.87 million cycles per second) about the $z$-axis, but in the $xy$-plane. If there were not in phase, but randomly precessing, their individual magnetic moments would cancel each other, and brain imaging would not be possible. Their joint in-phase precession, however, is now detectable by an RF receiver coil situated in the transverse plane.

Immediately after the net magnetization vector is rotated out of the $z$-plane

[3] While a quantum mechanical model provides the most accurate description of the behavior of protons in a magnetic field, the classical model of physics using magnetic vectors is applicable and generally more easily understood.

into the $xy$-plane (time = 0) the strength of the signal received by the RF coils is directly proportional to the number of mobile protons, N(H), in the volume of tissue and aligned with $B_0$ prior to the tipping pulse. The detectable signal, however, declines exponentially, due to dephasing of precessions (the result of T2 or spin-spin relaxation). The exponential decline in signal is referred to as the free induction decay. Small local variations in the magnetization of the tissue cause some protons to begin to precess slightly faster and some slightly slower. In fact, these local variations soon cause all the protons to dephase and cancel each other, reducing the net transverse magnetization vector to zero. Since there is no net transverse magnetization, there can be no detectable signal. The time it takes for the magnitude of $M_{xy}$ to decline by 63.2% is defined as T2. The value of T2 is always shorter than T1 in biological tissues, averaging about 10% of T1 (T1 and T2 are equivalent only in pure water). It is described by the Block equation

$$M_{xy} = M_z[e^{-TE/T2}], \tag{2}$$

where TE is the time between the 90° RF pulse and recording of the signal in the xy plane (the time to echo); $M_z$ is the net magnetization of the longitudinal vector immediately prior to rotation into the transverse plane (it is a function of $B_0$, tissue proton density, TR, and the tissue's T1); and $M_{xy}$ is the net magnetization in the $xy$ plane at time TE. When TE = T2, transverse magnetization will have lost 63.2% of its initial value.

Unfortunately, the superconducting magnetic, which is the source of the static external field, is not perfectly homogenous, leading to nonrandom variations in tissue magnetization. In practice, therefore, signal will be lost much more rapidly than would be suggested by the tissue's T2 time. The actual rate of signal decay is described by T2* ("star"). Fortunately, the source of dephasing is nonrandom, and thus, at some discrete time interval after the 90° pulse, T2* is compensated for by application of a 180° reversing pulse. It is this second pulse which then gives rise to the "spin echo" signal. The decay of a series of spin echo signals is accurately described by true T2.

In summary, there are three NMR parameters which determine signal intensity from tissue (and thus gray tissue–white tissue contrast): the T1 and T2 relaxation times and the density of mobile hydrogen protons in the tissue. Proton density, T1, and T2 vary across tissue type. Table 1 presents data on the percentage of water in different brain tissues. Gray matter has about 10% more water than white matter; this is helpful in creating tissue contrast. As stated earlier, signal in brain MRI arises for all practical purposes exclusively from brain $H_2O$. While the brain is rich in fats (another source of $^1H$), with 15% of the white matter and 6% of the gray matter composed of lipids (Suzuki, 1972), brain lipids are almost entirely bound in rigid membranes. Only fairly mobile protons contribute to MR signal. Immobile $^1H$ has extremely short T2 and very long T1. Therefore, these protons produce a very weak signal, too low to rise above the noise level (Wehrli, MacFall, Shutts, Breger, & Herfkens, 1984). One form of lipid, free fatty triglycerides (repositories of chemical energy), is mobile and thus contributes to signal. Although free fatty triglycerides are abundant outside the brain, they are sparse within the central nervous system (Agranoff, 1984) and compose, at most, only a few tenths of 1% of the total brain volume (Suzuki, 1972). *In vivo* brain NMR spectroscopy confirms these analytic chemical data, as

TABLE 1. PERCENTAGE OF WATER IN BRAIN TISSUES

| Tissue | % Water[a] | Reference |
|---|---|---|
| CSF | 99 | Documenta Geigy, 1978 |
| Gray matter | 81.9 | Suzuki, 1980 |
| | 84.3 | Documenta Geigy, 1978 |
| Average: | 83.1 | |
| White matter | 71.6 | Suzuki, 1980 |
| | 70.6 | Documenta Geigy, 1978 |
| | 70.0 | Kamman *et al.*, 1984 |
| Average: | 70.7 | |

[a]The literature provides surprisingly few estimates of brain tissue hydration. While estimation of between-subject variation is not provided, Suzuki (personal communication, November 1990) suggests that the typical range for white and gray matter water content is approximately 2 to 3%.

there is virtually no detectable lipid peak in brain tissue (Bottomley *et al.*, 1984; Kamman *et al.*, 1984).

Table 2 lists representative T1 and T2 times for white matter, gray matter, and cerebrospinal fluid (CSF). Variability of reported times is due to a combination of factors, including individual differences in tissue biochemistry (a little

TABLE 2. REPRESENTATIVE T1 AND T2 RELAXATION TIMES FOR BRAIN TISSUES

| Tissue | T1 (SD)[a] | T2 (SD) | N | Reference |
|---|---|---|---|---|
| CSF | | | | |
| Lateral ventricles | 1900[b] (353) | 250[b] (3) | 3 | Wehrli *et al.* (1984) |
| Gray matter | | | | |
| Thalamus | 945 (102) | 76 (4) | 5 | MacFall *et al.* (1987) |
| Thalamus | 855 (140) | 73 (4) | 27 | Breger *et al.* (1986)[c] |
| Thalamus | 985 (80) | 69 (5) | 27 | Breger *et al.* (1986)[c] |
| Caudate | 831 (177) | 84 (5) | 5 | MacFall *et al.* (1987) |
| Caudate | 850 (190) | 78 (5) | 27 | Breger *et al.* (1986)[c] |
| Caudate | 1170 (75) | 71 (4) | 27 | Breger *et al.* (1986)[c] |
| Putamen | 835 (155) | 70 (3) | 27 | Breger *et al.* (1986)[c] |
| Putamen | 1010 (50) | 65 (2) | 27 | Breger *et al.* (1986)[c] |
| Average: | 935 | 73.3 | | |
| White matter | | | | |
| Cortical | 687 (55) | 82 (5) | 5 | MacFall *et al.* (1987) |
| Centrum semiovale | 675 (70) | 79 (4) | 27 | Breger *et al.* (1986)[c] |
| Centrum semiovale | 740 (45) | 71 (3) | 27 | Breger *et al.* (1986)[c] |
| Corpus callosum | 610 (120) | 68 (4) | 27 | Breger *et al.* (1986) |
| Corpus callosum | 665 (20) | 66 (2) | 27 | Breger *et al.* (1986) |
| Internal capsule | 593 (57) | 69 (3) | 5 | MacFall *et al.* (1987) |
| Internal capsule | 655 (90) | 67 (3) | 27 | Breger *et al.* (1986)[c] |
| Internal capsule | 760 (60) | 64 (5) | 27 | Breger *et al.* (1986)[c] |
| Average: | 673 | 70.8 | | |

[a]Human *in vivo* relaxation times in milliseconds for brain tissue recorded at 1.5 Tesla (with the exception of CSF, which was measured at 1.4 Tesla). Estimates of cortical gray matter relaxation time in the literature are lacking presumably because of the difficulty in segmenting gray matter from sulcal CSF.

[b]This is probably an underestimate for CSF. We find pure CSF to have a T2 greater than 500 msec. Even small amounts of volume averaging dramatically decrease obtained T1 and T2 for CSF, because these parameters do not average linearly.

[c]Mean of reported right and left hemisphere values. Breger *et al.* used two separate instruments on the same subjects; as in their report, instrument C is reported before instrument A.

understood phenomenon), differences in scanner performance between studies, and measurement error within a study. Within a study, tissue volume averaging can be a significant source of measurement error. Attempting to measure relaxation times of cortical gray matter, for example, is problematic, as most voxels reported in the current literature are 5 mm thick, and thus gray matter is averaged with sulcal CSF in many pixels. The much longer relaxation time for CSF increases the estimate of gray matter relaxation parameters. Other tissues, such as the white matter of the centrum semiovale, are less susceptible to volume averaging effects because of their large volume.



We have avoided a discussion of fast imaging techniques, such as "spoiled grass" (SPGR), which achieve T1 weighted images in a shorter period of time than conventional spin echo imaging by reducing the flip angle to less than 90°. Tip angles of less than 90° cause only a fraction of longitudinally aligned protons to be rotated into the transverse plane. Since not all protons are rotated, some remain aligned with the $z$-axis, and one does not have to wait for full longitudinal recovery to flip additional protons. While this imaging procedure results in a time savings, it also produces images of a lower SNR, other things being equal.

### The Effect of Pulse Sequence on Contrast

The pulse sequence defines the timing of the 90° and 180° RF pulses. As an example in the discussion to follow, let us consider a commonly employed spin echo pulse sequence with the following parameters: TR = 2000 msec, TE = 30 msec and 80 msec. The time between 90° pulses is given by TR. Thus, in this example, every 2000 msec a new 90° pulse will excite the protons. Those protons that have not realigned with the longitudinal axis within 2000 msec are said to be saturated and will not contribute to pixel intensity. The time to echo is designated by TE. The 180° refocusing pulses are made at 15 msec (i.e., TE = 30/2) and 55 msec (i.e., [80 − 30]/2 + 30). The readers should refer to Figures 2 and 3 throughout the following discussion, as they are graphic representations of the effect of the pulse sequence on the contrast between cortical gray and white matter and CSF.

From equation 1 and using the average T1 times from Table 2 (i.e., T1 white = 673 msec, T1 cortical gray = 935 msec, T1 CSF = 3300 msec), the percentage of white and gray matter protons which have recovered at 2000 msec is 94.9% and 88.2%, respectively, while only about 48.5% of CSF protons will have returned to equilibrium (cf. Figure 2a). These percentages represent the relative fraction of mobile protons available for signal production at the beginning of each pulse cycle.

Cortical gray matter, white matter, and CSF also differ in their percentage of water, or proton density. Thus, when 94.9% of white and 88.2% of gray matter protons are recovered at 2000 msec, in absolute terms, there are not more white matter protons available for signal generation than gray, because white matter contains at least 10% less water. To determine contrast at any given TR, T1-governed percent proton recovery needs to be multiplied by N(H) (cf. Figure 2b). White matter, in fact, will be darker than gray at 2000 msec (see $y$-intercept at TR = 2000 msec, Figure 2b). Figure 2c shows the contrast expected at any time TR; contrast is the difference between the tissue relaxation curves from Figure 2b. Only at short TRs will white matter, because of its faster longitudinal relaxation rate (1/T1), be brighter than gray matter. With the T1 estimates

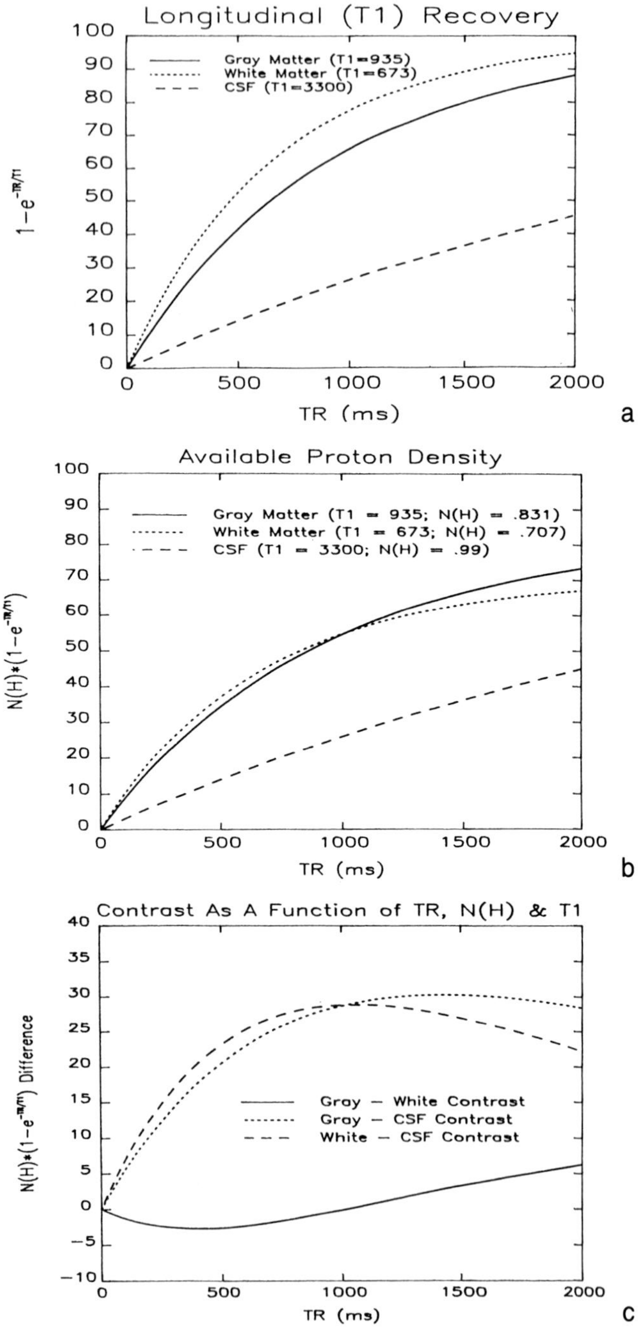


Figure 2. (a) Percent longitudinal proton recovery as a function of TR for three tissue types: gray matter, white matter (modeled using the average T1 relaxation value from the literature presented in Table 2) and cerebrospinal fluid (CSF) (T1 estimated as equal to free water). Note that gray and white matter protons are nearly fully recovered 2 seconds after RF excitation. (b) Proton density for CSF, gray matter, and white matter as a function of TR and T1. Proton density estimates (N(H)) are derived from the average tissue hydration values presented in Table 1. (c) Gray–white–CSF contrast as a function of TR, N(H), and tissue T1. Contrast is calculated by subtracting the curves from (b) and may be either positive or negative. Note that white matter is initially brighter than gray matter but that a contrast inversion occurs at about TR = 1000 msec with the tissue parameters modeled in this example. (d) T2 relaxation for each tissue type, employing the T1, T2, and N(H) values from

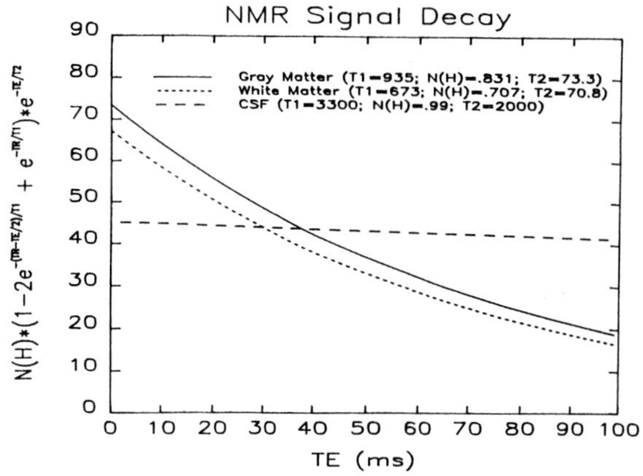


d

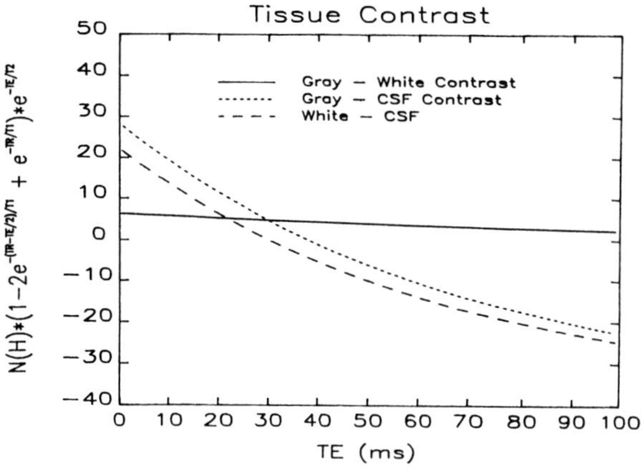


e

Figure 2. (*Continued*) Tables 1 and 2 and a TR of 2000 msec. Note that CSF, while initially darker than brain parenchyma, becomes brighter at about TE = 40 msec with the parameters modeled in this example. (e) Tissue contrast across echo time (TE) for a pulse sequence with TR = 2000 and T1, T2, and N(H) as in (a) through (d). Note that gray–white matter contrast decreases with time but remains positive, while there is a contrast inversion for brain parenchyma and CSF between TE = 30 and 40 msec.

employed in Figure 2b, Figure 2c shows that there will be a contrast inversion for TRs greater than about 1000 msec. Thus, in practice, TRs around 1000 msec are never used, since they result in minimal contrast. In summary, repetition time is a critical determinant of signal intensity and image contrast. Moreover, the shorter the TR, the fewer the available protons for image formation and the less the SNR.

Figure 2d shows the signal decay of gray matter, white matter, and CSF, for a TR of 2000 msec. Note that the initial positioning of the three decay curves along the *y*-axis is identical to their respective final positions in Figure 2C and is representative of the available proton density for each tissue type as a function of the TR and the tissue T1 and N(H). Gray matter signal decays at a slightly slower rate (1/T2) than white matter, and therefore, gray remains brighter at all echo times. On the other hand, CSF with its much longer T2, decays much more slowly than brain parenchyma. Therefore, with echo times greater than about 30 msec, CSF becomes brighter than white matter, and by the 80 msec echo CSF is also brighter than gray matter. Figure 2e shows how gray-white-CSF contrast

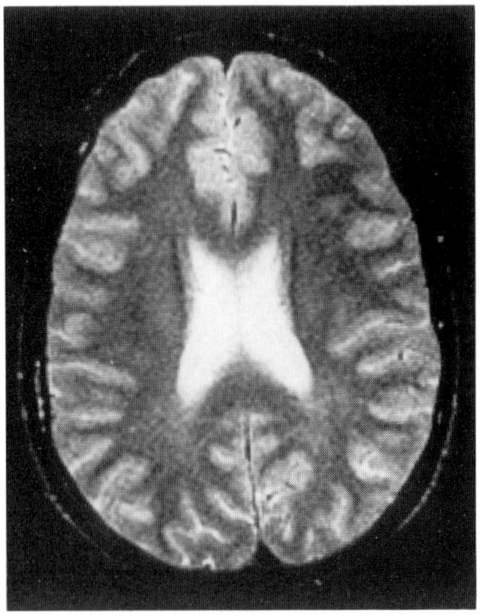

Figure 3. T2 weighted image (TR 2000 msec, TE = 80 msec, NEX = 1, voxel dimensions = $.9375^2 \times 5.0$ mm). Note the greater brightness of the ventricular CSF compared to brain parenchyma, as would be predicted from Figures 2d and 2e.

varies as a function of echo time. Note that gray–white contrast, defined as the difference in available phase-coherent proton density, declines linearly across time TE. Figure 3 presents an image formed at the 80 msec echo, which demonstrates the greater brightness of CSF compared to brain.

Images obtained with a TR of 2000 msec and a TE of 80 msec are frequently referred to as T2 weighted images (T2W). At TE-30 msec, images are usually called proton density weighted images (PDW). What do these labels mean? As described above, the density of the available hydrogen protons (i.e., the percentage recovered from the initial 90° pulse) at time TE = 0 is determined by N(H) and tissue T1. When TR is much greater than T1, T1 recovery approaches 100% and therefore T1 tissue characteristics cannot influence image contrast. When all or nearly all the protons recover to the longitudinal axis, the pixel intensities of early spin echoes are mostly a function of N(H) and are therefore called PDW. However, after 80 msec of signal decay, intensity is increasingly influenced by the T2 of the tissue, and thus the images formed at these later echoes are called T2W (cf. Figure 3).

In a spin echo pulse sequence, intensity[4] can now be defined with the complete equation

$$I = KN(H)(1 - e^{-TR/T1})e^{-TE/T2} \tag{3}$$

where the scaling constant $K$ is a function of the receiver coil, RF amplifier, receiver preamplifier, and the receiver gain (Breger, Wehrli, Charles, MacFall, & Haughton, 1986; Kjos, Ehman, & Brant-Zawadzki, 1985; Perman, Hilal, Simon, & Maudsley, 1984). Each time a subject is placed inside the MRI unit, a "prescanning procedure" is initiated to calibrate the machinery to the individual in order to assure that the RF pulse has the effect of tipping the net magnetization vector the desired degree (e.g., 90°) and to set the receiver gain properly.

[4] Note that Figures 2D and 2E us a slightly modified version of equation 3 which attempts to account for the small amount of T1 recovery that takes place during the time period between the 90° pulse and the measurement of a signal echo (e.g., Wehrli, MacFall, Shuts, Breger, & Herfkens, 1984).

## Pulse Sequence Suggestions

The discussion thus far has focused on the physical processes affecting tissue contrast. Earlier we described factors which affect the SNR and image resolution and the trade-offs between imaging time and image quality. Integrating this knowledge allows one to choose optimal pulse sequences for each measurement problem as framed by the study's hypotheses. We have systematically evaluated numerous pulse sequences for their ability to optimize the appearance of different structures to be measured, and we currently depend on three different sequences. These include two sagittal SPGR sequences and one traditional spin echo sequence acquired in the axial plane. These three are omnibus sequences, for they allow us to measure whole brain and all the individual structures. If one had a specific interest in only a single structure one might choose a different sequence to best illuminate those boundaries, though we doubt that alternative sequences would produce remarkably improved results over our suggested protocol. Technical parameters of the spin echo sequence are: TR = 2000 msec, TE = 30 and 80 msec, slice thickness (interleaved) = 4 mm, matrix size = 256 × 192, field of view (FOV) = 24 cm, NEX = 1. This results in images of fair resolution; voxel size = 4 × .9375 × .9375 $mm^3$. It requires 13.6 minutes of imaging time and provides 36 images covering 144 mm, thus allowing complete brain imaging. The second echo provides a T2 weighted image, optimal for detecting pathology. Moreover, as will be described below, we use the data from the two echoes to create calculated T2 images to assist with certain measurement problems (e.g., segmentation of CSF and CSF volume estimation).

The SPGR sequences are used for detailed anatomic measurement. Unfortunately, General Electric scanners do not allow slices thinner than 3 mm with spin echo sequences. But, SPGR images, acquired as a three-dimensional volume, have the significant advantage of permitting slices to be as thin as 1 mm. This limits partial volume effects and improves the ability to accurately measure structural volume, especially smaller structures or structures with irregular surfaces (e.g., the cortical surface, hippocampus, and basal ganglia), which are most susceptible to partial volume error. The strong T1 weighting of SPGR images also provides excellent gray–white tissue contrast. While we prefer a lengthy SPGR sequence (19.6 minutes) to acquire data with very good resolution and a high SNR, we will use an SPGR sequence of half that time (9.8 minutes) if we suspect patient movement might be a problem for the sample population. The shorter sequence maintains the same high SNR, but the resolution is less as voxel size is increased by 41%. Both sequences acquire the data as a three-dimensional volume and then format it in the sagittal orientation. Of the three dimensions, brain width (rarely greater than 135 mm) is less than brain length (no more than 150 mm) or height (less than 190 mm). Sagittal acquisition, therefore, has the advantage of allowing thinner slices while still covering the full extent of the brain volume in the minimal number of slices. In order to avoid blood flow artifact in sagittal images (e.g., from the carotid arteries) it is important to specify the frequency encoding direction as superior/inferior. Technical parameters of the longer SPGR are: TR = 24 msec, TE = 5 msec, flip angle = 45, slice thickness (contiguous) = 1.2 mm, matrix size = 256 × 192, FOV = 30 cm, NEX = 2. This sequence provides near isotropic voxels (1.17 × 1.17 × 1.2 $mm^3$), which is desirable if one plans on reslicing the data into other orientations with minimal distortion. The greater the departure from isotropy, the greater the distortions upon reslicing, especially if the data are reformatted at oblique angles. The second sequence reduces the imaging time by half by using a NEX of 1 instead of 2. To compensate for the reduction in the SNR we increase voxel size to $1.33 \times 1.33 \times 1.3$ $mm^3$. This is accomplished by using a slice thickness of 1.3 mm and an FOV of 34 cm. Both of these sequences provide 120 images. The longer sequence spans 149 mm, while the shorter one covers 156 mm. We have experimented with flip angles less than 45°, and while these result in a slightly better SNR, the trade-off is decreased gray matter–white matter contrast, and on balance we prefer the greater contrast.

### T1, T2, and N(H) as Biochemical Measures

The parameters T1 and T2 vary as a function of the average number of collisions and near collisions between molecules (Farrar & Becker, 1971). Collisions alter the motion of molecules and hence their magnetic field. The fluctuations in the magnetic field at the molecular level induce dephasing (T2) and return to the lower quantum energy state (T1). When collisions (and near collisions) occur at a rate equal to the resonant or Larmor frequency (e.g., 63.87 MHz with 1.5-tesla scanners), protons are stimulated to emit an RF signal and relax (longitudinally) to a lower energy state. Inhomogeneities in the tissue's chemical environment promote dephasing of spins (transverse relaxation). Therefore, T1 time is shortest in an environment where collisions occur with an average rate equal to the Larmor frequency. The T2 rate, on the other hand, varies inversely with frequency; protons are more effectively dephased when the average time between collisions is long, since very rapidly occurring molecular interactions tend to distribute the magnetic field fluctuations evenly, leaving the protons in precessional phase.

In pure water, collisions occur at a rate many times greater than the Larmor frequency, and thus T1 time is relatively long, and it is approximately equal to T2 time. As the collision rate decreases and approaches the Larmor frequency, T1 and T2 times decrease, but T2 decreases more rapidly. When the average rate of collision falls below the Larmor frequency, T1 time begins to lengthen again, while T2 time continues to decrease. Thus T1 time shows a "V-shaped" relationship to average collision time, with long T1 at both short and long average collision time, while T2 time drops off continuously as collision time decreases.

In biologic tissue, viscosity is a principal determinant of collision frequency. As tissue viscosity increases from that of pure water, the average collision rate decreases and T1 and T2 diverge. In CSF, T1 and T2 are nearly equal to that of free water, but considerably longer than that of brain parenchyma, which is more viscous. The hydrogen protons of lipids within the cell membranes have a very restricted motion and a low collision rate. Therefore, lipid protons in the brain, which exist in bound membranes, have a very long T1 and a very short T2. This is why structured fat gives off no detectable signal. After a TR of 2000 msec, virtually no longitudinal relaxation has taken place (i.e., the protons are saturated), and that which has dephases in the transverse plane after each 90° pulse in a few milliseconds and is therefore not observed.

Several models have been proposed to explain the relaxation behavior of water in solution. However, all are variations of a simple two-state model, which proposes that water is either free or bound in a hydration layer by electrochemical forces to larger molecules, such as the proteins and lipids which compose biological membranes. Bound water has restricted motion and therefore a much shorter T2 relaxation time than free water, whose motion and collision with other molecules is very frequent.

The simplest model suggests that these two states are in fast exchange (the fast exchange two-state model or FETS). Consider T1 relaxation under this model, as there is consensus about its observed relaxation. In the time frame of T1 relaxation, free and bound molecules will have exchanged and the observed T1 *rate* (1/T1) of a tissue will therefore represent the average weighted longitudinal relaxation *rate* of protons in each state. The T1 relaxation rate is given by:

$$1/T1 = \%\text{bound}(1/T1_{\text{bound}}) + \%\text{free}(1/T1_{\text{free}}) \qquad (4)$$

More elaborate formulas include a measure of the exchange rate as indicated by the residence time in each of the two states of water. The important concept in equation 4 is that T1 averages as a function of its rate (1/T1), not its time, and therefore the average T1 *time* of a tissue will be disproportionately influenced by the bound fraction of water. For example, the T1 time for brain parenchyma is about 500 to 1000 msec; this is considerably shorter than the T1 estimate of 3000–4000 msec for free water, though the fraction of water bound at the membrane surface is estimated to be only 5–10% for most tissues (Finch, 1979).

## Measurement Reliability

The reliability of MRI brain morphometry is best conceived of as a problem of test-retest reliability of the same image data, either with one or more than one observer (operator). Test-retest reliability provides an estimate of the error variance due to the measurement procedure. Because brain morphometry often occurs in steps (e.g., thresholding CSF, followed by some combination of visually guided hand tracing and region-growing segmentation), estimation of the reliability of the measurement procedure must include reproduction of the entire sequence of steps, since error at each step may be compounded. Moreover, reliability estimates must be based on the variable which will be used in later group or correlational analyses. If brain size is the variable of interest, reliability must be estimated by examining whole brain reproducibility, not single slice area, because the latter can yield spuriously large correlations. For example, we find that correlating brain area across multiple anatomically matched axial slices, from the crown of the brain to the foramen magnum, of two unrelated individual yields Pearson $r$ values of .97 to .99! This is explained by the species-typical constraints on the size of brain areas at each anatomical level; all human brains conform to the same general shape.

Reliability estimates pertain to specific measurement instruments, applied to specific populations, at a specific time. They do not necessarily generalize across populations or across time. In order to maximize reliability, we always approach each new morphometric problem with an extensive measurement training period, where the operator becomes familiar with variations in structural boundaries across individuals and vagaries particular to the task at hand. Once the operator can blindly reproduce his or her measurements across a small sample of cases (e.g., five to ten subjects), then actual data collection begins.

Morphometric reliability is maximized by having a written document which specifies the operational rules and specific instructions for making the measurements. As much as possible, without sacrificing validity, it is desirable to automate the measurement process, since variation in human performance is the principal source of measurement error. We have also found that environmental lighting can influence perception of structural boundaries, and that it is best to work in a darkened room so as to enhance the perception of tissue contrast. Smaller structures (e.g., planum temporale) are often more difficult to measure reliably than larger structures (e.g., whole brain volume), not because of issues related to image contrast, magnification, or resolution, but probably because small variations in boundary definition from repeat measurements have a proportionally greater impact on the size of the smaller structures (Schultz *et al.*, 1994). Taken a step further, we would predict, other things being equal, that the more round or spherical the structure, the higher the reliability of the measured area, since a circle maximizes the area per unit of the circumference (boundary).

While it is obvious that unreliable measurement of brain structure precludes a valid test of the study's hypotheses, it is less clear how one determines what the minimum degree of measurement reliability should be during the initial training period before continuation of measurement across the entire sample. What constitutes adequate reliability is, of course, dependent on the estimated statistical power of one's experimental design, since random measurement error will decrease the magnitude of observed relationships. However, measurement error is not always random. For example, we have found that inter-rater reliability is sometimes quite high (e.g., Pearson $r > .98$), but the slope of the regression can be substantially different from 1.0, signifying that one operator's measurements are systematically greater than the other's. The same can occur with intra-rater reliability, when the measurement procedure extends over a period of time, and subtle changes in habit cause systematic drift in how conservative or aggressive the operator is during boundary definition. Such drift can occur as the operator has more experience with the task and begins to appreciate more fully different visual features in the image. Hence, we always continue our reliability studies for the duration of the study by interleaving repeat subjects to which the operator is blind. If drift in measurement occurs across time, it is important that the subjects from a given cell in the experimental design not be measured sequentially in time, (e.g., the majority of controls measured before the majority of clinical cases), for this could invalidate statistical inference of group differences. An index sensitive to mean differences in measurement reproducibility, such as the intraclass correlation (e.g., see Guilford & Fruchter, 1973, p. 263) is thus more appropriate than a simple Pearson correlation when assessing reliability.

## Image Processing Techniques

Many brain structures can be measured reliably by simple, low-technology procedures, such as manually guided tracings, be it with radiologic films and a computer digitizing tablet or with an electronic cursor on the monitor of a high-speed workstation using the original digital image data. While the latter is more elegant, in that there is no loss of information resulting from the process of translating the image information from the original digital format to analog film back to a digital tracing, and in the long run is a more efficient process if one is tackling a large data set, we know of no study in the literature which argues on the basis of empirical data that tracings from film are less valid.

However, there are reasons to suspect that the two approaches differ in important respects. For example, digitized tracings of films do not allow the human operator to adjust image contrast or image brightness, both of which can have a significant impact on the perception of boundaries between adjacent structures. The contrast and brightness of radiographs are dependent on the "window" and "level" settings used during filming. This process can be quite variable, and unless one has the resources to make new films with different settings, the variability in contrast and brightness will contribute to measurement error. Working with the original image data also requires consideration of image intensity. Average image brightness, for example, can change by more than 100% when data are collected across many months or years. Large changes can be brought about by the periodic updates in the manufacturer's software used in the image reconstruction process. Moreover, we have observed a significant negative correlation between body size and average image intensity (Schultz, 1991). With a sample of 40 healthy college students, mean pixel intensity from a T2 weighted sequence significantly correlated with body mass measures in a multiple regression. The T2 weighted intensity (TR = 2000 msec, TE = 80 msec, matrix = 256 × 192, FOV = 24 cm, NEX = 1) was best predicted by height, brain area, and weight, in that order (multiple $R = -.84$, $p < .001$). A similar relationship ($R = -.83$, $p < .001$) was observed with proton density weighted images (TE = 30 msec). Body size, especially the mass closest to the superconducting magnet, apparently loads the coil and thereby affects the receiver gain and average pixel intensity. Moreover, the body is diamagnetic, and it changes the scanner's magnetic field slightly, causing the proton resonance frequency to change slightly with different subjects. The RF coil is tuned to the correct proton resonance frequency for each new subject (during Auto Prescan on the GE Signa system). The magnitude of resonant RF power is also adjusted, because larger subjects require greater RF to tip their protons the required number of degrees.

To alleviate potential measurement biases introduced by variations between subjects in image brightness and contrast, we have sometimes applied a global histogram equalization to the image data set so as to maximize contrast and to equalize variation in brightness between subjects (e.g., Schultz *et al.*, 1994). Normally, raw image data intensity values for brain span a relatively short portion of the possible gray scale. Contrast is enhanced through the histogram equalization technique by redistributing the pixel values over the full range and by forcing the distribution to approximate a rectangular distribution. Software packages also usually allow the magnification of regions of interest, thereby allowing more accurate measurement of structure boundaries. It is our experience that one source of measurement error is simply fine motor errors during visually guided hand tracings with a mouse and electronic cursor. By magnifying the area of interest to the fullest possible extent, these tremulous motor variations about the subjectively perceived ideal path are minimized in real tissue space.

With the raw pixel intensity information from a spin echo pulse sequence, one can calculate the T2 relaxation times at each pixel location, though the accuracy of these estimates can be influenced by imperfect RF pulses and inhomogeneities in the main magnetic field (Majumdar, Orphanoudakis, Gmitro, O'Donnell, & Gore, 1986a, 1986b). The T2 relaxation estimates may be a variable of interest bearing on the study's hypotheses. Moreover, they can be used to help segment tissues. Estimating T2 is impossible, however, when working with radiographic films, since the original data are lost when filming. We can calculate T2 times from the following equation:

$$T2 = \Delta TE/\ln(I_{PDW}/I_{T2W}) \tag{5}$$

where $I_{PDW}$ and $I_{T2W}$ represent pixel intensity values observed at each signal echo and $\Delta TE$ represents the difference between echo times (i.e., $80 - 30 = 50$ msec for the example we have been using). Often, however, data from more than two signal echoes (usually about four) are fitted to an exponential decay curve when estimating T2 (e.g., Wehrli *et al.,* 1984b). While in principle, any two-point estimation from equation 5 is exact, in practice, using more than two points increases the accuracy of the T2 estimate since any individual pixel value may be unduly influenced by noise. Figure 4a is an example of a calculated T2 image formed from two spin echoes (Figures 4b and 4c) with equation 5. Pixel intensity values are estimates of T2 relaxation time for each tissue voxel. We sometimes use calculated T2 images to classify CSF and skull (see later in this chapter), since these tissues have T2 times that are very different from brain parenchyma.

## Image Thresholding

Quantitatively based computer-driven image processing procedures serve to limit the number of subjective judgments made by the operator and can also greatly reduce the time involved in the morphometric analysis. At the simplest level one often chooses to threshold an image or a series of images so as to remove a certain range of intensity values. For example, as a first step in isolating the brain from the surrounding CSF, background noise, skull, and portions of the scalp, we have used a set threshold to remove all pixels less than a certain value (e.g., Willerman, Schultz, Ruttledge, & Bigler, 1991). Once established, the threshold procedure can be run in batch, off line, on all images across all subjects, thus greatly enhancing the efficiency of the image processing. This simple step can remove most of the subjective boundary decisions that would otherwise have to be defined by the operator, and thus it will enhance measurement reliability. The operator's input is then limited to boundary decisions in those cases where nonbrain pixels had gray scale values greater than the threshold values, and in these cases the operator uses local contrast and texture cues to finish the task with a manually guided electronic cursor.

We caution that using a preset lower threshold on all images can be problematic. If the data are collected over an extended time period, then changes in the MRI system software used to reconstruct the images can have a profound effect on the average brightness of a pixel. Moreover, as described above, subject characteristics can also affect mean intensity. Such factors can lead to serious misclassification errors. An alternative is to set unique threshold values for each subject and for each slice. This is not only very time-consuming, but in our experience, the reliability of such a method is low (i.e., intraclass correlation coefficient $< .90$) relative to what can be achieved with alternative strategies. With some images we recently have been using a procedure whereby the lower

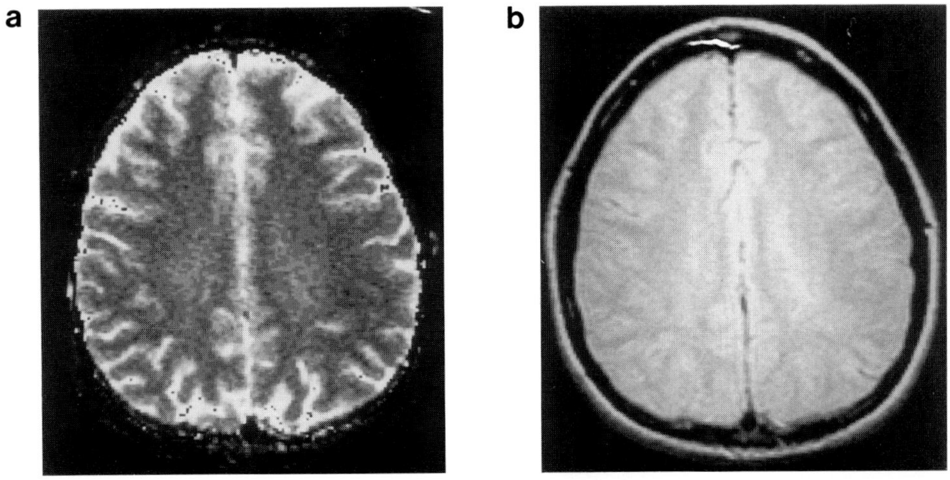


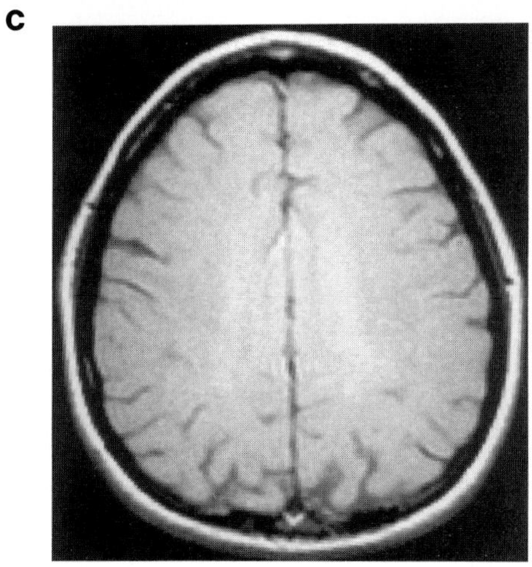


Figure 4. (a) Calculated T2 image, with pixel gray scale value equal to estimated T2 relaxation time of the tissue voxel. Note the very bright CSF pixels and dark scalp pixels. The calculated T2 image was formed using equation 5 for each pixel from two spin echo images, (b) and (c) (TR = 700 msec, TE = 30 and 80 msec, NEX = 2, voxel dimensions = $1.09^2 \times 5.0$ mm, imaging time = 9′ 3″). Neither (b) nor (c) show very good gray-white-CSF contrast, but the relationship between pixel intensity at two spin echo times accurately highlights the longer T2 of CSF and the shorter T2 of scalp and skull.

threshold value is estimated from the average brightness of the image—the brighter the image, the higher the threshold. The assumption is that the average brightness of the image data might vary between people but that this represents merely a shifting of the distribution toward darker or brighter values, and the relationship between brain pixel intensity and CSF pixel intensity at the outskirts of the brain and within sulci and ventricles would be a constant fraction of the overall brightness. Because gray matter and white matter differ in their average intensity, estimates of overall brain intensity are significantly influenced by the amount of gray and white in the region tested. To avoid this complication, we choose to sample only white matter, because it is much easier to sample a larger region of pure white matter than it is to sample a large region of fully volumed gray matter.

For example, in one study employing heavily T1 weighted images (CSF darker than brain parenchyma) we have determined that 40% of the average white matter intensity of the coronal slice immediately anterior to the lateral ventricles is consistently an excellent estimate of the threshold for excluding CSF at the brain's surface and within the ventricles. The rule is not completely rigid, as the operator is allowed to vary the CSF threshold by −2% or +1% in accordance with predefined criteria (e.g., all pure CSF pixels in middle of the lateral ventricles and in wide sulcil must be eliminated). Because the CSF threshold is near the upward sloping lower tail of the pixel gray scale distribution, more pixels are reclassified as nonbrain by a 1% increase in the threshold value than are reclassified by a 1% decrease in pixel values, hence the rule of 40% + 1 or −2%. In a separate study employing a slightly different T1 weighted image acquisition pulse sequence, the rule is 46% + 1 or −2%. The best algorithm needs to be determined separately for every pulse sequence. Once established, such procedures greatly improve the speed at which threshold decisions can be made and the morphometric reliability (e.g., for whole brain we have been achieving intraclass correlation coefficients > .95).

Magnetic resonance images usually contain an intensity gradient within each image and/or across images in the data set such that certain spatial locations are systematically brighter than others. Magnetic field inhomogeneities and imperfect RF pulses can create such intensity gradients or "shading" artifacts, especially for pixels located near the border of the field of view. Thus, pixel intensity can be influenced by spatial location as well as T1, T2, and N(H). Thresholding images with region-dependent intensity yields systematic tissue classification errors. Such artifacts can be usually be corrected with a high-pass filter or by eliminating selected low frequencies with a fast Fourier transform. Filtering is most effective when the brain is already isolated from the surroundings, since the intensity of nonbrain pixels influences the result. This presents a dilemma, since the accuracy of whole brain tissue segmentation is partially dependent on the magnitude of the intensity gradient artifact. Thus, an iterative process is required. First, one can threshold the image in a conservative fashion which eliminates some but not all of the CSF and other nonbrain pixels, but which preserves all brain pixels. This creates a gap between brain and skull, allowing elimination of most pixels outside the brain boundary by manual cursor tracing or region growing (see below). Next the image data can be filtered and then subjected to a less conservative threshold to remove the remaining nonbrain pixels. Unfortunately, the filtering also results in the loss of some informative low-frequency data, yielding images of reduced gray–white contrast. Reduced contrast will impede subsequent efforts to accurately segment certain brain regions. To circumvent this problem, one can then restore the original image characteristics by applying a binary mask of the filtered and correctly thresheld image to the original data. This may not be appropriate, however, for images which would then be subjected to overall gray–white matter segmentation, since the intensity gradient would have been reintroduced, and this could influence the accuracy of the tissue classification.

Figure 5a shows the gray scale pixel distribution for white matter, gray matter, and tissue representing a combination of pure sulcal CSF and sulcal CSF volume averaged with gray matter for proton density weighted image (TR = 2000 msec, TE = 30 msec, thickness = 5.0 mm, skip = 2.5 mm, FOV = 24 cm, matrix = 192 × 256, NEX = 1). The image used in Figure 5a is at the level of the centrum semiovale and contains no ventricular CSF. In these data, the diagonal from the right frontal to left posterior quadrant was darker. Other investigators have shown a front-to-back intensity gradient (e.g., Lim & Pfefferbaum, 1989). Figures 5b and 5c show the same pixel frequency distributions for the T2 weighted image (TR = 2000 msec, TE = 80 msec) and for the calculated T2 image for the three tissue types. Note the lack of a clean trimodal distribution in

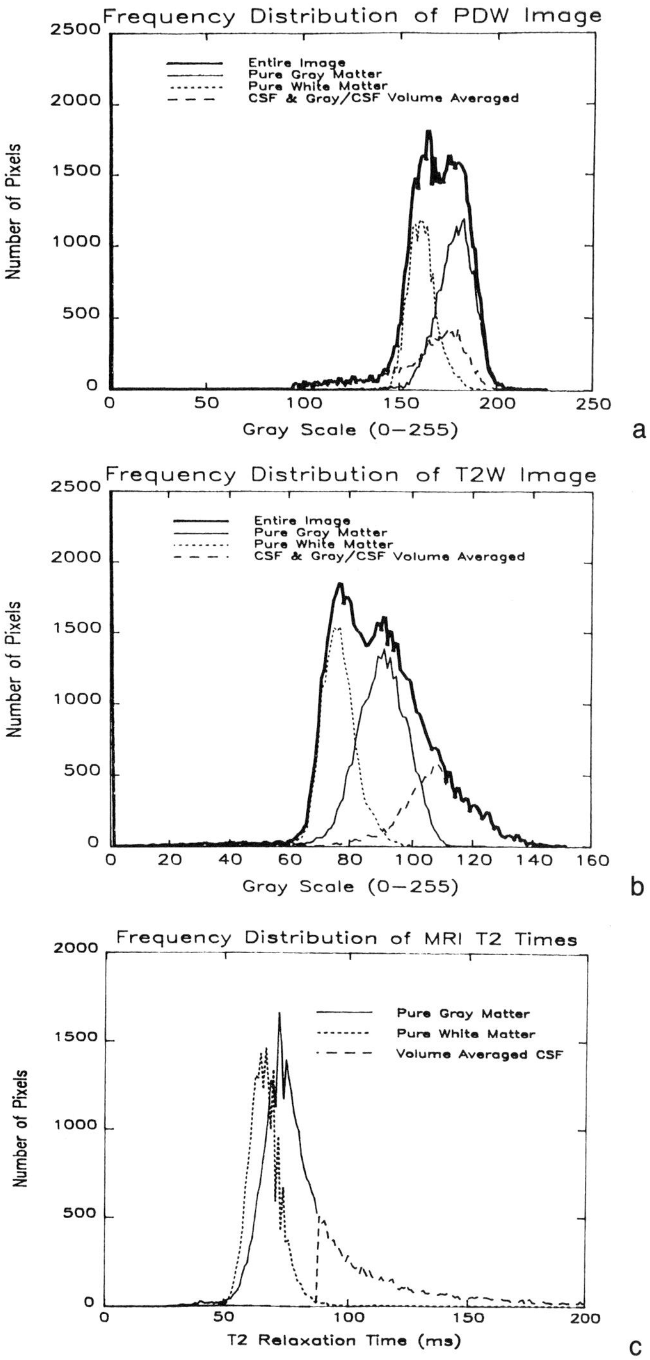


Figure 5. Frequency distribution of a proton density weighted image (a), a T2 weighted image (b), and a calculated T2 image (c) from images at the level of the centrum semiovale with no ventricular CSF (TR = 2000 msec, TE = 30 and 80 msec, NEX = 1, FOC = 24 cm, matrix = 192 × 256, slice thickness = 5.0, skip = 2.5 mm). Note the lack of a clear trimodal distribution in each image type, showing the difficulty of using simple linear thresholds for tissue classification. In (c) CSF was somewhat arbitrarily classified as pixels with T2 times greater than 90 msec, leaving only rather pure gray matter pixels in the cortical sulci. Gray and white matter were classified using a Laplacian of Gaussian segmentation strategy. The three image types were coregistered, so corresponding pixels were similarly classified.

either PDW or T2W pixel intensity, due in part to the shading artifact in the images. Attempts to segment CSF from brain on the basis of a single intensity value can result in considerable misclassification. With data such as these, one must adopt a very conservative thresholding to avoid classifying brain with CSF. Doing do, however, leaves many nonbrain pixels which must be reclassified by other methods.

### Calculated T2 Images

There does not appear to be an intensity gradient to calculated T2 images, suggesting that the image intensity gradient is identical at both spin echoes. Calculation of the ratio of pixel values in solving the Block equation for T2 relaxation apparently removes these spatially dependent intensity variations. If distortions in image intensity were not equivalent at each spin echo, then T2 images would continue to show regionally dependent T2 relaxation, which could be either linear or nonlinear functions of magnetic field and/or RF coil inhomogeneity.

Calculated T2 images show rather poor gray–white contrast, since the mean gray and white matter T2 is quite similar (cf. Figure 4a) and the T2 distributions for these tissues are greatly overlapping (cf. Figure 5c). However, calculated T2 images are excellent for discriminating brain from CSF, scalp, and skull. The thresholding methods for segmenting CSF discussed above relied on a trial and error method of determining the optimal threshold value. With T2 times, one has a theoretical basis for classifying CSF, since it is known to have a much longer T2 than brain parenchyma.[5] Thus pure CSF (i.e., voxels of CSF which are not volume averaged with other tissue types) might be classified as all pixels with T2 times greater than 250 msec. However, it is the CSF within the cortical sulci which are the most difficult pixels to classify accurately, since the cerebral sulci of many subjects are fairly tight, resulting in volume averaging of CSF and gray matter. The T2 times average as a function of their rate (1/T2), not their time. For example, if CSF T2 = 250 msec and gray matter T2 = 70 msec, a pixel representing equal parts of gray and CSF will not have a T2 time of 160 msec ([250 + 70]/2) but rather will be disproportionately biased toward the lower gray matter T2 time. Averaging of their rates (i.e., [1/250 + 1/70]/2) yields an expected T2 time (1/T2 rate) of 110 msec in this example. Figure 6 shows how one can threshold the calculated T2 image from Figure 4A to classify CSF and exterior nonbrain pixels (e.g., skull, scalp with short T2 because of low proton density and/or a large fraction of bound water). In Figure 6 pixels with T2 values less than 50 msec were classified as nonbrain, while pixels with T2 greater than 100 msec were deemed to represent largely CSF (in both cases pixel values were converted to zero for display). Figure 7b shows this classification applied to a T1 weighted image (Figure 7a) at the same anatomical level as the spin echo images used to generate the calculated T2 image. Using pixel T2 times to segment brain from CSF and nonbrain could be run off line in a fully automated fashion across all images and all subjects, resulting in tremendous time savings over alternative methods. Calculated T2 images also provide an excellent means of estimating

[5] It should be noted that in order to obtain accurate T2 relaxation times and to use these for image segmentation, it is imperative that the images be acquired either in an interleaved fashion or with a gap between slices which is equal to 50% of the slice thickness. Otherwise, "cross-talk" between the slices will distort the T2 estimates.

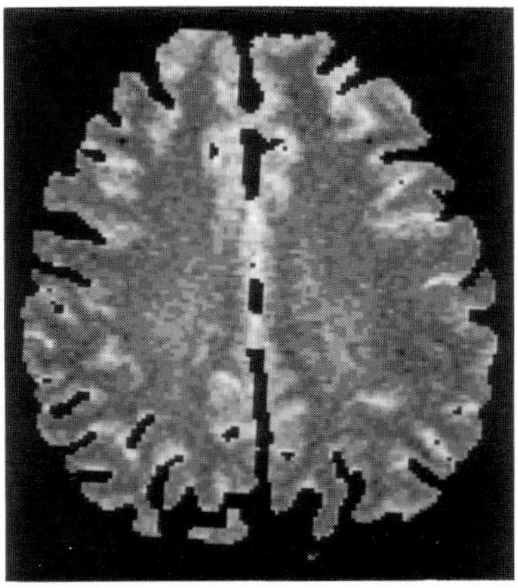

Figure 6. An example of CSF and nonbrain pixel classification with quantitative criteria using the calculated T2 image from Figure 4. All pixels with T2 less than 50 msec (scalp and skull) and greater than 100 msec (CSF) have been set to a gray scale value of zero. Note that a small cluster of scalp pixels remains at the bottom of the image. In this example, CSF classification is rather aggressive; CSF T2 thresholds between 100 and 150 msec, for example, would create less pronounced sulci and more gray matter and CSF partial voluming.

the volume of the CSF in the ventricles and/or sulci, a variables of considerable interest in patient populations.

Synthetic and Multispectral Approaches

It is common for researchers to use tissue T1 and T2 relaxation properties to segment image data, without acknowledging the fundamental physical basis of their approach. Let us consider two examples. First, researchers frequently create a weighted combination of T1 and T2 weighted images to enhance gray–white and parenchyma–CSF contrast for subsequent segmentation (e.g., Jernigan, Press, & Hesselkin, 1990; Lim & Pfefferbaum, 1989). Combinations which emphasize T2 weighted images show good parenchyma–CSF contrast because these are indirect methods of exploiting the large T2 relaxation time difference between the tissues (cf. Figures 2–6). For example, one might subtract a T2 weighted image from a proton density image to better resolve CSF (e.g., Lim & Pfefferbaum, 1989). Combinations of T1 or T2 weighted images can enhance gray–white contrast, though the contrast is less strong compared to parenchyma–CSF contrast because gray and white matter are more similar in T1, T2 and N(H). While fairly effective in practice, such methods are in principle less precise and flexible than a method which determines the T1, T2, and N(H) for each pixel and then bases segmentation on some optimal combination of these fundamental image contrast parameters. There are no other contributors to image contrast, and thus the ability to weight the contribution of each of these factors in synthetic images would provide unsurpassed tissue contrast.

While calculating voxel T2 and using these relaxation estimates for segmentation is not difficult, T1 estimation is more problematic. Calculated T1 images require fitting the exponential function (cf. equation 1) to several anatomically yoked image sets obtained with different TR values, a time-inefficient procedure which is sensitive to patient movement. The value of N(H) could be derived from the three-parameter equation, once two parameters were known, or by extrapolating the T2 relaxation curve for each tissue back to TE = 0 msec (the signal that would be observed immediately following the 90° excitation), a procedure that yields pseudodensity values. Pseudodensity is estimated by $Se^{\mathrm{TE}/\mathrm{T2}}$, where $S$ is the observed signal at time TE. It is devoid of any T2 effects (as long as the tissue T2 is not multiexponential) and therefore influenced only by N(H), T1,

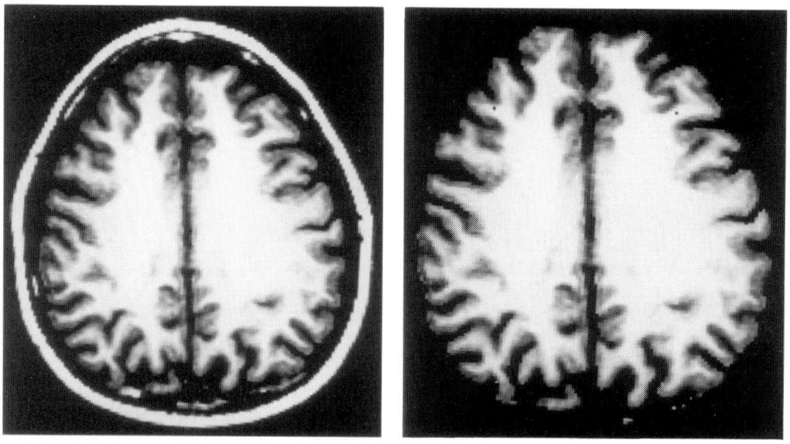

Figure 7. (a) T1 weighted image (TR = 317 msec, TE = 17 msec, NEX = 2, thickness = 5.0 mm with a 2.5 mm skip, matrix = 192 × 256, FOV = 28 cm; contrast enhanced via histogram equalization). (b) Segmentation of brain parenchyma from CSF and nonbrain skull and scalp pixels achieved by multiplying the image with a binary mask of the calculated T2 image from Figure 6 (1 = brain, 0 = nonbrain or CSF). This resultant image, which has gray-white contrast superior to the spin echo images (cf. Figures 4b and 4c) can then be further processed to achieve cortical gray-white classification (e.g., Figure 11). Also note two types of erroneous classification. First, as in Figure 6, there is a small cluster of scalp pixels at the bottom of the image. Also note that one left hemisphere and one right hemisphere white matter pixel was removed (toward the top of the image) because their T2 times were < 50 msec.

and K. These are referred to as pseudodensity values because when TR > T1, the estimates are largely a function of proton density. However, any estimation of N(H) requires that the scanning settings (e.g., receiver gain) which establish the scaling constant K (cf. equation 3) be identical between subjects. To date there are no published reports attempting comprehensive characterization of contrast parameters for tissue segmentation, due in part to the time-consuming nature of the problem. However, as was shown above, one can calculate pure T2 images from any spin echo sequence, and this relatively easy procedure allows quantitative and less ambiguous criteria for classifying CSF. Moreover, calculated T2 images could themselves be combined with PDW or T1W images to highlight gray–white contrast, though it is not clear that such an approach would have great utility, given the relatively low T2 contrast of these tissues (cf. Figures 2e, 4, and 5c).

A second method of MRI segmentation that relies on the three fundamental intensity parameters is multispectral analysis (e.g., Vannier *et al.*, 1985; Kohn *et al.*, 1991; Bonar, Schaper, Anderson, Rottenberg, & Strother, 1993). While the approach was developed initially for analyzing multiple satellite images of the same geographic terrain, the method can be extended to any type of image data where there is more than one image of the same object, with each image providing unique information. For the case of brain MRI segmentation, multispectral analysis relies on the relationship between two or more pixel intensities from different images of the same anatomy which are mapped graphically along orthogonal axes, creating a "feature space" histogram of pixel intensity.

For example, Kohn and colleagues (1991) describe a "bifeature space" method, where the relationship between pixel intensities at two spin echoes is dis-

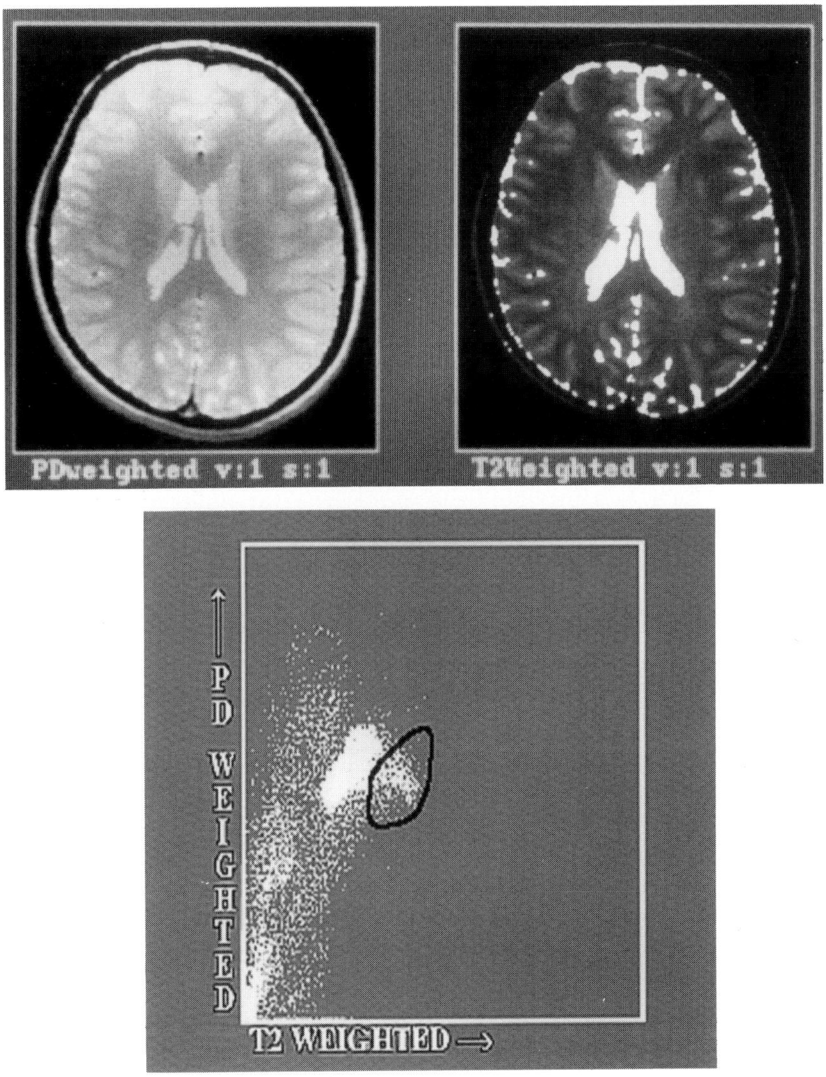


Figure 8. A multispectral approach to CSF-brain parenchyma segmentation. The proton density weighted (PDW) (TR = 2000 msec, TE = 30 msec) and T2 weighted (T2W) (TE = 80 msec) images are mapped in feature space. Note the clustering of pixels in two linear arrays emanating from the origin. The operator has encircled the region on the two-dimensional histogram believed to represent CSF. The bright white pixels superimposed on the PDW and T2W images show the result of this interactive CSF classification procedure.

played graphically (cf. Figure 8). Pixel brightness at the first echo can be displayed along the ordinate, while brightness at the second echo is coded along the abscissa. When the approach is used with images formed from different spin echoes of the same pulse sequence, the relationship between the data points is perfectly described by the T2 relaxation of the tissue voxels. Thus, there are not two features as the term "bifeature" implies, but only one, the decay rate in signal intensity across time. Collapsing the data toward the origin, one obtains a one-dimensional distribution of T2 relaxation times. Kohn *et al.* (1991) show that brain pixels cluster in two fairly distinct regions (cf. Figure 8), one representing

CSF, the other brain parenchyma. In their procedure, the operator used an electronic cursor to encircle the CSF cluster in order to segment brain parenchyma from CSF. Such a method is much more subjective and time-consuming than using calculated T2 images with quantitative threshold criteria for tissue segmentation. While coregistration of images from pulse sequences with different TR values would provide information about T1 relaxation, such an approach is highly susceptible to patient movement during or between image acquisitions. Thus, in the current literature, the multispectral approach has been applied only to images acquired at different spin echoes within the same pulse sequence.

Both a calculated T2 image (Figure 4A) and a two-dimensional mapping of intensity at different echoes (Figure 8) show that gray and white matter are not sufficiently differentiated to allow accurate classification (cf. Figure 5C). The elongated shape of the pixel clusters in Figure 8 is largely a result of spatial variability in pixel brightness (i.e., the shading gradient) and the presence of tissues with different T1, T2, and N(H). Spread in cluster dimension orthogonal to the cluster's long axis probably represent true variability in T2 relaxation times and the effects of volume averaging.

## Gray–White Matter Image Segmentation

Nearly all problems of brain pixel classification involve differentiating one or more gray matter structures from adjacent white matter and/or CSF. In fact, the juxtaposition of the three classes of tissue simplifies many image processing tasks. While sometimes it is desirable to separate adjacent gray matter structures from one another (e.g., the amygdala from the hippocampus, or the putamen and caudate nucleus from the nucleus accumbens), in practice there is very little, if any, contrast in these cases, making it difficult to accurately assess the true size of the structures. In these cases, one must use operational definitions of structure boundaries based on other local anatomical landmarks. The majority of image quantification problems, however, do involve some degree of gray-white-CSF segmentation.

There are two general approaches to the problem of automated or semi-automated gray–white segmentation, each with unique advantages and disadvantages. The first approach is based on region-growing segmentation methods (Geman & Geman, 1984; Leahy, Herbert, & Lee, 1989). This is a class of related methods which rely on the gray scale homogeneity of spatially localized features. The other approach is based on the methods of edge detection (Marr & Hildreth, 1980; Canny, 1986) or boundary finding (Staib & Duncan, 1992; Kass, Witkin, & Terzopoulus, 1988). These use the gradient features in an image (i.e., sharp changes in pixel intensity) at a subset of the spatial positions (near an object boundary). There are a number of issues pertaining to each of these approaches which make the segmentation problem difficult.

1. As discussed above, MR image quality is in part a function of the SNR. While image noise is an obstacle to accurate pixel classification for both region growing and edge detection, gradient-based techniques are more sensitive to noisy images.
2. While sometimes the contrast between adjacent tissues is quite good, the high-frequency information (the sharp change in pixel intensities at anatomical borders) is often distorted or blurred as the result of partial

volume effects. This problem affects gradient-based methods more than region-growing techniques.

3. Across the images in a single data set, the shape of a structure of interest varies, but often this variation is around some average shape. Thus, edge detection, which can incorporate shape priors, is at an advantage compared to region-based segmentation. On the other hand, edge detection for very convoluted or complex shapes which do not have a well-defined border in all places may lead to either false or discontinuous edges. In addition, tissue such as the cortical white matter is frequently present in multiple noncontiguous areas, posing the problem for edge detection routines of grouping identical regions into the same tissue class. Postprocessing is required to identify true edges and to identify relevant structures by grouping together identical regions.
4. MR images frequently have an intensity gradient both within and between images. Boundary finding, being a gradient-based method, is less prone to error due to such systematic spatial variations in pixel intensity, because it relies on the change in gray level rather than absolute values. On the other hand, uncorrected systematic inhomogeneity in mean pixel brightness across the image can pose a significant challenge for region-growing techniques.

Thus, region-growing methods for image segmentation are best suited when the task is to measure structures of complex shape with noncontiguous components with a suboptimal SNR. This describes fairly well many T1 weighted image data sets and the task and cortical gray matter–white matter segmentation. On the other hand, when there are significant changes in mean tissue brightness at different spatial locations which cannot be adequately corrected by spatial filtering, the gradient-based edge detection methodologies may be preferable.

EDGE DETECTION TECHNIQUES. The contrast between gray and white matter is a function of the difference between the mean gray and white matter intensity. The difference in intensity at the border between tissues is exploited by edge detection techniques, which look for instances of change in pixel intensity.

*The Laplacian of Gaussian.* One method of isolating gray and white matter is by using a Laplacian of Gaussian segmentation strategy (Marr & Hildreth, 1980). With this technique, the digital image data are first smoothed with a Gaussian filter to eliminate much of the high-frequency noise which would otherwise contribute to the formation of unwanted edges within like tissue. After image smoothing, edge detection is achieved by a "zero crossing" operator, the Laplacian. It is an isotropic second-degree differential operator that responds well to edges of any orientation, yielding a single pixel thick "edge map" with closed contours which can be superimposed on the original image (Figure 9a, b). However, because it is a second-degree derivative operator, it is quite sensitive to noise. This results in many unwanted edges. The Laplacian does not distinguish between more and less probable edges. False edges, therefore, can only be discarded with operator input. For example, the operator can manually correct edges with an electronic cursor while referring to the original image (e.g., Schultz, 1991). The corrected edge map is then used to isolate the gray and white

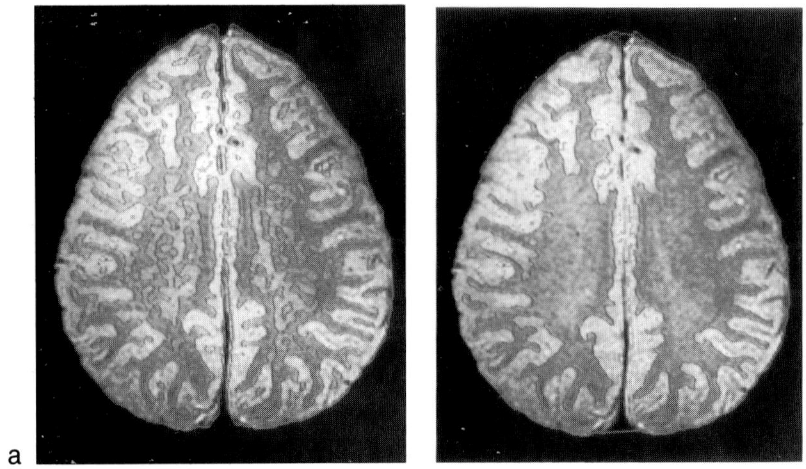


Figure 9. (a) Laplacian of Gaussian edge map superimposed on the input image (TR = 2000 msec, TE = 30 msec, NEX = 1, 192 × 256 matrix, 24 cm FOV, 5 mm thick slice). Note the large number of false edges. (b) False edges have been removed, leaving only an edge at the true gray matter–white matter boundary and another at the surface of the brain.

matter. This process is very time-consuming, especially if one segments numerous slices per subject.

*Canny Edge Detection Techniques.* The Canny edge detector (Canny, 1986) has certain advantages compared to the Laplacian of Gaussian. Mathematically, Canny seeks to optimize by making the best trade-off between detection and localization of edges, the main yet opposing goals of edge detection. In its simplest form it consists of first convolving the image with a Gaussian of appropriate size to get a smooth version of the image and then taking the local maxima of the gradient. Like the Marr-Hildreth operator, it also gives a single pixel thick edge map. However, an advantage of Canny is that it also provides information on "edge strength", the magnitude of the gradient of the Gaussian smoothed image. Each point of the edge map is coded by pixel intensity, with higher values representing a steeper gradient and greater likelihood of a true boundary between two tissue classes. Unwanted edges can be rejected by simple thresholding, yielding an edge map of only the most prominent gradient features (Figure 10). A disadvantage of Canny is that under all but ideal circumstances it results in broken edges. Thus, identifying objects requires grouping appropriate edges, either interactively or by using some additional computer algorithm. An ideal application for Canny might be region-specific edge detection (e.g., the planum temporale).

REGION-BASED SEGMENTATION. Region growing or region-based segmentation is a technique for classifying an image into several regions of uniform or smoothly varying intensity. Each region represents a different class of tissue. A good example of this method is provided by the ANALYZE software package and its "Auto Trace" tool. Auto Trace involves region growing within a specific gray scale range from a seed placed in a tissue of interest. The routine classifies pixels by thresholding all those outside a regionally specified contour. For noisy

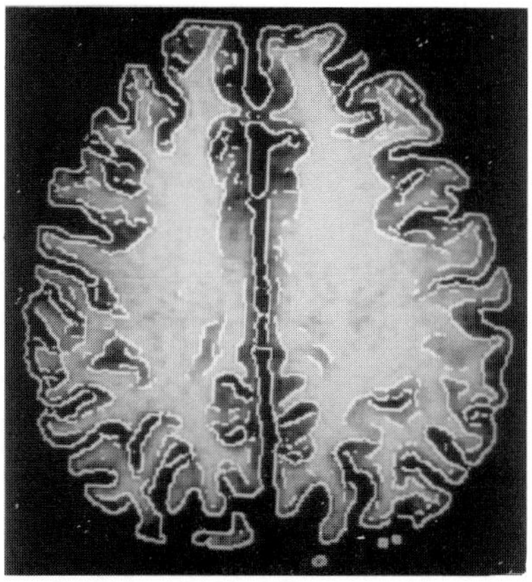

Figure 10. A Canny edge map superimposed on the input image (TR = 317 msec, TE = 17 msec, FOV = 28 cm, 192 × 256 matrix, thickness = 5.0 mm, NEX = 1). Note that Canny results in broken edges, unlike the Laplacian of Gaussian. However, by thresholding the Canny edge map, one can remove the majority of weak and unwanted edges.

images, where it is difficult to maintain the contour within the structure boundary, it is sometimes helpful to use this tool subsequent to some sort of smoothing function to reduce image noise. A disadvantage of this method is that it doesn't classify like tissue which is separated in space by intervening tissue classes. The user must separately seed each tissue cluster.

*Bayesian Analysis with Markov Priors.* To segment gray matter from white matter we have adopted a region-growing method utilizing a Bayesian analysis with Markov priors (Figure 11). The chapter appendix provides a detailed mathematical description of the method. The method looks for similarity within the whole image in order to identify all pixels of a given tissue class, even when these pixels are interspersed among pixels of other tissue types. It is an ideal approach for the problem of gray–white semiautomated segmentation across multiple images, where the two tissue classes often divide each other into several spatially localized clusters.

The method assumes that the image can be modeled as several regions of smoothly varying intensities, plus some noise. Theoretically, if the noise is eliminated, one would be able to classify pixels with 100% accuracy. For each pixel in the image one must decide to which class it belongs. A very trivial way to do this is to compare the gray scale value of each pixel with some threshold range and classify the pixel on the basis of only gray scale value. Let us take a hypothetical example. For images with 256 levels of gray, one might classify all pixels with values $\geq 128$ as white matter and values $< 128$ as gray matter (where CSF has already been converted to zero by way of a separate step). However, application of simple thresholding schemes to MR images usually results in a number of classification errors; the better the contrast to noise ratio of the image, the fewer the errors of classification. If the noise in the image can be approximated by a Gaussian distribution with a mean of zero and a nonzero variance, this can explain many of the classification errors, especially for pixels with values near 128. However, since a Gaussian distribution has infinite tails, theoretically, misclassification could occur (albeit with a small probability) with pixels whose true gray scale value is far removed from the threshold value of 128. Thus, images segmented by simple thresholding will have islands of misclassified tissues embedded in accurately classified tissues because of the influence of noise.

In the one-dimensional case, one typically eliminates the effects of noise by

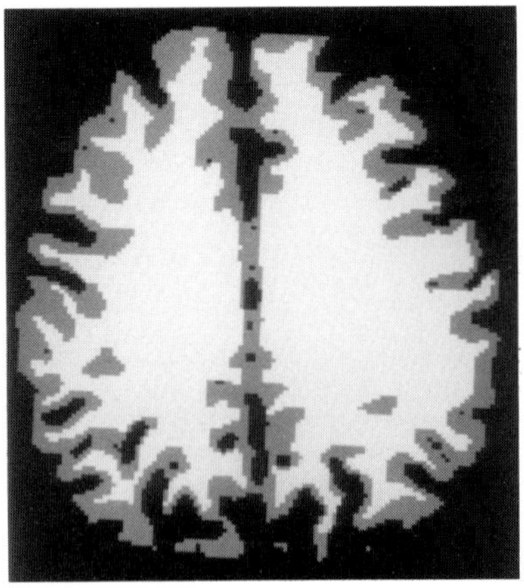

Figure 11. Gray matter–white matter segmentation map showing the result of tissue classification using a Bayesian analysis with Markov priors.

using low-pass filters. The islands of misclassified tissue are instances of high-frequency effects, which in a one-dimensional case are characterized by large variations in signal level. A low-pass filter keeps the low-frequency components and eliminates all high-frequency components. For two-dimensional image data, the simplest equivalent operation is to smooth the data by local averaging, where each pixel value is replaced by the average of a group of pixels within its local neighborhood.

This procedure has very interesting probabilistic connotations. In an image, we know that pixel values are not independent of each other but rather are correlated with others in the immediate vicinity. Except at the boundaries, the pixel values should not change very rapidly. Since the noise is random, the likelihood of a particular pixel being distorted could be high, but it is unlikely that all the surrounding pixels will be distorted in a similar way. Thus, smoothing is based on the assumption that it is more likely that a pixel with a deviant gray value relative to its neighbors has been corrupted by noise than it is that the pixel actually belongs to a different tissue class. Such a model emphasizes similarity between neighboring pixels at the expense of the less frequent event of a single pixel belonging to a different class than those in the neighborhood with which it is embedded.

One way to incorporate this spatial smoothness assumption is to model the image as a Markov Random Field (MRF) so that the prior probability of a particular pixel belonging to a particular class, given the classification of its neighbors, is given by the so-called Gibbs distribution (Geman & Geman, 1984). Like the Gaussian distribution it also belongs to the exponential family of probability density functions. The exponent is a function of the difference in the labels (labels characterize the class, with each class having a separate label) between the pixel in consideration and its neighbors. The simplest neighborhood consists of the four closest pixels (left, right, top, bottom). If we know the classification of the neighbors and if all or the majority of them belong to one class, the prior probability of the center pixel belonging to the same class as its neighbors is much higher compared to the prior probability of it belonging to a different class. This argument is easily extended to include more than two classes and higher-order neighborhoods.

One of the most commonly used and efficient class estimation procedures of this model is the maximization of the *a posteriori* probability. Given the intensity distribution of the observed image, we define the *a posteriori* probability density function for the classification of the image (see appendix). The density function has two components: one constrains the region intensity to be close to the observed gray scale data (likelihood term) and the other imposes spatial continuity (prior term). The spatial continuity term is derived from the MRF model as described above. Our method is similar to the one used by Leahy, Herbert, and Lee (1989).

The classification is posed as a function maximization (minimization) over the whole image. Typically, this is a combinatorial problem, where there are numerous combinations of all the different pixels belonging to the different classes. Out of all these combinations, the one which optimizes the *a posteriori* probability function is chosen. The only proven way of achieving global optima under such circumstances is to use the method of simulated annealing (Geman & Geman, 1984). This method is, however, extremely time-consuming. Hence, for our purposes, we have used the Besag's Iterated Conditional Maximizing (minimizing) method (Besag, 1986). With this method, at every iteration, the probability of a particular pixel being assigned to different classes is computed, and then the pixel is assigned to the tissue type with the highest probability. The procedure stops when there are no changes between iterations. This method is guaranteed to converge to a local minimum because at each step the function that is optimized moves in a single direction. Even though there is no guarantee of achieving global optima, for most cases this procedure gives good results. Figure 11 shows the result of this method applied to the T1 weighted brain image previously segmented from CSF and nonbrain exterior pixels on the basis of T2 relaxation times (Figure 7b).

## Parcellation of the Cerebral Cortex

Measurement of the area or volume of specific cerebral cortical regions is particularly challenging. While many noncortical structures have rather clear boundaries, the landmarks of the cerebral cortex are notoriously variable (Ono, Kubik, & Abernathey, 1990). For example, while a few major sulci, such as the sylvian fissure and the central sulcus rarely show discontinuities in their path, almost all other sulci which might be used as cortical landmarks (e.g., precentral, superior frontal, and intraparietal sulci) show discontinuities (path interruptions) in more than 50% of individual (Ono *et al.,* 1990). In fact, the medial surface of the frontal lobe shows such diversity between individuals that some sulci have yet to be named. The presumption underlying the desire to measure regional cortical area or volume is that there is a relationship among gross anatomical divisions, cytoarchitectonic divisions, and neuropsychological function and that the variability in cortical morphology might predict individual or group differences in behavior.

In the literature, there are currently two general approaches to the problem of cortical morphometry. The first is based on direct measurement of cortical sulci, for detailed anatomic comparison between individuals (e.g., Jouandet *et al.,* 1989). This is most easily accomplished by rendering the brain in three dimensions and then interactively identifying the limiting sulci of interest. Often the goal is to parcel the cortical volume (which may include the underlying white matter), not to assess linear dimensions of sulci. In such cases, identified sulci can form the basis for erecting dividing boundaries for regional segmentation. This technique might be used profitably in combination with a gray–white segmentation strategy. Even though direct measurement of all brain sulci affords unmatched measurement precision, it can be extraordinarily time-consuming, and it may not be practical for application to large numbers of subjects. In fact, in the literature to date the approach has been applied to just one pair of twins (Jouandet *et al.*, 1989). It is also not clear whether the method has acceptable reliability, an issue which supersedes the apparent face validity of direct measurement. Moreover, in order to analyze such data statistically in the context of a between-group or correlational design, an explicit system of criteria is needed for grouping sulci that may be variably split into more than one segment in some subjects but not others and that have a different pattern of supplemental branching between people.

The second approach involves indirect localization of cerebral regions via comparison of each brain to a normative reference system (e.g., Tailarach & Tournoux, 1988; Damasio & Damasio, 1989; Steinmetz, Fürst, & Freund, 1989, 1990). The most widely used indirect system is the proportional grid method of Tailarach (e.g., Tailarach & Tournoux, 1988; Steinmetz, Fürst, & Freund, 1989, 1990). After first rotating the plane which contains the anterior and posterior commissures to a horizontal position, the method places each brain into a three-dimensional box and erects a Cartesian scaffolding with dimensions defined by brain height, width, and length. Every major structure can be referenced by its normative Cartesian coordinates, providing a time-efficient method of localization. The Tailarach atlas is predicated on the existence of a general scaling factor for brain structure size. That is, if one region or structure is large, others will be proportionally large so that a proportional distance between structures is maintained. If the sizes of different brain components were completely independent of one another, a proportional grid system would fail. Because it is a proportional system, the size of each internal subdivision is dependent on the size of the overall brain, thus adding no regionally specific information with regard to size. Although the proportional grid system does not aid in region-specific volumetric comparisons between individuals, it could be used to localize a set of sulci bounding a region of interest which is then accurately measured by some additional sequence of steps.

Rademacher, Galaburda, Kennedy, Filipek, and Caviness (1992) have proposed a method which combines aspects of the proportional grid method and the method of direct measurement of the principal limiting fissures of each lobe. Thus, it maintains some of the precision of direct anatomic measurement while having the advantage of somewhat greater time efficiency. Moreover, one can choose to focus on select cortical regions and not devote time to measuring all the principal sulci catalogued in the system. Rademacher and colleagues first divide the brain into two planes in order to establish a Cartesian reference system for comparisons between subjects. One plane (defined here as the $z$-plane) passes through the interhemispheric fissure. Like the Tailarach system, the $y$-plane is established so that it passes through both the anterior and posterior commissures (the AC-PC plane) as seen from the midsagittal orientation. Next, they establish a set of 16 coronal planes perpendicular to $y$ and $z$, 15 of which are keyed to anatomic landmarks. However, unlike a proportional grid system, the establishment of coronal planes with reference to actual anatomy provides regionally specific size information. The cortex may then be further segmented by establishing dividing lines along hand-traced sulci or with further anatomically keyed *y*- and *z*-planes.

The accuracy of indirect localization systems, such as those of Tailarach and Rademacher and colleagues, is dependent on the anatomic location of the AC and PC relative to the position of the cortical region of interest. The AC-PC plane was originally chosen as the base plane for Tailarach's indirect measurement system during the 1950s in order to localize subcortical ganglia for stereotactic neurosurgery. It was not chosen for its unvarying relationship to cortical regions. In fact, while between-subject variability in subcortical localization with the AC-PC based Tailarach system is relatively small (Tailarach & Tournoux, 1988), in part because of the proximity of the AC and PC to the subcortex, variation in the topography of the major cortical sulci (e.g., the central sulcus) is relatively greater. Steinmetz *et al.* (1989) report that the typical range in the location of major cortical sulci is about 2 cm, the equivalent of 20 high-resolution (1 $mm^3$) MR brain images. Thus, the choice of a base *y*-plane needs to be carefully considered, for it will have very important consequences for measurements throughout the brain.

While many workers in this field establish *y*- and *z*-axes to realign brains to a standard orientation, not all have chosen to use the AC-PC plane. In some cases, images are of suboptimal quality, and these small structures are not reliably identifiable. For example, Jernigan and colleagues have established a *y*-axis by projecting a plane through the anterior-most point of the genu of the corpus callosum and the posterior-most point of the splenium (e.g., Jernigan *et al.*, 1990). Both this method and the AC-PC establish a reference *y*-axis from which perpendicular coronal planes can be erected to further subdivide the brain. To our knowledge, no one has systematically tested whether one such *y*-plane is more valid for brain parcellation than another, and while Steinmetz has reported on the variability in the location of several major cortical sulci using the Tailarach system, neither the statistical significance of this variation nor the possibility that other scaffoldings might reduce this variability has been evaluated quantitatively. If cortical regions of interest were bounded on all sides by reliably identified and invariant sulci, one would not need an indirect system of cortical volume definition. In current practice, however, cortical region definition borrows on some combination of actual anatomy and operationally defined boundaries. It is the validity of the artificial boundaries and their impact on the volume of cortical regions which need to be better understood.

When adopting an indirect reference system for measurement, the accuracy of brain segmentation is highly dependent on the consistency between people of the underlying Cartesian coordinate system. If the landmark coordinates used to establish the Cartesian scaffolding are inherently variable between people and *not related* to variability in cortical regions of interest, then the scaffolding will introduce error variance into all measurements. Establishing the best *y*-plane is critical, because it will have a cascading influence on the size of all brain regions later measured from the Cartesian system. For example, if there is a systematic difference in the position of the PC between two groups and the AC and PC landmarks are used as the basis for cortical parcellation, one may invalidly infer that there are real differences between the two groups in the volume of various cortical regions (e.g., prefrontal cortical volume), when the true difference lies at the base of the measurement system (see Figure 12).

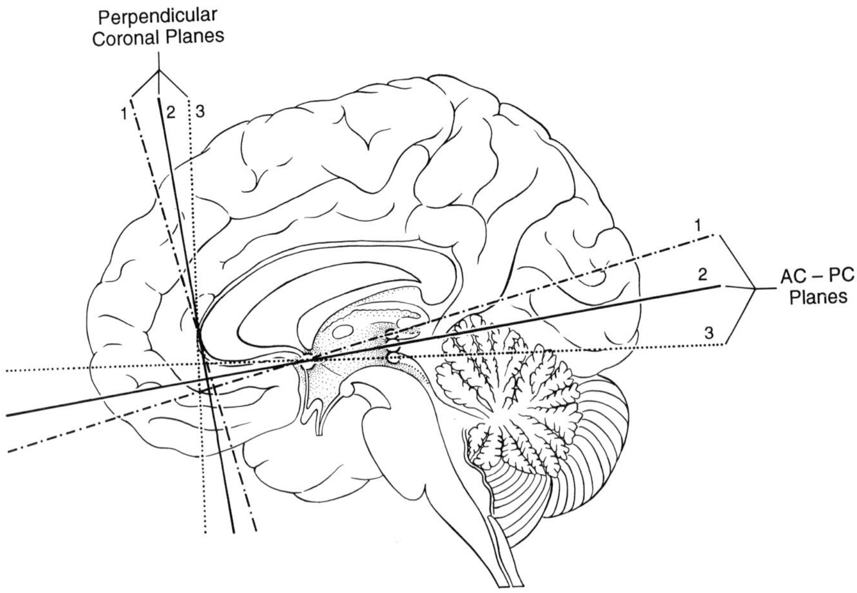


Figure 12. Midsagittal schematic of the brain with three hypothetical positions for the posterior commissure (PC). Small differences in the real or the operator-identified PC position create important differences in the angle of orientation of the AC-PC plane and of the perpendicular coronal segmentation planes erected from each of the AC-PC planes. These differences result in variable definition of the prefrontal region. Coronal plane 1 would result in the smallest midsagittal prefrontal area, while plane 3 would result in the largest. Although the size differences appear small, they may be compounded as the planes are extended toward the lateral surface of the cortex to define a complete volume.

Geometrically, the AC and PC landmarks are not ideal because of their proximity within the brain. Small variations in the real or operator-defined position of these landmarks can exert strong influence on the angle of the segmentation plane and hence the definition of cortical areas (see Figure 12). Steinmetz and colleagues (1989) noted that an error of only 5° in the angle of the AC-PC base plane could create as much as 14 mm of positional variability in the central sulcus. If one is interested in regional cortical volume, it might prove possible to find landmarks which are more widely spaced than the AC and PC and which are anatomically related to the cortical region of interest, possibly enabling less variable localization. The problem is really twofold. The first issue concerns the orientation angle of the $y$-plane. If it is variable between subjects, then perpendicular coronal planes for regional brain segmentation and definition will be variable. The second issue concerns the relative elevation of the plane within the sagittal orientation. If this varies, then segmentation using it or parallel planes will produce variable definitions.

To understand the impact of the choice of base $y$-plane, the issue of whole brain and regional brain volume variability *between subjects* must first be addressed. There is some amount of true variance between subjects in the size of the brain and in the size of individual brain structures. Ideally one wants the method of parceling the brain to represent this accurately without introducing additional variability due to measurement error. Brain size, however, is a compound variable, being composed of numerous different structures. Because substructure sizes are not perfectly correlated within the population, the sum of the Standard deviation (SD) of each component of whole brain will exceed the SD of whole brain size, even if the segmentation plane is unvarying between people relative to all other brain locations. While not perfectly correlated, the sizes of brain subunits are moderately correlated to each other, suggesting that some general scaling factor exists for components of overall brain (e.g., Schultz *et al.*, 1994). Borrowing from standard psychometric theory, one can model the population variability in regional brain size as a function of variability in the general scaling factor, specific region variability (independent of the general scaling factor), and measurement error. Every measurement procedure will introduce error. By definition error is random and thus can only serve to increase the variability of regional brain size measurements for a sample of subjects. The *y*-plane parcellation schemes can therefore be pitted against one another within the same sample, and differences in regional volume variance can be attributed to differences in error variance. These differences can be tested for statistical significance.

After realigning the interhemispheric fissure to a vertical orientation, the first step would be to divide the brain in half with each *y*-plane and then assess the amount of variability between subjects for both the superior and inferior half of the brain for each method in the same sample of subjects. Between-subject variation in the proportion of tissue above and below the dividing plane would largely be a function of differences in the relative elevation of the AC-PC and alternative planes within the sagital orientation, since angle variability would tend to produce trade-offs in anterior and posterior volume and less net volume change[6]. Using the F-test of variance for correlated observations, one can test the statistical significance of differences in variability introduced by different methods (e.g., Kanji, 1993, p. 38). Next, each method might then be used to measure a specific cortical region of interest. For example, one might operationally define the prefrontal region as anterior to a coronal plane at the anterior-most aspect of the genu of the corpus callosum. The more variable the angle of orientation of the *y*-plane, the more variable the perpendicular coronal and the greater the difference between subjects in the region defined as prefrontal. To obtain a statistically significant difference in prefrontal variance between alternative methods the varying angle of the coronal plane would need to produce net changes in prefrontal volume. This might occur because the superior surface of the frontal lobe slopes upward toward the central sulcus, while the inferior boundary is relatively flat. From Figure 12 it can be seen that coronal plane 3 would produce a slightly larger operationally defined prefrontal region than plane 1. But the problem is really three-dimensional, because regional volume variance is also a function of the changing shape of the lateral convexities in reference to the angle of the coronal perpendicular. Thus, size variability of alternative methods might be greater in one slice than in others. The approach of comparing method variance that we are advocating might even be more sensitive if the perpendicular coronal which forms the posterior boundary of the prefrontal region is used in conjunction with other anatomic boundaries, such as cortical sulci, to define smaller subregions, since the imperfect correla-

[6] Variability should be expressed as the fluctuation in the percentage of tissue above or below the dividing plane, since the variance of volumes is partially dependent on the relationships between the sizes of brain regions and the relative amount of tissue above or below the segmentation plane.

tion between the size of adjacent subunits within a larger area, such as the prefrontal cortex, could mask real method differences.

If a method such as the AC-PC dividing plane does introduce significant measurement variance, it would be important to explore alternative anatomically keyed dividing planes to find an approach that minimizes the volumetric variability between individuals. One might choose anatomical points for forming the plane that are more greatly separated in tissue space, since small differences between people in the relative position of two points which are close together, like the AC and PC, create greater differences in the orientation angle of dividing planes than that of two more greatly separated points. Alternatively, one might base the $y$-plane on a least squares fit of a number of landmarks spread across the midsagittal plane. Of course, it may be that all possible dividing planes introduce significant variability simply because individual differences in the spatial relationship between regions are inherent and that the popular AC-PC plane is the most satisfactory solution. Nevertheless, this is an important unanswered empirical question, one that we have begun to study in our laboratory.

## Appendix

Here we try to present a detailed mathematical development of the algorithm. Let us consider a $M * M$ image.
Hence,

$$S = \{1, 2, 3, \ldots, M^2\}$$

corresponds to the set of pixels or the sites.
Let

$$x_i = \text{random variable associated with site i}$$

Thus the image is

$$X = \{x_1, x_2, \ldots, x_{M^2}\} \in R^{M^2}$$

Now, we make the important assumption that the spatial relationships in the image can be modeled adequately by a Markov Random Field. Thus,

$$p(x_i|X_{S/i}) = p(x_i|X_{N/i})$$

where

$$X_{S/i} = \{x_1, \ldots, x_{i-1}, x_{i+1}, \ldots, x_{M^2}\}$$
$$X_{N/i} = \{\text{neighbors of } x_i \text{ forming the MRF}\}$$

The requirements for the probability distribution function are:

(i) $p(x_i) > 0 \ \forall i$
(ii) $p(x_1, x_2, \ldots) > 0$

Then it can be shown that $x$ is a Markov Random Field and

$$p(X) = \frac{1}{K} \exp(-\ U(x)/\beta)$$

where β is a constant. When β goes to infinity, it approaches the uniform distribution and

$$U(x) = \sum_{c} V_c(x)$$

corresponds to the potential function with $c$ defining the neighborhoods.

Now, let us assume that there are $L$ classes and there is only one random variable of each class. Hence,

$$X = \{x_i\};\ x_i = l, \quad l = \{1, 2, \ldots, L\}$$

where $l$ corresponds to the class to which it belongs.

The contribution of $U(x)$ is to define the priors and this eventually results in the smoothing operation. Thus we need to make sure that the choice of the function $f$ in the following expression is such that the neighbors of similar types are rewarded (see Leahy, Herbert, & Lee, 1989).

$$U(x) = \sum_{j \in S} \left\{ \sum_{i \in N_i} (-\ f(x_i - x_j)) \right\}$$

Let the data set be represented by

$$y = \{y_1, y_2, \ldots, y_{m^2}\}$$

Hence, by the MRF assumption,

$$p(y_i | Y_{S/i}, x_i = l, X_{S/i}) = p(y_i | Y_{N/i}, x_i = l, X_{N/i})$$

Now, if we further assume that we have a Gaussian Markov Random Field, then

$$\begin{aligned} p(y_i | Y_{N/i}, x_i = l, X_{N/i}) &= \frac{p(y_i, x_i = l, X_{N/i})}{p(Y_{N/i}, x_i = l, X_{N/i})} \\ &= \frac{1}{\sqrt{2\pi}\sigma_1} \exp\{-\ D\} \end{aligned}$$

where

$$D = \left\{ \frac{1}{2\sigma_l^2} \left((y_i - \mu_l)^2 - \sum_{j \in N} \alpha_l (y_j - \mu_l)^2\right) \right\}$$

Thus for the particular problem, we assume *$y$ corresponds to the data set or the image* and *$x_i$ corresponds to the data type.*
*We need to maximize*

$$p(x_i|Y, X_{S/i}) = p(x_i|Y, X_{N/i})$$

$$\text{or } p(x_i|Y, X_{S/i}) = \frac{p(Y, x_i, X_{N/i})}{p(Y, X_{Ni})}$$

$$\text{or } p(x_i|Y, X_{S/i}) = \frac{p(Y|x_i, X_{N/i})\,p(x_i, X_{N/i})}{p(Y, X_{Ni})}$$

$$\text{or } p(x_i|Y, X_{S/i}) = \frac{p(Y|x_i, X_{N/i})\,p(x_i|X_{N/i})\,p(X_{N/i})}{p(Y, X_{Ni})}$$

$$\text{or } p(x_i|Y, X_{S/i}) = \frac{p(Y|x_i, X_{N/i})\,p(x_i|X_{N/i})}{p(Y, X_{Ni})}$$

$$\text{or } p(x_i|Y, X_{S/i}) \propto p(Y|x_i, X_{N/i})\,p(x_i|X_{N/i})$$

The last step is valid because the denominator is basically a normalizing term.

Now,

$$p(Y|x_i, X_{S/i}) = p((y_i Y_{S/i})|x_i, X_{S/i})$$

$$= \frac{p(y_i, Y_{S/i}, x_i, X_{S/i})}{p(x_i, X_{S/i})}$$

$$= \frac{p(y_i|Y_{S/i}, x_i, X_{S/i})\,p(Y_{S/i}, x_i, X_{S/i})}{p(x_i, X_{S/i})}$$

$$= p(y_i|Y_{S/i}, x_i, X_{S/i})\,p(Y_{S/i}|X)$$

We must note that the second term on the right-hand side is independent of the label at $i$.
Thus,

$$p(Y|x_i, X_{S/i} \propto p(y_i|Y_{S/i}, x_i, X_{S/i})$$

Hence,

$$p(_{xi}|Y, X_{S/i} \propto p(y_i|Y_{S/i}, x_i, X_{S/i})\; p(x_i|X_{S/i})$$

Thus, our aim is

$$\max_{\{i=1,2,\ldots,L\}} p(y_i|Y_{S/i}, x_i, X_{S/i})\ p(x_i|X_{S/i})$$

Thus, under the above-mentioned assumptions, we want

$$\max_{\{x_i \in l\}} \exp(-D - \sum_{j \in N} f(x_i - x_j))$$

where $f$ and $D$ are defined as before.

## References

Agranoff, B. W. (1984). Lipids. In G. J. Siegel, B. W. Agranoff, R. W. Albers, & P. B. Molioff (Eds.), *Basic neurochemistry: Molecular, cellular and medical aspects* (4th ed., pp. 91–107). New York: Raven Press.

Allen, L. S., Richey, M. F., Chai, Y. M., & Gorski, R. A. (1991). Sex differences in the corpus callosum of the living human being. *Journal of Neuroscience, 11,* 933–942.

Besag, J. (1986). On the statistical analysis of dirty pictures. *Journal of the Royal Statistical Society, 48,* 259–302.

Bigler, E. D. (1994). Neuroimaging and neuropsychological assessment. In C. R. Reynolds (Ed.), *Cognitive assessment: A multidisciplinary perspective* (pp. 1–34). New York: Plenum Press.

Bird, C. R., Hedberg, M., Drayer, B. P., Keller, P. J., Flom, R. A., & Hodak, J. A. (1989). MR assessment of myelination in infants and children: Usefulness of marker sites. *American Journal of Neuroradiology, 10,* 731-740.

Bonar, D. C. Schaper, K. A., Anderson, J. R., Rottenberg, D. A., & Strother, S. C. (1993). Graphical analysis of MR feature space for measurement of CSF, gray matter, and white matter volumes. *Journal of Computer Assisted Tomography, 17,* 461–470.

Bottomley, P. A., Hart, H. R., Edelstein, W. A., Schenck, J. F., Smith, L. S., Leue, W. M., Otward, M. M., & Redington, R. W. (1984). Anatomy and metabolism of the normal human brain studied by magnetic resonance at 1.5 tesla. *Radiology, 150,* 441–446.

Breger, R. K., Wehrli, F. W., Charles, H. C., MacFall, J. R., & Haughton, V. M. (1986). Reproducibility of relaxation and spin-density parameters in phantoms and the human brain measured by MR imaging at 1.5 T. *Magnetic Resonance in Medicine, 3,* 649–662.

Canny (1986). A computational approach to edge detection. *IEEE Transactions on Pattern Analysis and Machine Intelligence, 8,* 679–698.

Courschesne, E., Yeung-Courschesne, R., & Egaas, B. (1994). Methodology in neuroanatomic measurement. *Neurology, 44,* 203–208.

Damasio, H., & Damasio, A. R. (1989). *Lesion analysis in neuropsychology.* New York: Oxford University Press.

Denenberg, V. H., Kertesz, A., and Cowell, P. E. (1991). A factor analysis of the human's corpus callosum. *Brain Research, 548,* 126–132.

Dietrich, R. B., Bradley, W. G., Zaragoza, E. J., Otto, R. J., Taira, R. K., Wilson, G. H., & Kangarloo, H. (1988). MR evaluation of early myelination patterns in normal and developmentally delayed infants. *American Journal of Radiology, 150,* 889–896.

Dobbing, J., & Sands, J. (1973). Quantitative growth and development of human brain. *Archives of Diseases in Childhood, 48,* 757–767.

Documenta Geigy (1970). *Scientific tables* (7th ed.). Basel, Switzerland: J. R. Geigy S.A.

Farrar, T. C., & Becker, E. D. (1971). *Pulse and Fourier transform NMR: Introduction to theory and methods.* New York: Academic Press.

Finch, E. D. (1979). Nuclear magnetic resonance studies of the state of water and ions in aqueous cytoplasm. In A. D. Keith (Ed.), *The aqueous cytoplasm* (pp. 60–90). New York: Marcel Dekker, Inc.

Geman, S., & Geman, D. (1984). Stochastic relaxation, Gibbs distribution and the Bayesian restoration of images. *IEEE Transactions on Pattern Analysis and Machine Intelligence, 6,* 721–741.

Gore, J. C., O'Donnell, M., Pope, C. F., Majumdar, S., & Adams, W. J. (1988). Quantitative NMR tissue characterization using calculated images and automated image segmentation. In C. L. Partain, R. R. Price, J. A. Patton, M. V. Kularni, & A. E. James (Eds.), *Magnetic resonance imaging: Vol. 1* (pp. 117–132). Philadelphia: W. B. Saunders.

Guilford, J. P., & Fruchter, B. (1973). *Fundamental statistics in psychology and education* (5th ed.). New York: McGraw-Hill Book Company.

Holland, B. A., Haas, D. K., Norman, D., Brant-Zawadzki, M., & Newton, T. H. (1986). MRI of normal brain maturation. *American Journal of Neuroradiology, 7*, 201–208.

Interactive Data Language [computer software] (1987). Denver, Co: Research Systems Incorporated.

Jack, C. R., Bentley, M. D., Twomey, C. K., & Zinmeister, A. R. (1990). MR imaging-based volume measurements of the hippocampal formation and anterior temporal lobe: Validation studies. *Radiology, 176*, 205–209.

Jernigan, T. L., Press, G. A., & Hesselink, J. R. (1990). Methods for measuring brain morphologic features on magnetic resonance images: Validation and normal aging. *Archives of Neurology, 47*, 27–32.

Jernigan, T. L., Zisook, S., Heaton, R. K., Moranville, J. T., Hesselink, J. R., & Braff, D. L. (1991). Magnetic resonance imaging abnormalities in lenticular nuclei and cerebral cortex in schizophrenia. *Archives of General Psychiatry, 48*, 881–890.

Jouandet, M. L., Tramo, M. J., Herron, D. M., Hermann, A., Loftus, W. C., Bazeell, J ., & Gazzaniga, M. S. (1989). Brain prints: Computer-generated two dimensional maps of the human cerebral cortex in vivo. *Journal of Cognitive Neuroscience, 1*, 88–117.

Kamman, R. L., Go, K. G., Muskiet, F. A. J., Stomp, G. P., Van Dijk, P., & Berendsen, H. J. C. (1984). Proton spin relaxation studies of fatty tissue and cerebral white matter. *Magnetic Resonance Imaging, 2*, 211–220.

Kanji, G. K. (1993). *100 statistical tests.* London: Sage Publications.

Kass, M ., Witkins, A., & Terzopoulus, D. (1988). Snakes: Active contour models. *International Journal of Computer Vision, 1*, 321–331.

Kjos, B. O., Ehman, R. L., & Brant-Zawadzki, M. (1985). Reproducibility of T1 and T2 relaxation times calculated from routine MR imaging sequences: Phantom study. *American Journal of Neuroradiology, 6*, 277–283.

Kohn, M. I., Tanna, N. K., Herman, G. T., Resnick, S. M., Mozely, P. D., Gur, R. E., Alavi, A., Zimmerman, R. A., & Gur, R. C. (1991). Analysis of brain and cerebrospinal fluid volumes with MR imaging. *Radiology, 178*, 115–122.

Kulynych, J. J., Vladar, K., Jones, D. W., & Weinberger, D. R. (1993). Three-dimensional surface rendering in MRI morphometry: A study of the planum temporale. *Journal of Computer Assisted Tomography, 17*, 529–535.

Leahy, R., Herbert, T., & Lee, R. (1989). Application of Markov random fields in medical imaging. In D. Orthendahl & J. Lacer (Eds.), *Proceedings of information processing in medical imaging* (pp. 1–14). Wiley-Liss.

Leonard, C. M., Voeller, K. K. S., Lombardino, L. J., Morris, M. K., Hynd, G. W., Alexander, A. W., Andersen, H. G., Garofalakis, M., Honeyman, J. C., Mao, J. Agee, F., & Staab, E. V. (1993). Anomalous cerebral morphology in dyslexia revealed with MR imaging. *Archives of Neurology, 50*, 461–469.

Lim, K. O., & Pfefferbaum, A. (1989). Segmentation of MR brain images into cerebrospinal fluid spaces, white and gray matter. *Journal of Computer Assisted Tomography, 13*, 588–593.

MacFall, J. R., Wehrli, F. W., Boreger, R. K., and Johnson, G. A. (1987). Methodology for measurement and analysis of relaxation times in proton imaging. *Magnetic Resonance Imaging, 5*, 209–220.

Majumdar, S., Orphanoudakis, S. C., Gmitro, A., O'Donnell, M., & Gore, J. C. (1986a). Errors in the measurement of T2 using multiple-echo MRI techniques: I. Effects of radiofrequency pulse imperfections. *Magnetic Resonance in Medicine, 3*, 397–414.

Majumdar, S., Orphanoudakis, S. C., Gmitro, A., O'Donnell, M., & Gore, J. C. (1986b). Errors in the measurement of T2 using multiple-echo MRI techniques: II. Effects of static field inhomogeneity. *Magnetic Resonance in Medicine, 3*, 562–574.

Marr, D., & Hildreth, E. (1980). Theory of edge detection. *Proceedings of the Royal Society of London, 207*, 187–217.

Ono, M., Kubik, S., & Abernathey, C. D. (1990). *Atlas of the cerebral sulci.* New York: Thieme Medical Publishers, Inc.

Perman, W. H., Hilal, S. K., Simon, H. E., & Maudsley, A. A. (1984). Contrast manipulation in NMR imaging. *Magnetic Resonance Imaging, 2*, 23–32.

Peterson, B., Riddle, M. A., Cohen, D. J., Katz, L. D., Smith, J. C., Hardin, M. T., & Leckman, J. F. (1993). Reduced basal ganglia volumes in Tourette's syndrome using three-dimensional reconstruction techniques from magnetic resonance images. *Neurology, 43*, 941–949.

Press, G. A., Amarad, D. G., & Squire, L. R. (1989). Hippocampal abnormalities in amnesic patients revealed by high resolution magnetic resonance imaging. *Nature, 341,* 54–57.
Rademacher, J., Galaburda, A. M., Kennedy, D. N., Filipek, P. A., & Caviness, V. S. (1992). Human cerebral cortex: Localization, parcellation, and morphometry with magnetic resonance imaging. *Journal of Cognitive Neuroscience, 4,* 352–374.
Raz, N., Torres, I. J., Spencer, W. D., White, K., & Acker, J. D. (1992). Age-related regional differences in cerebellar vermis observed in vivo. *Archives of Neurology, 49,* 412–416.
Robb, R. A., & Hansen, D. P. (1990). ANALYZE: A software system for biomedical image analysis. *Proceedings from the first conference on visualization in biomedical computing* (pp. 507–518). Los Alamitos, CA: IEEE Press.
Runge, V. M. (1994). *Magnetic resonance imaging of the brain.* Philadelphia: Lippincott.
Schultz, R. T. (1991). *The relationship between intelligence and gray-white image contrast: An MRI study of college students.* Unpublished doctoral dissertation, University of Texas at Austin.
Schultz, R. T., Cho, N. K., Staib, L. H., Kier, L. E., Fletcher, J. M., Shaywitz, S. E., Shankweiler, D. P., Katz, L., Gore, J. C., Duncan, J. S., & Shaywitz, B. A. (1994). Brain morphology in normal and dyslexic children: The influence of sex and age. *Annals of Neurology, 35,* 732–742.
Singer, H. S., Reiss, A. L., Brown, J . E., Aylward, E. H., Shih, B., Chee, E., Harris, E. L., Reader, M. J., Chase, G. A., Bryan, R. N., & Denckla, M. B. (1993). Volumetric MRI changes in basal ganglia of children with Tourette's syndrome. *Neurology, 43,* 950–956.
Staib, L. H., & Duncan, J. S. (1992). Boundary finding with parametrically deformable models. *IEEE Transactions on Pattern Analysis and Machine Intelligence, 14,* 1061–1075.
Steinmetz, H., Furst, G., & Freund, H. J. (1989). Cerebral cortical localization: Application and validation of the proportional grid system in MR imaging. *Journal of Computer Assisted Tomography, 13,* 10–19.
Steinmetz, H., Fürst, G., & Freund, H. J. (1990). Validation of perisylvian and calcarine anatomic landmarks within stereotaxic proportional coordinates. *American Journal of Neuroradiology, 11,* 1123–1130.
Steinmetz, H., Rademacher, J., Huang, Y., Zilles, K., Thron, A., & Freund, H. J. (1990). Cerebral asymmetry: MR planimetry of the human planum temporale. *Journal of Computer Assisted Tomography, 13,* 996–1005.
Suzuki, K. (1972). Chemistry and metabolism of brain lipids. In R. W. Albers, G. J. Siegel, R. Katzman, & B. W. Agranoff (Eds.), *Basic neurochemistry* (1st ed., pp. 207–225). Boston: Little, Brown and Co.
Tailarach, J., & Tournoux, P. (1988). *Co-planar stereotaxic atlas of the human brain.* New York: Thieme Medical Publishers, Inc.
Van Balen, R., Kate, T., Koelma, D., Smeulders, A., & Groen, F. [computer software] (1991). Scil-image: A multi-image interactive image processing environment. Amsterdam: University of Amsterdam.
Vannier, M. W., Butterfield, R. L., Jordan, D., Murphy, W. A., Levitt, R. G., & Gado, M. (1985). Multispectral analysis of magnetic resonance images. *Radiology, 154,* 221–224.
VoxBlast [computer software] (1993). Fairfield, IA: VayTek, Inc.
Watson, C., Andermann, F., Gloor, P., Jones-Gotman, M., Peters, T., Olivier, A., Melanson, D., & Leroux, R. T. (1992). Anatomic basis of amygdaloid and hippocampal volume measurement by magnetic resonance imaging. *Neurology, 42,* 1743–1750.
Wehrli, F. W., MacFall, M. R., Glover, G. H., Grigsby, N., Haughton, V., & Johanson, J. (1984). The dependence of nuclear magnetic resonance (NMR) image contrast on intrinsic and pulse timing parameters. *Magnetic Resonance Imaging, 2,* 3–16.
Wehrli, F. W., MacFall, J. R. Shutts, D., Breger, R. & Herfkens, R. J. (1984). Mechanisms of contrast in NMR imaging. *Journal of Computer Assisted Tomography, 8,* 369–380.
Willerman, L., Schultz, R., Rutledge, N., & Bigler, E. (1991). *In vivo* brain size and intelligence. *Intelligence, 15,* 223–228.
Wood, M. L., & Henkelman, R. M. (1985). Truncation artifacts in magnetic resonance imaging. *Magnetic Resonance in Medicine, 2,* 517–526.

3

# Imaging the Developing Human Brain

ELIZABETH R. SOWELL AND TERRY L. JERNIGAN

## INTRODUCTION

Questions about the development of the human nervous system have been addressed using a number of different experimental approaches. One method has been the examination of neurogenesis in autopsy studies of fetuses, infants, and young children. In autopsy studies, information has been recorded about the shape, size, cellular structure, and chemical makeup of the human brain at different stages in development. However, in postmortem studies individual subjects can be examined at only one point in development, leaving unanswered questions about individual differences in the course of maturation. In reviewing autopsy studies, careful assessment must be made as to the normality of the subjects examined. For example, in Conel's (1939–1967) studies of the postnatal development of the human cerebral cortex in the six-month-old, his seven subjects died from various causes ranging from asphyxia to malnutrition. It is unclear what effect such maladies had on the brains of these infants.

ELIZABETH R. SOWELL San Diego State University/University of California, San Diego, Joint Doctoral Program in Clinical Psychology, University of California, San Diego, School of Medicine, La Jolla, California 92093. TERRY L. JERNIGAN Veterans Affairs Medical Center, San Diego, and Departments of Psychiatry and Radiology, University of California, San Diego, La Jolla, California 92093.

*Neuroimaging I: Basic Science*, edited by Erin D. Bigler. Plenum Press, New York, 1996.

Animal studies represent another important approach to questions about human maturation. Advantages of these studies are that the animals can be raised in controlled environments and be studied at specific, predetermined stages of development to address may of the same questions asked by neuropathologists. True experiments can be conducted by randomly assigning subjects to groups to determine what factors contribute differentially to neurodevelopment. The problem of nonrepresentative samples present in human autopsy studies is effectively eliminated, and relatively large groups of subjects can be studied. However, the animals are still generally evaluated at only one point in their development, and it is unclear to what extent findings from animal studies can be generalized to human neurodevelopment. Nevertheless, it is likely that the principles that govern neurodevelopment in human and nonhuman species are sufficiently general that much insight into human maturation can be gained from animal studies.

Fortunately, new technology that has emerged in the last few decades permits the study of the human brain during life. Such neuroimaging techniques as x-ray computed tomography (CT), magnetic resonance imaging (MRI), positron emission tomography (PET), and single photon emission computed tomograph (SPECT) provide promising alternatives to the more traditional approaches discussed above. While these methodologies have their limitations, larger, more representative human samples can be examined with these *in vivo* techniques. Many questions can be answered that cannot be answered via autopsy studies. For example, the same subject can be studied repeatedly during his or her development to determine within-subject maturational changes. Prospective studies can also be designed in which detailed behavioral assessments can be collected concurrently and correlated with changes observed with the brain imaging examinations.

Just as some questions are better answered with *in vivo* methods, others are difficult to address without detailed postmortem examinations. For example, the resolution of *in vivo* imaging systems is unlikely to provide information at the single-cell level in the near future. The early embryonic stages of CNS development are also best examined by means of autopsy studies because of the small size of the nervous system in these stages. Finally, it should be noted that it is often unclear what is being observed with *in vivo* methodologies. Changes in signal characteristics are measured with all imaging techniques, but the biophysical basis of the change in signal with maturation can only be inferred from what is currently known about the human brain based on autopsy studies. Some of the most exciting possibilities for the future of developmental neurobiology lie in the combination of the above-mentioned scientific methods: human autopsy, animal, and neuroimaging studies.

This chapter will review the research to date that has focused on neuroimaging techniques and the developing human brain. The information will be organized into sections on the progression of myelin deposition observed with MRI, morphological changes observed with MRI, a brief section on changes observed with functional imaging techniques such as PET and regional cerebral blood flow (rCBF), and finally, a section discussing methodological concerns in the study of normal neurodevelopment *in vivo*. The research summarized in this chapter will focus on normal development in infancy, childhood, and adolescence.

## Progressing Myelin

### Evidence from Postmortem Studies

From postmortem studies, we know that myelination begins near the end of the second trimester of fetal development and extends beyond the second decade of life (Yakovlev & Lecours, 1967). Autopsy studies consistently reveal that myelination occurs in a systematic sequence with different brain regions myelinating at different times. Generally, myelination is thought to progress from inferior to superior brain regions and from posterior to anterior; that is, brain stem and cerebellar regions myelinate prior to the cerebral hemispheres, and the occipital lobes myelinate prior to the frontal lobes (Yakovlev & Lecours, 1967; Brody, Kinney, Kloman, & Gilles, 1987). This process is thought to reflect the regional pattern of functional maturation of the brain (Yakovlev & Lecours, 1967).

In the first column of Table 1, the findings of Yakovlev and Lecours (1967) are summarized for a number of different brain regions. For each region, the age at which myelination first appeared and the age at which it was apparently completed are listed. It is clear that different regions have different patterns of myelination. For example, the optic radiations start to myelinate at 10 fetal months and attain the adult appearance roughly four months later at 4 postnatal months. Other brain regions show a more protracted myelinogenesis. The white matter in the cerebral hemispheres, for example, begins myelinating at birth and continues myelinating into adulthood.

It should be noted that other researchers have also studied patterns of myelination in postmortem material. For example, Brody *et al.* (1987) found the same patterns of myelination as Yakovlev and Lecours, but the onset of myelination was consistently later across all brain regions in the latter study. The discrepancies between these two studies are most likely due to either differences in the subject populations or differences in the methods used to measure the process of myelination (Holmes, 1986). The study by Yakovlev and Lecours (1967) is used here as a standard for assessing the findings in MR studies because later postmortem studies generally confirm these earlier findings.

### Biochemical Basis

Before reviewing the results of the MR studies, it is useful to consider what is known about biochemical changes associated with myelin deposition. The stimulus that evokes the progression of myelination in the central nervous system has not yet been determined. However, one hypothesis is that the emission of neurotransmitters from the neuron is the initiating stimulus needed to set the myelination process into action (Valk & van der Knaap, 1989). Once the stimulating event has occurred, there is a rapid proliferation of oligodendrocytes, dramatically increasing cell density in the region to be myelinated. This process, known as myelination gliosis, precedes the process of myelination *per se,* during which the cell membranes of the oligodendrocytes wrap around the axons and form the myelin sheath (Friede, 1989). At birth the gray and white matter have comparable water and lipid concentration. However, myelination gliosis and the formation of the myelin sheath are associated with increasing lipid content and

TABLE 1. AGE AT FIRST APPEARANCE AND COMPLETION OF MYELINATION[a]

| Anatomic region | MR protocol[b] A | B | C | D | E | F |
|---|---|---|---|---|---|---|
| Brain stem | | | | | | |
| Medial leminiscus | 5 fm (12 pm) | | | birth | | |
| Reticular formation | birth (adolescence) | | | | | |
| Cerebellar white | | birth (4 pm) | 3–5 pm | | | < 1 pm [vermis] |
| Superior cerebellar peduncle | 8 fm (8 pm) | birth | birth | birth | | |
| Middle cerebellar peduncle | birth (48 pm) | birth | birth (18 pm) | 3 pm | | |
| Inferior cerebellar peduncle | 6 fm (8 fm) | | | birth | birth | < 1 pm |
| Forebrain | | | | | | |
| Inferior colliculi | 7 fm (4 pm) | | | | | |
| Superior colliculi | 9 fm (4 pm) | | | | | |
| Optic radiations | 10 fm (4 pm) | < 1 pm | 1pm | 3 pm | 3 pm | 3 pm |
| Corpus callosum | 4 pm (adolescence) | 3–6 pm | 4–8 pm | 5 pm | | 5 pm (> 12 pm) |

| | | | | | | |
|---|---|---|---|---|---|---|
| Fornix | 3 pm (24 pm) | | | | | |
| Striatum | 3 pm (24 pm) | | | | | |
| Internal capsule | | | | | | |
| Posterior limb | 9 fm (4 pm) | birth | birth–7 pm | 3 pm | > 1 pm | < 1 pm |
| Anterior limb | 4 pm (8 pm) | 2–3 pm | 7 pm (11 pm) | 5 pm | 6 pm | < 1 pm (adulthood) |
| Centrum semiovale | | 2–4 pm | 7–11 pm | birth [corona radiata] | | < 1 pm [central portion] |
| Cerebral hemispheres | | | | | | 6–7 pm |
| Frontal white | birth (adulthood) | 3–11 pm | 11–18 pm | | 8 pm | 6–7 pm |
| Parietal white | birth (adulthood) | | | | 8 pm | 5 pm |
| Lateral temporal white | birth (adulthood) | | | | 12 pm | |
| Limbic white | 2 pm (adolescence) | | | | | |
| Occipital white | | 3 pm (7 pm) | 9–15 pm | | 3 pm | |

[a] Ages are given in fetal months (fm) and postnatal months (pm). Age at completion of myelination is given in parentheses.
[b] MR protocol conditions: (A) Postmortem, n > 200 (Yakovlev & Lecours, 1967); (B) T1, SE: TR 600/TE 20, n = 82 (Barkovich *et al.*, 1988); (C) T2, SE: TR 2500/TE 70, n = 82 (Barkovich *et al.*, 1988); (D) T1, IR: TR 3000/TE 30, T1 800, n = 51 (Christophe *et al.*, 1990); (E) T2, TR 2000/TE 80 or 84, n = 64 (Dietrich *et al.*, 1988); (F) T1, TR 350 or 450/TE 35, n = 36 (Hayakawa *et al.*, 1990).

decreasing water content in the white matter. This process leaves the gray matter with a higher water content relative to the myelinated white matter, which has a higher lipid concentration (Girard, Raybaud, & du Lac, 1991).

The myelin sheath does not contain mobile protons and therefore does not itself emit an MR signal. Myelin is a fatty substance which is hydrophobic, and thus the white matter has lower water content than gray matter. Some authors have suggested that signal changes on T1 weighted (T1W) images are induced during the early process of myelination gliosis in association with rapid increases in brain cholesterol, glycolipids, or protein. These MR signal changes may be due to the effects of the lipids or proteins on the T1 of water. The T2W images may be more sensitive to the longer-lasting decreases in water content that occur as the myelin sheath matures. For a more detailed discussion of the changes that occur prior to and during the myelination process and their effects on the signal values observed on MRI, see Barkovich, Kjos, Jackson, and Norman (1988) and Girard *et al.* (1991).

### Evidence from MR Studies

Developmental changes during infancy and early childhood have been noted in a number of brain imaging studies, and most of these studies have focused on myelination. Magnetic resonance imaging is a particularly useful *in vivo* tool for assessing the deposition of myelin because of its sensitivity to differences in water and lipid concentrations of the tissues. The literature on postmortem observations (e.g., Conel, 1939–1967; Yakovlev & Lecours, 1967; Brody *et al.*, 1987) provides evidence upon which the interpretation of *in vivo* studies of myelination can be based.

As Table 1 reveals, the different studies targeted different brain regions in their assessment of myelination, with only a few structures examined in multiple studies. It is also apparent that the postmortem study (Yakovlev & Lecours, 1967) examined the process in much greater detail, listing many relatively small neuroanatomical structures. Unfortunately, the areas discussed in the MR studies only roughly correspond to those of Yakovlev and Lecours, partly because the resolution of current *in vivo* techniques does not yet approach that of postmortem methods. It should be noted in reviewing Table 1 that the results related to different anatomical structures within the same general brain region are given under a common general label, and thus, the exact anatomical terms used by the individual authors in their studies may not be those used in the table.

### Changes in White Matter Appearance

Most neuroimaging studies of myelination survey the brain changes that occur in the earliest postnatal days of life when much of the myelination process occurs. Barkovich *et al.* (1988) conducted a study of normal brain maturation in 82 infants ages 4 days to 2 years. The subjects were referred for nonspecific neurological complications, suspected child abuse, and possible macrocephaly. Three of the authors subjectively rated the relative intensity of gray to white matter in each of 14 anatomic areas. These regions were chosen because they varied in the age of onset of myelination according to previous autopsy studies. The subjects were assigned a score of +2 to −2 for each of the 14 regions; +2 was given if the relative intensity was completely adultlike[1] and −2 if the contrast appeared very immature. A summary of the chronological changes in myelination observed in this study is given in Table 1. In general, the table shows that the timing and regional pattern of myelination in the brain is similar in MR and autopsy studies. One can also conclude that myelination appears consistently earlier on MR with T1 weighted images than with T2 weighted images. For example, the white matter in the central frontal region shows signs of myelination by age 3–6 months on T1W images, whereas on T2W images, evidence of myelination does not occur until 11 to 16 months. In contrast, myelination in the middle cerebellar peduncle has a similar time course on T1W and T2W images. Barkovich and colleagues (1988) speculate that such discrepancies may relate to differing time courses for increasing lipid and protein content of white matter on the one hand, and decreasing water content of white matter on the other.

Martin *et al.* (1988) also conducted an MR study of normal brain maturation in 50 infants and young children. They developed a subjective staging system for assessing the extent of myelin deposition in four locations: in subcortical and periventricular white matter of the 1) frontal, 2) temporal, and 3) occipital lobes, and 4) in the internal capsule. Typical ages at which each stage of myelination occurred for each structure were described. Similar to Barkovich *et al.* (1988), Martin *et al.* (1988) found discrepancies between data collected with T1W and T2W images, with T1W images showing evidence of myelination earlier than T2W images. Like others, these authors noted that myelination appeared to proceed from inferior to superior brain regions and from posterior to anterior regions.

Hayakawa, Konishi, Kuriyama, Konishi, and Matsuda (1990) examined normal cerebral maturation on T1W and T2W images in 90 subjects from 6 months to 60 years of age. They conducted qualitative and quantitative analyses of the pattern of myelination observed in their subjects. The T2W protocol was used to quantify the myelination process in the centrum semiovale region of the brain. The authors selected one 10-mm section of brain from each subject at the level of the centrum semiovale and traced the outer boundaries of one of the cerebral hemispheres as well as the border between low- and high-intensity areas in that hemisphere (presumably the border between myelinated an unmyelinated tissue). With these measures, the authors computed a ratio of the hypointense area to the area of the whole hemisphere to create a percentage measure of myelination in the centrum semiovale. They found that the percent hypointensity (presumed myelination) in the brain ranges from roughly 23% at ages 6 months to 1 year to roughly 55% by age 60. The most dramatic changes occurred in the first three years of life; however, definite gradual increases continued into adulthood. Visually, this increase in myelination has the appearance of a thinning cortical rim. Yakovlev and Lecours (1967) described continuing myelination in supralimbic regions (which include the centrum semiovale) well into the third decade of life. These findings parallel those of Hayakawa *et al.* (1990), that is, an increasing percentage of hypointensity (presumably myelination) in their older subjects.

Hayakawa *et al.* (1990) also used qualitative methods to examine changes in

[1] Gray matter is less intense than white on T1 weighted images, and gray matter is more intense than white on T2 weighted images.

myelination with both T1W and T2W images in 36 subjects ranging in age from 3 weeks to 2 years. Based on their observations, the authors described the pattern of myelination observed at three stages of development. On T1W images progressing myelination was reflected in an increase in the regional extent of high signal (i.e., higher than the gray matter of cortex and subcortical nuclei). In stage 1 development (less than 4 weeks), large areas of the unmyelinated white matter actually had signal values lower than the gray matter. The high signal associated with myelination was observed only in some brain stem and cerebellar structures and in portions of the internal capsule and centrum semiovale (see Table 1). In stage 2 development (age 1–6 months) the less myelinated regions of white matter could no longer be distinguished from the gray matter, that is, they were isointense, and the high signal associated with myelination was now visible in corpus callosum, optic radiations, and some cortical gyri adjacent to the central sulcus. Finally, in stage 3 (7 months and older), the authors observed that myelination proceeded rostrally and peripherally into the cortical gyri. The authors noted that on T2W images unmyelinated regions within the white matter had a high signal intensity relative to gray matter until about 6 months of age. The low signal areas associated with myelination were visible at this stage in dorsal pons, midbrain, and cerebellar vermis and peduncles. The relative signal intensities of gray and white matter were similar throughout much of the cerebrum after 6 months of age and continuing until about 18 months. After this, the myelinating white matter had a low signal value relative to the gray matter.

It should be noted that in their qualitative analyses, Hayakawa *et al.* (1990) used a different T1W protocol for infants in stage 1 of maturation (SE 350/35) and for those in stages 2 and 3 (SE 450/35). It is unclear how this change in protocol would affect the qualitative observations made by the authors. Thus, the differences observed between stage 1 and stage 2 maturation on the T1W images may have been due, to some extent, to the shift in protocol.

McArdle *et al.* (1987) conducted an MR study of changes in gray matter–white matter differentiation and the onset of myelination in the perinatal period. They obtained T1W images of the brains of 51 premature and full-term infants ranging in age from 29 weeks to 42 weeks postconception. On the basis of a subjective rating of gray–white differentiation in these essentially unmyelinated brains, McArdle *et al.* (1987) divided the subjects into five groups ranging from high contrast between gray and white in stage 1 to little contrast in stage 5. In the youngest subjects, white matter signal was generally low relative to gray matter signal (contrast was high). In older infants white matter had generally increased in signal intensity relative to gray matter (contrast was low). In their description of characteristic images of the five stages, they also mention the infolding of the cortical surface, from lissencephaly in stage 1 to deep gyral infolding in stage 5. Because cortical surface development is such a salient visual characteristic, it is unclear how the appearance of the cortex may have influenced the subjective rating of the tissue contrast. However, the authors also used a less subjective measure of the changing signal values of white matter by taking sample signal values in the white matter as well as in the vitreous fluid of the globes (of the eyes) in selected subjects from each group. A ratio of the two measures was computed as an index of white matter intensity. They chose vitreous fluid in the globes as a control for overall signal fluctuation because the chemical composition of the vitreous fluid presumably does not change in the first four to five years of life, unlike the chemical composition of brain tissue. With this measure, McArdle *et al.* (1987) were able to establish that the differences between their staged groups, observed visually, were indeed associated with increasing signal values of the white matter.

The second parameter McArdle *et al.* (1987) used to stage their subjects was the rostro-caudal level at which myelination was observable. They classified the subjects into four groups based on the most superior level in the brain at which the high signal intensity of myelination was observed. In stage 1, only the dorsal aspect of the pons exhibited myelination, whereas in stage 5, myelin could be observed up to the centrum semiovale region of the brain. The authors graphically display the distribution of myelination stages by age and note that there is much variability. Several subjects between 29 and 34 weeks of age postconception show stage 3 myelination, and several infants in the older age groups show stage 2 myelination. However, overall, the authors observe that there is some correlation between myelination stage and postconception age because none of the 25 infants in their sample who were less than 36 postconception weeks exhibited stage 4 myelination, and none of the 15 term infants had stage 1 myelination. McArdle *et al.* (1987) are among the very few researchers who mention a longitudinal component in their study. In eight of their 51 neonates serial studies were performed at least two weeks apart. Although the results are anecdotal, they are consistent with the pattern emerging from the cross-sectional analyses.

Dietrich *et al.* (1988) used a T2W protocol to classify 34 normal infants, ranging in age from 4 days to 36 months, into three groups. Although these authors also examined a group of developmentally delayed infants, only the findings from the normal subjects will be discussed here. Two independent raters subjectively graded the signal intensity of both gray and white matter in several predefined brain regions using a 10-point scale (0 = air, lowest signal, to 10 = subcutaneous fat, highest signal). Using these subjective gradings as a basis, the authors defined three patterns of signal differentiation. The patterns were based on the relative ratings of gray and white matter for the centrum semiovale region. The first pattern (infantile pattern), which was observed in all infants less than 8 months of age, was characterized by white matter rated at least two grades higher in signal intensity than gray matter. The second pattern (transient pattern), exhibited in most children between 8 and 12 months of age, was characterized by isointense gray and white matter, that is, gray and white matter ratings that differed by less than two grades. Finally, an "early adult" pattern, with white matter ratings of two or more grades lower than gray matter, was seen in all infants older than 12 months of age.

Christophe *et al.* (1990) used phase-sensitive inversion recovery (PS IR) MR to assess brain maturation. The authors argue that this protocol is particularly sensitive to the progression of myelination; however, the evidence presented is entirely subjective. The IR sequences take considerably longer to perform, which can result in motion artifact with young subjects. This may be the reason that this protocol is seldom used in MR studies of children.

Finally, Girard *et al.* (1991) also conducted a study of the progression of myelination. These authors described qualitative changes and how they differed on T1W versus T2W images. Fifty-one young infants, ages 40 to 62 weeks postmenstrual, were studied, and the results paralleled those described previously. These authors provided a particularly thoughtful discussion of possible biological bases for the observed changes.

## Estimation of T1 and T2 during Maturation

One of the earliest reports in the literature examining normal brain maturation was conducted by Holland, Haas, Norman, Brant-Zawadzki, and Newton (1986). In their study, the authors examined 59 subjects ranging in age from term to 16 years using an imaging protocol that permits estimation of T1 and T2 (two TR sequences each with two echoes: SE 2000/28 and 2000/56 and SE 500/28 and 500/56). Signal values were sampled in several gray and white matter regions of interest in corresponding locations on each of the four images for each subject. The authors calculated estimates of T1 and T2 relaxation times for gray and white matter averaged across the several regions of interest for each age group. Holland *et al.* (1986) also examined the T2W images qualitatively for changes in appearance of the developing brain.

Quantitatively, they found that T1 and T2 relaxation times for gray and white matter were much longer in neonates than in the older subjects. Newborns had an average white matter T1 relaxation time of 1615 msec with a relatively rapid decrease in relaxation times to 580 msec by 1 year of age. Between 1.5 and 14 years of age, the T1 relaxation time decreased from an average of 570 msec to 487 msec, a relatively slow decline when compared with the changes seen in the first year of life. The gray matter T1 relaxation times followed a similar course, rapidly decreasing from 1590 msec at birth to 890 msec at 1 year of age. At age 14, relaxation times for gray matter averaged 805 msec.

Holland *et al.* (1986) found that T2 relaxation values for gray and white matter were within 3 msec of each other from birth to 6 months of age, with a relatively fast decline in that age range from about 88 msec to about 67 msec for gray matter and from about 91 msec to about 64 msec for white matter. By 1 year of age, there was considerable difference between white and gray matter T2 relaxation times: white matter averaged 57 msec and gray matter 68 msec. At age 14, the average T2 white matter relaxation time was 49 msec, and the average gray matter T2 relaxation times was 59 msec. Estimates of T2 from multiple-echo protocols are not very robust against variations in the TE times samples. However, the relative changes in relaxation rates between early infancy and the late childhood are of interest and help to explain the appearance of the tissues on different imaging protocols.

In their qualitative analysis of T2W images, Holland *et al.* (1986) note the emergence of visible myelination in their subjects. Again, their findings are consistent with both neuropathological data as well as other neuroimaging studies conducted subsequently. Maturation in the appearance of gray and white matter signal intensities was observed to occur first in inferior brain regions and to progress to superior and anterior regions with increasing age. Holland *et al.* (1986) noted the reversal in relative signal value of gray and white matter in the first 6 months of life, consistent with several studies reviewed above.

Hassink, Hiltbrunner, Muller, and Lutschg (1992) conducted a similar study with a more restricted age range. They examined maturational changes in estimated T2 relaxation times (SE 2500/90) in a group of 17 children and adults; five were 8 years old, four were 10 years old, and eight were either 24 or 25 years of age. The T2 relaxation times were calculated for samples from dorsal frontal white matter, lateral frontal white matter, orbital frontal white matter, caudate nucleus, internal capsule, corpus callosum, and frontal cortex. In their sample, only the splenium of the corpus callosum did not exhibit a significant decrease in T2 between the ages of 8 and 25. Significant decreases were reported for all other regional measures. A significant decrease in T2 between the ages of 8 and 10 was observed only in the internal capsule. In the study by Holland *et al.* (1986), all white matter regions sampled were said to have reached "adult signal values" by the age of 3 years. It is understandable why in this study the later changes are minimized in comparison to the dramatic changes that occur in the first year of life. However, as the study by Hassink *et al.* (1992) demonstrates, small but definite reductions in T2 relaxation times continue beyond this age into late childhood and early adulthood. Relatively little attention has been focused on these later changes in T2.

### Summary of MR Studies of Tissue Characteristics

When the different MR protocols and diverse quantitative methods are taken into account, the findings in the studies reviewed are highly consistent. For example, it can be confidently concluded from the studies reviewed that myelination does progress from more inferior to more superior brain regions and that the changes before 18 months postnatal are best observed on T1W images, while subsequent changes are best observed on T2W images. In general, MR studies have focused on the perinatal and infancy periods when dramatic changes in myelination are taking place. Fewer MR studies have been directed toward detecting the more subtle changes that occur in childhood, adolescence, and early adulthood, even though postmortem studies (e.g., Yakovlev & Lecours, 1967) suggest that myelination continues into adulthood in many brain regions. Thus, the literature is more informative about the normal limits for myelination in infants than in older children and adolescents.

## Morphological Changes on Magnetic Resonance Imaging

### Evidence from Postmortem Studies

The perinatal development of the cerebral cortex has been examined in detail in human autopsy studies. One of the earlier studies, by Hesdorffer and Scammon (1935), examined cortical surface area in the brains of 20 subjects ranging in age from 4 months postconception to 2 years. Their methods consisted of sectioning formalin-fixed brains into 2- to 3.5-mm sections, measuring the outline of each section with a "chartometer" and multiplying the reading by the section thickness. Results of this study indicated that there is a rapid growth spurt in the total volume and surface area of the cerebrum between the sixth postconception month and 2 years of age.

Starting in 1939, and for several decades to follow, J. LeRoy Conel published eight volumes on the postnatal development of the human cerebral cortex (Conel, 1939–1967). In these volumes, he provided qualitative observations as well as meticulous quantitative data regarding the gross morphological features of the human brain and the microscopic structure throughout the entire cortex at five different ages between birth and 6 years. While the morphological changes are of greatest interest here, it is clear in reviewing Conel's work that regional changes in cell size and in number of axonal and dendritic processes are largely responsible for the observed changes in cortical thickness and in gyral formation. Conel reported a large number of regional cortical thickness measures across the different ages. We have summarized these by computing averages across the 37 regions examined for each age between birth and 6 years of age. These averages were 1.70 mm at birth, 1.79 mm at 1 month, 1.97 mm at 3 months, 2.16 mm at 6 months, 2.49 mm at 15 months, 2.27 mm at 24 months, 2.64 mm at 4 years, and 2.49 mm at 6 years (Conel, 1939–1967).

In a more recent study, Dobbing and Sands (1973) examined 139 autopsy specimens ranging in age of death from 10 weeks gestation to 7 years. They measured overall brain weight, brain weight relative to body weight, total cell count, and cellularity (cell density per gram of tissue), among other measures. To estimate cell count they used an indirect measure, DNA content, the value of which is known to correlate with the number of diploid cell nuclei in the area measured. Dobbing and Sands (1973) found that the growth spurt in the human brain, previously only well described in animal brains, begins after the adult neuronal number has already been nearly attained. They thus conclude that the change in weight and size of the human brain between midgestation and 2 years of age is due to a great increase in glial cell populations.

### Evidence from *in Vivo* Studies

Several neuroimaging studies have examined gross morphological changes during neurodevelopment, both qualitatively and quantitatively. Results of studies using *in vivo* methods have generally been consistent with Conel (1939–1967) and Dobbing and Sands (1973).

Martin *et al.* (1988) proposed a system of staging cortical maturation *in vivo* with MRI. The authors suggested that their standards could be used by clinicians as a "brain age" chart, much like "bone age" charts are used by pediatric endocrinologists. Their rating system defines four levels of brain surface gyration from the most lissencephalic to the fully gyrated adult brain. Stage 1 is characterized by a smooth brain surface. In stage 2 the primary gyri are identifiable and are separated by shallow fluid-filled sulci. By stage 3 the sulci are more deeply infolded but there is no visible cerebrospinal fluid (CSF) between the tightly compressed gyri. Finally the adult pattern, stage 4, is characterized by deep fluid-filled sulci with primary and secondary gyri visible.

The authors of this study noted considerable variation between subjects of similar age in the stage of cortical gyration; however, some age consistency was observed. For example, Martin *et al.* (1988) found that 25% of the infants in their sample had reached stage 4 (adult pattern) brain surface architecture by the postconception age (PCA) of 87 weeks, whereas by the PCA of 143 weeks, 75% had reached stage 4 development. Generally, these researchers maintain that neonates born at term displayed stage 2 pattern of gyration, while the adult pattern (stage 4) was usually not observed until 1 year of age. The authors reported that the pattern of gyration occurs in an occipital-rostral direction, that is, beginning with the occipital lobes and progressing upward and forward.

Barkovich and Truwit (1990) discuss qualitative developmental changes observed on MRI in their *Atlas of Neonatal Brain Development.* They provide sections illustrating developmental changes in myelination, corpus callosum size and morphology, bone marrow signal characteristics, and normal iron deposition. In the section on normal brain myelination, Barkovich and Truwit (1990) compiled T1W and/or T2W images of 12 infants between 31 weeks gestation and 24 months of age. For each image, the authors describe the observable myelination in structures visible in the plane of section, and for some images, they discuss the maturity of gyral formation. In the section on the maturation of the corpus callosum, the midline sagittal images of eight infants between 31 weeks of gestation and 10 months of age are provided as well as one image of a 15-year-old adolescent. This atlas is helpful in illustrating the gross morphological differences in the brains of infants of different ages as well as in showing the differences in appearance between T1W and T2W images.

Barkovich and Kjos (1988) conducted a quantitative analysis of changes in the corpus callosum in the first year of life. In this study, 63 patients between 3 days and 12 months of age were examined clinically for various reasons. Nearly half of the subjects were suspected or confirmed seizure disorder patients, and it is not clear what effect such a disorder would have on the developing brain. The authors obtained measurements of the length of the callosum and the length of the greatest anterior-posterior dimension of the brain to compute a ratio of the two measures, presumably to correct for differences in overall brain size. Finally, they collected qualitative information regarding changes in appearance of the callosum secondary to myelination. The first postnatal change documented in this study is the gradual thickening of the genu which occurs in the second and third months of life. The next change occurs in the splenium which "slowly enlarges until the fifth and sixth months, when a rapid increase in size occurs, resulting in the bulbous splenium seen in the adult brain" (Barkovich & Kjos, 1988, p. 489). The body of the corpus callosum grows slowly without any detectable growth spurts over the first 12 months. The authors comment on how their findings differ from the findings in a postmortem study by Rakic and Yakovlev (1968) in which the same measurements were obtained. Barkovich and Kjos (1988) obtained consistently larger measurements of the corpus callosum than Rakic and Yakovlev reported. However, it seems likely, as Barkovich and Kjos suggest, that the process of formalin fixation in the postmortem study resulted in smaller raw measurements.

### Morphological Changes in Older Children

Jernigan and Tallal (1990) reported that children ages 8 to 10 years had significantly more cortical gray matter relative to cerebral size than did young adults. However, the timing of the changes observed could not be assessed because no data were available for subjects in the adolescent age range. Jernigan, Trauner, Hesselink, and Tallal (1991) added to the original sample subjects between 11 and 24 years of age and examined age-related changes in regional cortical and subcortical gray matter and CSF. Magnetic resonance images from 39 normal children and young adults ranging in age from 8 years to 35 years were included in this study.

Detailed descriptions of the image-analytic approach used by Jernigan, Trauner, Hesselink, and Tallal (1991) can be found in several articles (Jernigan, Press, & Hesselink, 1990; Jernigan, Archibald, Berhow, Sowell, Foster, & Hesselink, 1991; Jernigan, Salmon, Butters, & Hesselink, 1991). Image data sets are assigned (random) numeric codes, and all analyses are conducted blind to any subject characteristics. Each pixel location within a section of the imaged brain is classified on the basis of its signal values in both original images (TR = 2000 msec; TE = 25 msec, TE = 70 msec) as most resembling CSF, gray matter, white matter, or signal hyperintensity (tissue abnormality). This is accomplished in two steps: first, two new linear combinations of the pixel values are computed to optimize tissue contrast (CSF/brain and gray/white). Then, classification criteria that are adjusted separately for each section are applied to these computed values. The full series of axial images is analyzed, beginning at the bottom of the cerebellar hemispheres and extending through the vertex.

To delineate specific gray matter and CSF structures, the operators circumscribe pixels classified as gray matter or CSF that are visually determined to be in the ventricles, cortical sulci, caudate nuclei, lenticular nuclei, lenticular nuclei, diencephalic gray matter (a region which includes but does not separately delineate mammillary bodies, hypothalamic gray, septal nuclei, and thalamus), and cerebral cortex (including the contiguous amygdala). The operators do not trace the edges of the structures, but, for each region, they define polygons that include all gray matter or CSF pixels within the structure and exclude those associated with other structures.

Visual determination of specific cortical structures on MRI depends upon the presence of visible gross morphological features relative to which the boundaries of the cortical regions can be defined. Standard regional divisions for the cortex are based to a large extent on cortical gyral patterns, but the accurate localization of particular gyri or sulci, throughout a series of images, is often impossible with lower-resolution imaging protocols like the one used by Jernigan, Trauner, Hesselink, and Tallal (1991). Also, even when attempts are made to standardize head positioning, rotation of the head (relative to the imaging plane) occurs in all three planes. Careful inspection indicates that relatively small rotations substantially change the appearance of brain structures in the image plane, further complicating their visual identification. Thus, manually tracing the structures in the sections where they are best visualized often leads to inaccurate boundary determinations and volume assessments. To address these problems, Jernigan, Trauner, Hesselink, and Tallal (1991) applied the stereotactic techniques described below.

To define anatomically consistent cortical regions, a method was adopted for making subdivisions of the cerebrum relative to the centromedial structural midline and two consistently identifiable (operator-designated) points: the most anterior midline point in the genu and the most posterior midline point in the splenium of the corpus callosum. By calculating estimated rotation angles using these landmarks, it was possible to perform a three-dimensional rotation of the images, thus correcting each individual's image data for rotation out of the optimal imaging plane. Regions could then be constructed which resulted in highly consistent placement of regional boundaries relative to gross anatomical landmarks.

Jernigan, Trauner, Hesselink and Tallal (1991) used these methods to divide the cerebrum into four zones. Based on an axial dividing plane (passing through the two corpus callosum landmarks and perpendicular to the midsagittal plane) the cerebral cranial vault was divided into a superior and an inferior zone. A second (coronal) dividing plane (perpendicular to the first and passing through the midpoint between the callosal points) divided each of these zones into an anterior and a posterior zone (see Figure 1). When the authors examined the relationship between age and the overall volume of these four regions of the

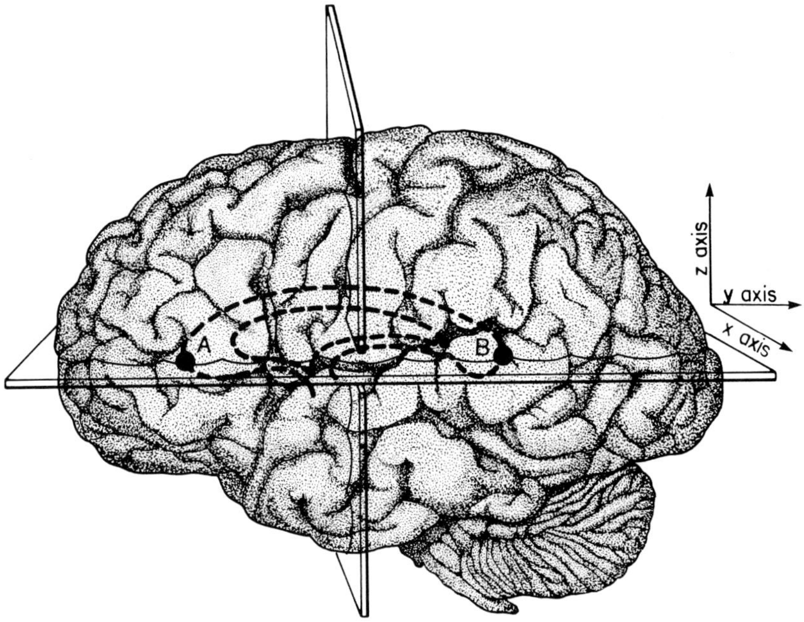


Figure 1. Cerebral regions are defined as follows: A and B in the corpus callosum are the most anterior midline point in the genu and the most posterior midline point in the splenium, respectively. An axial plane passing through these two points is defined, as shown, perpendicular to the midsagittal plane. A coronal plane is defined perpendicular to the axial plane and passing through the midpoint between points A and B. This coronal plane divides the cerebrum into an anterior and a posterior zone.

cranial vault, they found evidence for an increase in size of the superior cranial vault, particularly in the anterior region. Within the superior cranial vault, the cortical gray matter appeared to be decreasing with age while CSF in this region increased. The inferior cortical gray matter volumes (including orbitofrontal, temporal, and occipital regions) did not appear to change across the age range.

It is well known that during neurogenesis more neurons are produced than ultimately survive (Purves & Lichtman, 1980). Such regressive events are also evidenced in the reduction in synaptic density that occurs during childhood in the human cortex (Huttenlocher, 1979; Huttenlocher & de Courten, 1987). Rakic and colleagues have reported reductions in synaptic density in the primate brain during a period coinciding with the onset of puberty in this species (Zecevic & Rakic, 1991). Jernigan, Trauner, Hesselink, and Tallal (1991) proposed that their observation of a "thinning" cortex could be related to the processes by which the earlier reported reductions in synaptic density occur.

Jernigan, Trauner, Hesselink, and Tallal (1991) also examined subcortical gray matter and ventricular CSF volumes. In simultaneous regression analyses predicting regional cortical volumes from measures of overall cerebral volume, gender, and age, they found that the caudate nucleus, the lenticular nucleus (including putamen, claustrum, and some globus pallidus), and the thalamus all significantly decreased in volume with increasing age. As would be expected, a ventricular volume increase paralleled the caudate and lenticular volume decreases. Interestingly, the only remaining subcortical gray matter region actually appeared to be increasing in volume with age. This region was the anterior diencephalon which included hypothalamus and septal gray matter structures. It was hypothesized that this change could be related to peripubertal changes in gonadal steroid hormones. Animal studies have shown that hypothalamic and septal structures have high concentrations of receptors for these hormones (Stumph & Sar, 1978; Rainbow, Parsons, MacLusky, & McEwen, 1982).

The anomalous finding by Jernigan, Trauner, Hesselink, and Tallal (1991) of age-related increases in the volume of certain subcortical limbic structures raised questions about effects that might be occurring in the related limbic structures on the mesial surface of the temporal lobe. These structures had been combined with neocortical structures in the inferior cortical zones that exhibited no change in volume. We extended these observations by examining separately the mesial temporal lobes and the remaining cortical structures in the inferior zone (Sowell & Jernigan, submitted). The sample was also expanded to include 57 subjects. The mesial surface of the temporal lobe (MTL, including the uncus, hippocampus, parahippocampal gyrus, and amygdala) was isolated from the rest of the inferior cortical region using a combination of visually guided and stereotactic methods. Thus, three cortical regions were analyzed in this study: the MTL, ventral lateral cortex (remaining inferior zone), and dorsal cortex (superior zone). Four subcortical regions were also examined in this and the previous study: caudate nucleus, lenticular nucleus, thalamus, and anterior diencephalon (including hypothalamus and septal nuclei).

We found that the MTL region actually significantly increased in volume with age, whereas the more lateral ventral cortices significantly decreased in volume with age (Sowell and Jernigan, submitted). As in the earlier study (Jernigan, Trauner, Hesselink, and Tallal, 1991), the anterior diencephalic region significantly increased in volume with age, while the other subcortical structures (caudate, lenticular nucleus, and thalamus) decreased in volume. Also consistent with the earlier study, the dorsal cortex continued to show a significant decrease in volume with age.

We noted that the only two brain regions measured in these studies that appeared to increase with age were those brain regions that have been shown in animals to have high concentrations of gonadal steroid hormone receptors (Stumph & Sar, 1978; Rainbow *et al.*, 1982). Unfortunately age-related changes in MTL and anterior diencephalic structures have not been examined in human postmortem studies. However, animal studies in the literature discuss regionally specific trophic effects induced by the introduction of gonadal steroid hormones (for a review, see Jones, 1988). Perhaps the increase in circulating hormones that occurs at puberty (see Rosenfield, 1991; Styne, 1991) is associated with organizational effects on brain morphology.

The findings from this study provide further evidence that dynamic morphological changes are occurring in diverse regions of the brain between the ages of 8 years and 35 years. These findings should be substantiated with further *in vivo* studies of morphological changes occurring in this age range, preferably with larger sample sizes, longitudinal methods, and concurrent somatic, hormonal, and behavioral measures. This and other studies reviewed here demonstrate that *in vivo* methods now provide an important alternative to postmortem examination for elucidating gross morphological changes during brain development.

To date, very few functional neuroimaging studies involving positron emission tomography or single photon emission tomography have been conducted on populations of normal children because of the risks involved in exposure to radiation. The few studies of age-related changes on functional imaging in children have examined symptomatic children referred for functional imaging who were later determined to have little or no neurologic component to their disorders. However, as Chugani points out, "ontogenetic changes of various biochemical and physiologic processes in the normally developing brain must first be established before widespread application of functional neuroimaging to improve management and treatment of childhood diseases" (Chugani, 1992, p. 778). Therefore, the incentive is strong to attempt to estimate normal values using clinical material in spite of the limitations inherent in this method.

Chugani and Phelps (1986) were the first to report normal maturational changes in local cerebral metabolic rates for glucose (lCMRGlc) in human infants using 2-deoxy-2 [$^{18}$F]fluoro-D-glucose (FDG) and PET. From the 60 or more infants studied at their institution, nine were determined to be nearly normal at the time of imaging. These children were between 5 days and 17 months of age and most of them were referred for seizure disorders or suspected Sturge-Weber syndrome. Six of the nine infants were medicated with phenobarbital on the day of the PET study, but none showed signs of oversedation or any other side effects.

Briefly, Chugani and Phelps (1986) found that in the four infants 5 weeks of age or younger lCMRGlc was "highest in the sensorimotor cortex, thalamus, midbrain–brain stem, and cerebellum" (Chugani & Phelps, 1986, p. 840). By 3 months of age, the lCMRGlc increased in much of the cerebral cortex, and levels in the striatum approached that of the thalamus. However, only after 7.5 months of age did the pattern of glucose utilization resemble that of an adult in frontal and association cortices. The authors note the heterogeneity with which different brain regions attain adult patterns of glucose utilization.

In a second study, Chugani, Phelps, and Mazziotta (1987) examined the functional development of the human brain in subjects ranging from 5 days to 15.1 years of age. They used the same method and rationale described above to select the 29 subjects. Eighteen of the subjects were taking anticonvulsant medications, and these subjects were fairly well distributed across the age range. Overall, lCMRGlc in the cortex appeared to rise to values well above those observed in adults and then to decline gradually.

The findings for the nine subjects 1.5 years and younger were consistent with the previous findings by this group. The regional cortical distribution of lCMRGlc by 1 year of age was similar to the anatomical distribution observed in adults, however, the absolute values of lCMRGlc were not yet at adult rates. The sensorimotor cortex had reached 86% of the adult value by 1 year of age, and this region was the earliest to exhibit this maturational rise. Between 3 and 8 years of age, a dramatic increase in glucose utilization occurred in most cortical regions, to roughly 200% of the corresponding adult values. In the ensuing years up to age 15 there was a consistent decline in lCMRGlc in all cortical regions; however, the decreasing values were not yet comparable to adult values at this age.

Of the subcortical structures, the thalamus displayed maturational increases in lCMRGlc earlier than the caudate or lenticular nucleus, and all three structures seemed to show the maturational rise more quickly than the cortical regions. The increase in glucose utilization over adult values that occurred between 3 and 8 years of age was less extreme for the subcortical structures, peaking at about 160% of adult values.

The age changes in lCMRGlc of the brain stem and cerebellum displayed a somewhat different maturational pattern from the cortical and subcortical gray matter regions discussed earlier. In these structures, lCMRGlc reached adult values earlier, by approximately 1 year of age. The peak values for these regions were approximately 175% of adult values. The cerebellum lCMRGlc seemed to mature in a phylogenetic manner, with the "older" vermis exhibiting the maturational rise earlier than the phylogenetically younger cerebellar hemispheres.

Chugani and colleagues (1986, 1987) note that the regional pattern of functional brain maturation seems consistent with the development of different behaviors in infants and young children. The brain stem, cerebellum, sensorimotor cortex, and thalamus mature early, whereas association cortex, thought to be involved in higher-order processing, matures later. Chugani *et al.* (1987) discuss the possible effects of anticonvulsant medication on their findings. The authors point to studies describing the effects of anticonvulsant medications on cerebral glucose metabolism. One of the studies documents a 37% increase in mean metabolic rate upon withdrawal from phenobarbital (Theodore *et al.,* 1986). Thus, in their study of infants, it seems likely that some of the regional increase in lCMRGlc between 5 days and 17 months of age could be due to the fact that only one of the four infants over age 13.5 weeks was on phenobarbital, whereas all five infants between 5 days and 11 weeks of age were on the medication. Nonetheless, most of the change in lCMRGlc seems to be attributable to maturation because the increase in glucose utilization is much greater than would be expected from the medication alone.

Chugani and his associates note that the early proliferation of dendritic processes and synapses in the postnatal period, and the later gradual decreases in synaptic density, parallel the time course observed in their studies of cortical metabolic rate. They hypothesize that changes in cortical metabolic rate may be largely in response to the metabolic demands of cortical synapses.

Rubinstein *et al.* (1989) studied regional cerebral blood flow (rCBF) in 30 infants using SPECT with [$^{133}$I] Iodoamphetamine (IMP). Twenty-three of the subjects were newborns (36 to 44 weeks gestation). The remaining 7 subjects were between 2 months and 2 years of age and all 30 were selected retrospectively. In this study, four observers, blind to subject age, interpreted the SPECT results, and different patterns of rCBF were recorded. Rubinstein and his colleagues noted that all of the infants could be categorized into six groups based on the patterns observed. The first pattern was observed in newborn infants between 36 and 38 weeks gestation and was characterized by prominent thalamic rCBF with low activity in parietal, occipital, and frontal cortical areas. Pattern 2, observed in newborn infants between 39 and 42 weeks gestation, was characterized by increased activity in the parietal cortex. By 44 weeks gestation, newborn infants still had prominent thalamic and parietal rCBF with increasing activity in the occipital regions (pattern 3). Infants with pattern 4, aged 2 to 3 months, exhibited predominant activity in parietal and occipital cortex. Between 6 and 10 months of age, pattern 5 rCBF was observed and was characterized by high parietal and occipital activity and the first appearance of activity in the frontal cortex. Finally, by about 1 year of age, infants displayed pattern 6 in which high activity was observed in all cortical regions. Like Chugani and Phelps (1986), Rubinstein *et al.* (1989) believe that these findings are consistent with the behavioral evolution which occurs during infancy.

In a similar study measuring regional cerebral blood flow using SPECT, Chiron *et al.* (1992) studied brain maturation in 42 subjects ages 2 days to 19 years. The subjects, like the subjects in the Chugani and Rubinstein studies, were chosen *a posteriori.* Chiron *et al.* reported both absolute and relative rCBF[2] values in an attempt to describe global rCBF changes as well as changes in the regional pattern of rCBF.

Chiron *et al.* (1992) obtained estimates of global CBF at birth of 50 ml/min/100 g, a value very similar to adult values. Much like with lCMRGlc (Chugani, Phelps, & Mazziotta 1987), there was an increase in CBF to considerably above the adult level, up to 71 ml/min/100 g, by about 5 years of age, after which CBF declined gradually to reach adult levels at approximately 19 years of age. Absolute CBF values follow a similar pattern for all cortical regions, being slightly below the adult value at birth, increasing to maximum by 5 or 6 years (70% higher than adult levels), and decreasing slowly to adult levels after age 15.

Relative CBF values revealed that within the cortex, primary sensory and motor areas appeared to mature earlier than association areas. Furthermore, while absolute cortical CBF values rise to above adult values and then decline, relative cortical CBF rises to adult values during the first two years of life and remains stable. The authors suggest that if this is true, and if the fall in cortical CBF is due to synaptic depletion, as Chugani *et al.* (1987) suggest, then such synaptic depletion must be occurring in subcortical structures as well as the cortex. This would also be consistent with the brain morphological findings of Jernigan, Trauner, Hesselink, and Tallal (1991) and Sowell and Jernigan (submitted) of similar age-related volume losses in cortical and subcortical structures.

## Logistical and Methodological Issues

Several practical problems arise in the MRI study of young subjects. Perhaps the most significant one is motion during the imaging session and the resulting degradation of image quality. For clinical imaging, children are often safety sedated with, for example, oral chloral hydrate (Barkovich, 1990); however, many investigators and parents are reluctant to sedate normal children for research purposes. Experience in our laboratory suggests that children younger than 8 years of age are frequently unable to remain motionless for the periods of time needed to acquire high-resolution images (around 20 minutes). However, technical advances in MR are dramatically decreasing the time required for some protocols, so this issue may become less important in the future. It is worth noting that some MR examinations of normal children become available because they are obtained, usually with sedation, for clinical purposes, that is, to rule out neurological conditions or to assess other organs such as the heart. If standardized imaging conditions can be implemented, and if careful post hoc screening of such cases is performed, such material can be used to great advantage to elucidate brain maturation in young children.

[2] Relative rCBF was expressed as the percentage (%) of global CBF for each age group.

Another practical problem involves the standardization of head position in the imager's field of view. Many investigators attempt to standardize the plane of section in their brain images by aligning the head in a standard position based on external cranial landmarks. In our experience, it is virtually impossible to accomplish this across subjects of different ages. Variation in the relationship of head to body size and in the flexibility of the neck result in systematic age differences in the range of tolerable head positions during imaging sessions. Children are generally most comfortable in positions in which the chin is tilted in toward the chest, while older subjects assume positions with the chin rotated upward and the head extended further back. Unless great care is taken, this results in considerable change in the plane of section. This is most problematic when structures are measured within specific sections from the original set of images. "Standard" sections identified based on midline structures will often contain quite different cortical structures in children and young adults. Thus spurious age-related differences in the sizes or asymmetries of structures may occur due to age-related variation in head rotation. The best solutions to this and other problems associated with head tilt are to construct structural boundaries stereotactically (based on standard anatomical landmarks) or to perform region definition on sections resliced relative to a standard anatomical coordinate system. Current methods provide very thin sections so that resliced images lose little or none of the high in-plane resolution of the original images.

The preceding review highlights the importance of the choice of imaging protocol in MRI of young subjects. The most obvious example is provided by the contrast between the appearance of early myelination on T1W and T2W images. However, subtler differences in the sensitivity of different protocols at different ages almost certainly exist. Thus careful standardization of the imaging protocol across the age range of the subjects examined is important in any developmental study. Furthermore, it is probably important to examine a variety of different protocols across the full age range to ensure that important information about the timing of different tissue alterations is not missed. The lack of standardization of imaging protocols across studies complicates the interpretation of apparently disparate findings. Unfortunately, the technology is evolving so rapidly that it is probably unrealistic to expect strict standardization in the near future. However, as experience in the field evolves, a few roughly comparable protocols are becoming more widely employed than others. For example most investigators now acquire a dual-echo series, providing both high T2 sensitivity and good tissue contrast, as well as a high-resolution volume acquisition with T1 weighting.

Methods for analyzing and quantifying imaging results are even less well standardized than imaging protocols. A variety of techniques, ranging from subjective ratings to semiautomated morphometrics have been applied to brain images. It is difficult to compare the results obtained with such disparate methods, especially when subjective or global assessments fail to reveal changes observed with quantitative or local measurements. Trends in the field suggest that the latter are considerably more sensitive and could be highly informative if they were applied more frequently in developmental imaging studies.

Finally, the virtual absence of longitudinal MRI studies of brain development is a significant methodological weakness in this research area. As a noninvasive method with virtually no known contraindictions, MRI provides one of the first opportunities to monitor brain development within individuals. Longitudinal studies would provide not only stronger evidence regarding the timing and sequencing of different changes in tissue and morphology, but also an opportunity for functional correlations. This is, it should be possible to establish the relationship between behavioral indices of rate of functional maturation and concomitant brain changes within systems putatively linked to the functions under examination.

## An Exciting Development for Future Neurodevelopmental Studies

Magnetic resonance research groups in a number of centers have recently demonstrated that MRI can be used to measure regional changes in blood flow. This is because the MR signal is affected by the local concentration of deoxyhemoglobin. Apparently when blood flow increases to an activated area in the brain, the flow increase is disproportionately greater than the increase in oxygen extraction by the tissue. Therefore, the blood becomes relatively oxygenated. Since deoxyhemoglobin reduces the MR signal, and since the concentration of deoxyhemoglobin is reduced in the oxygenated blood, a local MR signal increase results. The measurement of blood flow with MRI is referred to as functional MRI (fMRI). Most of the early experiments demonstrating flow changes in response to brain activation were conducted in centers with specialized high field strength magnets and ultrafast imaging capabilities. However, the technique is becoming more generally available as methods are developed for use on conventional imagers and as instruments and their software are upgraded (for a review, see Cohen & Bookheimer, 1994).

This technique represents a very exciting opportunity for neurodevelopmental researchers, because the method is safe for children and can be repeated throughout development. It is likely that functional specialization of brain subsystems evolves during development. The results of fMRI mapping studies may differ between children and adults. If so, it may be possible to use fMRI to track alterations in patterns of activation as cerebral systems mature functionally. It is hoped that in the near future, as behavioral activation paradigms for fMRI are validated in adults, investigators will begin to apply these paradigms in children.

*Acknowledgments*

Investigations presented in this paper were supported by funds from the Medical Research Service of the Department of Veterans Affairs to Dr. Terry Jernigan, and grants P01 DC01289 and P50 NS22343 from the National Institutes of Health for the Center for Developmental Cognitive Neuroscience.

## References

Barkovich, A. J. (1990). Techniques and methods in pediatric imaging. In *Contemporary neuroimaging: Vol. I. Pediatric neuroimaging* (pp. 1–4). New York: Raven Press.

Barkovich, A. J., & Kjos, B. O. (1988). Normal postnatal development of the corpus callosum as demonstrated by MR imaging. *American Journal of Neuroradiology, 9,* 487–491

Barkovich, A. J., Kjos, B. O., Jackson, D. E., & Norman, D. (1988). Normal maturation of the neonatal and infant brain: MR imaging at 1.5 T. *Neuroradiology, 166,* 173–180.

Barkovich, A. J., & Truwit, C. L. (1990). *Practical MRI atlas of neonatal brain development.* New York: Raven Press.

Bradley, W. G. (1989, January/February). Magnetization: Physical basis for T1 and T2, part 2. *MRI Decisions,* 17–25.

Brody, B. A., Kinney, H. C., Kloman, A. S., & Gilles, F. H. (1987). Sequence of central nervous system myelination in human infancy. I. An autopsy study. *Journal of Neuropathology and Experimental Neurology, 46*(3), 283–301.

Chiron, C., Raynaud, C., Maziere, B., Zilbovicius, M., La Flamme, L., Masure, M., Dulac, O., Bourguignon, M., & Syrota, A. (1992). Changes in regional cerebral blood flow during brain maturation in children and adolescents. *Journal of Nuclear Medicine, 33*(5), 696–703.

Christophe, C., Muller, M. F., Baleriaux, D., Kahn, A., Pardou, A., Perlmutter, N., Szliwowski, H., & Segebarth, C. (1990). Mapping of normal brain maturation in infants on phase-sensitive inversion-recovery MR images. *Neuroradiology, 32,* 173–178.

Chugani, H. T. (1992). Functional brain imaging in pediatrics. *Pediatric Clinics of North America, 39*(4), 777–799.

Chugani, H. T., & Phelps, M. E. (1986). Maturational changes in cerebral function in infants determined by $^{18}$FDG positron emission tomography. *Science, 231,* 840–843.

Chugani, H. T., Phelps, M.E., & Mazziotta, J. C. (1987). Positron emission tomography study of human brain functional development. *Annals of Neurology, 22*(4), 487–497.

Cohen, M. S., & Bookheimer, S. Y. (1994). Localization of brain function using magnetic resonance imaging. *Trends in Neuroscience, 17*(7), 268–277.

Conel, J. L. (1939–1967). *The Postnatal development of the human cerebral cortex* (Vols. 1–8). Cambridge, MA: Harvard University Press.

Dietrich, R. B., Bradley, W. G., Zaragoza, E. J., Otto, R. J., Taira, R. K., Wilson, G. H., & Kangarloo, H. (1988). MR evaluation of early myelination patterns in normal and developmentally delayed infants. *American Journal of Roentgenology, 150,* 889–896.

Dobbing, J., & Sands, J. (1973). Quantitative growth and development of human brain. *Archives of Disease in Childhood, 48,* 757-767.

Friede, R. L. (1989). Gross and microscopic development of the central nervous system. In R. L. Friede, (Ed.), *Developmental Neuropathology* (2nd ed., pp. 2–20). Berlin: Springer.

Girard, N., Raybaud, C., & du Lac, P. (1991). MRI study of brain myelination. *Journal of Neuroradiology, 18,* 291–307.

Hassink, R. I., Hiltbrunner, B., Muller, S., & Lutschg, J. (1992). Assessment of brain maturation by T2-weighted MRI. *Neuropediatrics, 23,* 72–74.

Hayakawa, K., Konishi, Y., Kuriyama, M., Konishi, K., & Matsuda, T. (1990). Normal brain maturation in MRI. *European Journal of Radiology, 12,* 208–215.

Hesdorffer, M. B., & Scammon, R. E. (1935). Growth of human nervous system. I. Growth of cerebral surface. *Proceedings of the Society for Experimental Biology and Medicine, 33,* 415–418.

Holland, B. A., Haas, D. K., Norman, D., Brant-Zawadzki, M., & Newton, T. H. (1986). MRI of normal brain maturation. *American Journal of Neuroradiology, 7,* 201–208.

Holmes, G. L. (1986). Morphological and physiological maturation of the brain in the neonate and young child. *Journal of Clinical Neurophysiology, 3*(3) 209–238.

Huttenlocher, P. R. (1979). Synaptic density in human frontal cortex: Developmental changes and effects of aging. *Brain Research, 163,* 195–205.

Huttenlocher, P. R., & de Courten, C. (1987). The development of synapses in striate cortex of man. *Human Neurobiology, 6,* 1–9.

Jernigan, T. L., Archibald, S. L., Berhow, M. T., Sowell, E. R., Foster, D. S., & Hesselink, J. R. (1991). Cerebral structure on MRI, I: Localization of age-related changes. *Biological Psychiatry, 29,* 55–67.

Jernigan, T. L., Press, G. A., & Hesselink, J. R. (1990). Methods for measuring brain morphologic features on magnetic resonance images: Validation and normal aging. *Archives of Neurology, 47,* 27–32.

Jernigan, T. L., Salmon, D. P., Butters, N., & Hesselink, J. R. (1991). Cerebral structure on MRI, II: Specific changes in Alzheimer's and Huntington's diseases. *Biological Psychiatry, 29,* 68–81.

Jernigan, T. L., & Tallal, P. (1990). Late childhood changes in brain morphology observable with MRI. *Developmental Medicine and Child Neurology, 32,* 379–385.

Jernigan, T. L., Trauner, D. A., Hesselink, J. R., & Tallal, P. A. (1991). Maturation of the human cerebrum observed *in vivo* during adolescence. *Brain, 114,* 2037–2049.

Jones, K. J. (1988). Steroid hormones and neurotrophism: Relationship to nerve injury. *Metabolic Brain Disease, 3*(1), 1–18.
Martin, E., Kikinis, R., Zuerrer, M., Boesch, C., Briner, J., Kewitz, G., & Kaelin, P. (1988). Developmental stages of human brain: An MR study. *Journal of Computer Assisted Tomography, 12*(6), 917–922.
McArdle, C. B., Richardson, C. J., Nicholas, D. A., Mirfakhraee, M., Hayden, C. K., & Amparo, E. G. (1987). Developmental features of the neonatal brain: MR imaging, Part I. Gray-white matter differentiation and myelination. *Radiology, 162,* 223–229.
Purves, D., & Lichtman, J. W. (1980). Elimination of synapses in the developing nervous system. *Science, 210,* 153–157.
Rainbow, T. C., Parsons, B., MacLusky, N. J., & McEwen, B. S. (1982). Estradiol receptor levels in rat hypothalamic and limbic nuclei. *Journal of Neuroscience, 2*(10), 1439–1445.
Rakic, P ., & Yakovlev, P. I. (1968). Development of the corpus callosum and cavum septi in man. *Journal of Comparative Neurology, 132,* 45–72.
Rosenfield, R. L. (1991). Puberty and its disorders in girls. *Endocrinology and Metabolism Clinics of North America, 20*(1), 15–42.
Rubinstein, M., Denays, R., Ham, H. R., Piepsz, A., VanPachterbeke, T., Haumont, D., & Noel, P. (1989). Functional imaging of brain maturation in humans using iodine 123 iodoamphetamine and SPECT. *Journal of Nuclear Medicine, 30*(12), 1982–1985.
Sowell, E. R., & Jernigan, T. L. (submitted). Further MRI evidence of late brain maturation: Limbic volume increases and changing asymmetries during childhood and adolescence.
Stumph, W. E., & Sar, M. (1978). Anatomical distribution of estrogen, androgen, progestin, corticosteroid and thyroid hormone target sites in the brain of mammals: Phylogeny and ontogeny. *American Zoology, 18,* 435–445.
Styne, D. M. (1991). Puberty and its disorders in boys. *Endocrinology and Metabolism Clinics of North America, 20*(1), 43–69.
Theodore, W. H., DiChiro, G., Margolin, R., Fishbein, D., Porter, R. J., & Brooks, R. A. (1986). Barbituates reduce human cerebral glucose metabolism. *Neurology, 36,* 60–63.
Valk, J., & van der Knaap, M. S. (1989). *Magnetic resonance of myelin, myelination, and myelin disorders.* Berlin: Springer.
Yakovlev, P. I., & Lecours, A. R. (1967). The myelogenetic cycles of regional maturation of the brain. In A. Minkowski (Ed.), *Regional development of the brain in early life* (pp. 3–70). Oxford: Blackwell Scientific.
Zecevic, N., & Rakic, P. (1991). Synaptogenesis in monkey somatosensory cortex. *Cerebral Cortex, 1*(6), 510–523.

# II

# Basic Methods and Techniques

# 4

# A Normative Database from Magnetic Resonance Imaging

DUANE D. BLATTER, ERIN D. BIGLER, C. STERLING JOHNSON, CAROL ANDERSON, AND SHAWN D. GALE

Without controversy, magnetic resonance (MR) imaging achieves exquisite approximation of gross anatomy. This has been appreciated from the onset of the se of MR technology to image the brain. However, what has been clinically lacking in the first decade of MR use is a rapid yet automated means to quantify various neuroanatomic structures. As such, much of the interpretation of MR imaging has been based primarily on a qualitative rather than quantitative analysis (see Osborn, 1994).

Recently, a number of researchers have provided various types of normative MR imaging data (see Jernigan, Press, & Hesselink, 1990; Seab *et al.*, 1988; Kesslak, Nalcioglu & Cotman, 1991; Pearlson *et al.*, 1992; Kesole, Cadet, Pickar, & Weinberg, 1988; Lim, Zipursky, Murphy, & Pfefferbaum, 1990; Coffey *et al.*, 1992; Watson *et al.*, 1992; Kohn *et al.*, 1991; Jack, 1991; Filipek *et al.*, 1989; Pfefferbaum *et al.*, 1994; Gur *et al.*, 1991). However, none of this research has approached this in a fully systematic, volume-based, morphometric quantification of cerebral structures over the life span. To that end, we have examined over 200 "normal" individuals with MR analysis of the brain, evaluated by semiautomated and rapid quantitative assessment of target cerebral structures (see Blatter *et al.*, 1995). This chapter represents a distillation of aspects of the Blatter *et*

DUANE D. BLATTER Division of Neuroradiology, Department of Radiology, LDS Hospital, Salt Lake City, Utah 84143. ERIN D. BIGLER, C. STERLING JOHNSON, CAROL ANDERSON, AND SHAWN D. GALE Department of Psychology, Brigham Young University, Provo, Utah 84602-5543.

*Neuroimaging I: Basic Science*, edited by Erin D. Bigler. Plenum Press, New York, 1996.

*al.* (1995) study and extends that work by the addition of a medical control group. The original Blatter *et al.* study did not include some medical controls and had a slightly smaller sample size. Aspects of this study are reproduced with permission of the *American Journal of Neuroradiology*. For this chapter both normal nonmedical controls as well as a combined normal and medical control group will be presented. The "normal and medical" control group is included, because often research, as well as clinical studies, incorporate medical controls in a normative data set. However, "medical" controls may not be entirely representative of a "normal" sample. Thus, for the current chapter, we present the data from a "normal" and a "mixed normal and medical control" group, separately.

## Quantitative Neuroimaging Methods

Detailed methodologic procedures are given in Blatter *et al.* (1995), and the reader is referred to that article for elaboration of any methods.

TABLE 1. NORMATIVE VOLUMETRIC MEASURES OF INTRACRANIAL STRUCTURES OF FEMALES (COMBINED NORMAL AND MEDICAL CONTROL) FOR MRI BY AGE[a]

| | Age Decade | | | | | |
|---|---|---|---|---|---|---|
| | *16–25* n = 26; mean = 21 SD = 2.83 | | *26–35* n = 28; mean = 30.9 SD = 3.19 | | *36–45* n = 23; mean = 40.56 SD = 2.97 | |
| Structure | Uncorrected | Corrected | Uncorrected | Corrected | Uncorrected | Corrected |
| Total CSF | 87.85 (32.7) | 89.6 (30.48) | 91.57 (34.08) | 99.02 (34.46) | 111.58 (31.02) | 118.62 (30.53) |
| Subarachnoid CSF | 70.32 (27.74) | 71.74 (26.07) | 75.26 (31.19) | 81.34 (31.95) | 96.21 (29.22) | 102.33 (29.4) |
| Total ventricle | 17.53 (6.37) | 17.87 (5.89) | 16.31 (7.08) | 17.68 (7.27) | 15.37 (4.88) | 16.29 (4.73) |
| Lateral ventricle | 14.84 (6.02) | 15.12 (5.64) | 13.85 (6.75) | 14.97 (6.93) | 12.77 (4.63) | 13.52 (4.54) |
| Left temporal horn | 0.24 (0.14) | — | 0.18 (0.16) | — | 0.21 (0.23) | — |
| Right temporal horn | 0.21 (0.11) | — | 0.25 (0.31) | — | 0.17 (0.14) | — |
| Third ventricle | 0.7 (0.17) | 0.72 (0.16) | 0.62 (0.25) | 0.68 (0.27) | 0.71 (0.3) | 0.76 (0.3) |
| Fourth ventricle | 1.54 (0.57) | 1.58 (0.57) | 1.41 (0.42) | 1.54 (0.48) | 1.51 (0.37) | 1.62 (0.41) |
| Gray matter | 706.04 (104.6) | 724.22 (89.9) | 638.63 (98.7) | 694.03 (79.9) | 612.71 (103.5) | 652.51 (101) |
| White matter | 613.96 (96.97) | 632.46 (103.58) | 597.56 (83.4) | 653.24 (96.59) | 634.12 (109.5) | 675.17 (103.59) |
| Total brain volume | 1319.99 (78.49) | 1356.68 (30.48) | 1236.2 (95.49) | 1347.28 (34.46) | 1246.84 (105.3) | 1327.68 (30.53) |
| TICV[b] | 1407.8 (88.57) | — | 1327.77 (106.32) | — | 1358.42 (113.05) | — |
| VBR[c] | 1.33 (0.47) | — | 1.32 (0.56) | — | 1.23 (0.37) | — |

[a]Measurements are given in cubic centimeters with standard deviation in parentheses.
[b]Total intracranial volume.
[c]Ventricle-to-brain ratio.

SUBJECTS

MIXED NORMAL AND MEDICAL CONTROL GROUP. In the study, 206 individuals (98 males and 108 females) from age 16 to 65 were examined. The sample size and age breakdown for the five decades studied are provided in Tables 1 and 2. Twenty-nine of these subjects were medical controls who presented neurologic symptoms (i.e., headache, dizziness, rule out pituitary tumor) but in whom no identifiable neurologic abnormalities were found.

NORMAL GROUP. In this group, 194 normal volunteers (89 males and 105 females) from age 16 to 65 were examined.

IMAGING

The MR images were acquired at 1.5 tesla, with a quadrature head coil using standard clinical protocols. Sigittal T1 weighted (500/11/2; TR/TE/excitations) images were acquired and used for localization. Axial intermediate and T2 weighted (3000/31; 90/1) spin echo images were then acquired. The slice thick-

TABLE 1. (*Continued*)

| *46–55* n = 26; mean = 50.27 SD = 2.69 | | *56–65* n = 20; mean = 59.6 SD = 2.04 | | F score | | p value | |
|---|---|---|---|---|---|---|---|
| Uncorrected | Corrected | Uncorrected | Corrected | U | C | U | C |
| 115.43 (29.24) | 123.66 (28.62) | 147.91 (39.94) | 157.57 (42.99) | 11.63 | 14.01 | 0.0001 | 0.0001 |
| 99.89 (27.88) | 107.03 (27.76) | 124.18 (35.67) | 132.24 (37.83) | 11.69 | 13.82 | 0.0001 | 0.0001 |
| 15.53 (3.78) | 16.63 (3.52) | 23.73 (12) | 25.33 (12.98) | 5.03 | 5.46 | 0.0009 | 0.0005 |
| 13.19 (3.69) | 14.12 (3.57) | 20.68 (11.12) | 22.07 (11.98) | 4.89 | 5.32 | 0.001 | 0.0006 |
| 0.14 (0.11) | — | 0.22 (0.23) | — | 1.24 | — | 0.29 | — |
| 0.17 (0.11) | — | 0.28 (0.21) | — | 1.45 | — | 0.222 | — |
| 0.75 (0.23) | 0.81 (0.23) | 1.14 (0.54) | 1.22 (0.6) | 9.37 | 9.33 | 0.0001 | 0.0001 |
| 1.28 (0.47) | 1.37 (0.48) | 1.41 (0.53) | 1.51 (0.6) | 1.13 | 0.89 | 0.346 | 0.474 |
| 585.67 (86.51) | 627.41 (68.8) | 619.87 (72.88) | 657.81 (63.2) | 5.86 | 5.54 | 0.0002 | 0.0004 |
| 650.09 (100.1) | 695.22 (74.52) | 595.04 (99.14) | 630.92 (92.59) | 1.45 | 2.09 | 0.221 | 0.086 |
| 1235.76 (140.1) | 1322.64 (28.62) | 1214.91 (99.15) | 1288.73 (42.99) | 3.62 | 14.01 | 0.008 | 0.0001 |
| 1351.19 (147.12) | — | 1362.82 (92.53) | — | 1.8 | — | 0.133 | — |
| 1.26 (0.28) | — | 1.98 (1.05) | — | 6.12 | — | 0.0002 | — |

TABLE 2. NORMATIVE VOLUMETRIC MEASURES OF INTRACRANIAL STRUCTURES OF MALES (COMBINED NORMAL AND MEDICAL CONTROL) FOR MRI BY AGE[a]

| | *Age Decade* | | | | | |
|---|---|---|---|---|---|---|
| | *16–25* n = 26; mean = 23 SD = 2.1 | | *26–35* n = 20; mean = 31.05 SD = 3.32 | | *36–45* n = 17; mean = 40.88 SD = 2.8 | |
| Structure | Uncorrected | Corrected | Uncorrected | Corrected | Uncorrected | Corrected |
| Total CSF | 87.45 (31.7) | 80.53 (30.5) | 113.58 (36.96) | 105.93 (32.89) | 141.86 (39.14) | 132.61 (34.86) |
| Subarachnoid CSF | 69.34 (30.64) | 64.03 (29.51) | 95.3 (34.95) | 88.87 (31.28) | 122.19 (36.49) | 114.22 (32.7) |
| Total ventricle | 18.11 (6.46) | 16.49 (5.61) | 18.28 (4.37) | 17.06 (3.87) | 19.68 (6.26) | 18.39 (5.57) |
| Lateral ventricle | 15.33 (6.14) | 13.95 (5.35) | 15.15 (4.34) | 14.12 (3.85) | 16.16 (6.07) | 15.08 (5.4) |
| Left temporal horn | 0.17 (0.11) | — | 0.18 (0.15) | — | 0.23 (0.18) | — |
| Right temporal horn | 0.22 (0.25) | — | 0.27 (0.16) | — | 0.22 (0.19) | — |
| Third ventricle | 0.68 (0.21) | 0.63 (0.19) | 0.73 (0.2) | 0.68 (0.19) | 0.998 (0.38) | 0.93 (0.35) |
| Fourth ventricle | 1.71 (0.56) | 1.57 (0.51) | 1.95 (0.7) | 1.83 (0.68) | 2.07 (0.55) | 1.95 (0.56) |
| Gray matter | 757.81 (104.7) | 692.74 (76) | 747.12 (81.87) | 699.04 (77) | 740.19 (109.1) | 692.66 (91.9) |
| White matter | 735.26 (95.45) | 673.03 (77.58) | 686.96 (103.7) | 641.33 (86.56) | 664.09 (117) | 621.03 (98.41) |
| Total brain volume | 1493.08 (115.8) | 1365.77 (30.5) | 1434.08 (78.57) | 1340.37 (32.89) | 1404.28 (96.99) | 1313.69 (34.86) |
| TICV[b] | 1580.5 (109.1) | — | 1547.66 (80.11) | — | 1546.14 (100.64) | — |
| VBR[c] | 1.21 (0.42) | — | 1.28 (0.3) | — | 1.41 (0.44) | — |

[a]Measurements are given in cubic centimeters with standard deviation in parentheses.
[b]Total intracranial volume.
[c]Ventricle-to-brain ratio.

ness was 5 mm, with an interspace gap of 1.5 mm. A 22-cm field of view (FOV) was used with a 256 × 192 acquisition matrix. Flow compensation, an inferior saturation pulse, and variable bandwidth were used. This sequence was part of our standard clinical protocol.

## VOLUMETRIC IMAGE ANALYSIS

The axial intermediate and T2 weighted spin echo images were processed using ANALYZE (Biomedical Imaging Resource, Mayo Foundation, Rochester, Minnesota; Robb, 1990) running on SPARC 10 workstations (Sun Microsystems, Mountain View, California). The original 16-bit images were converted to 8-bit images in ANALYZE file format and then archived permanently on optical disc using a lossless compression algorithm. A multistep volume analysis was then performed using several image processing tools available in ANALYZE, including (1) multispectral tissue segmentation, (2) interactive image editing, and (3) region of interest pixel counting. The multispectral tissue segmentation was performed in a manner similar to that described in a previous report (Cline, Lorensen, Kikinis, & Jolesz, 1990). Regions of cerebro-spinal fluid (CSF), white matter, and gray matter were defined by the user and plotted in a two-dimen-

TABLE 2. (*Continued*)

| 46–55<br>n = 17; mean = 51.06<br>SD = 2.59 | | 56–65<br>n = 18; mean = 60.67<br>SD = 3.25 | | F score | | p value | |
|---|---|---|---|---|---|---|---|
| Uncorrected | Corrected | Uncorrected | Corrected | U | C | U | C |
| 137.35<br>(39.24) | 130.1<br>(41.02) | 189.63<br>(62.29) | 175.85<br>(53.19) | 16.75 | 17.65 | 0.0001 | 0.0001 |
| 116.32<br>(37.87) | 110.34<br>(38.94) | 160.51<br>(58.43) | 149.03<br>(51.31) | 15.03 | 15.52 | 0.0001 | 0.0001 |
| 21.02<br>(4.99) | 19.76<br>(5.02) | 29.12<br>(12.91) | 26.81<br>(10.59) | 6.97 | 8.02 | 0.0001 | 0.0001 |
| 18.08<br>(4.76) | 16.98<br>(4.66) | 25.75<br>(12.46) | 23.66<br>(10.24) | 7.06 | 8.05 | 0.0001 | 0.0001 |
| 0.17<br>(0.14) | — | 0.22<br>(0.19) | — | 0.75 | — | 0.56 | — |
| 0.19<br>(0.1) | — | 0.26<br>(0.19) | — | 0.6 | — | 0.665 | — |
| 1.11<br>(0.38) | 1.04<br>(0.37) | 1.37<br>(0.63) | 1.26<br>(0.53) | 11.11 | 12.4 | 0.0001 | 0.0001 |
| 1.48<br>(0.54) | 1.4<br>(0.54) | 1.51<br>(0.4) | 1.43<br>(0.48) | 3.86 | 3.45 | 0.006 | 0.011 |
| 711.4<br>(71.94) | 666.69<br>(61.2) | 702.73<br>(143.4) | 650.62<br>(103) | 1.02 | 1.19 | 0.399 | 0.319 |
| 696.78<br>(122.9) | 649.51<br>(82.92) | 663.64<br>(126.3) | 619.83<br>(117.7) | 1.54 | 1.23 | 0.197 | 0.303 |
| 1408.18<br>(133.6) | 1316.2<br>(41.02) | 1366.37<br>(120.1) | 1270.45<br>(53.19) | 4.01 | 17.65 | 0.0048 | 0.0001 |
| 1545.53<br>(112.02) | — | 1555.99<br>(126.03) | — | 0.45 | — | 0.772 | — |
| 1.51<br>(0.41) | — | 2.12<br>(0.88) | — | 9.5 | — | 0.0001 | — |

sional feature space where the pixel signal intensity in the T2 weighted sequence is the value on the *x*-axis and the pixel signal intensity in the intermediate weighted image is the *y*-axis (see Figure 1, Chapter 10, Volume II of this series). A *k*-nearest neighbor (k-NN) multispectral algorithm (Dudu & Hart, 1973; Clarke *et al.*, 1993) was then applied to the pixels of the entire slice. If the classified image did not accurately represent the three tissue types, the process was repeated including additional or differently defined regions. When a feature space map was obtained that accurately represented the three tissue types, using the original spin echo images as a reference, it was stored and then applied to the remaining slices in the study.

The classified images were edited using a manual trace tool to remove pixels representing the calvarium and extracranial soft tissues. The inner table of the skull was used as the landmark for separation of intracranial versus extracranial compartments. All of the pixels assigned to each segmented category (gray matter, white matter, CSF) were then summed over all of the classified, edited images from foramen magnum to vertex. In a second editing step, subregions that included the lateral ventricles, temporal horns of the lateral ventricles, third ventricle, and fourth ventricle were manually traced. Because the segmentation step of the process had already identified the interface between CSF and brain, precise tracing was not necessary as long as all of the desired CSF-containing structure was included along with some surrounding brain parenchyma. The pixels assigned to CSF in these subregions were again summed.

The total brain volume was determined by summing the white and gray matter pixels and then multiplying by the voxel dimension ($4.801 \times 10^{-3}$cm$^3$). Total ventricular volume was obtained by summing the measurements of the lateral ventricles and the third and fourth ventricles. The subarachnoid CSF volume was obtained by taking the difference between total CSF and total ventricle. The total intracranial volume was obtained by summing the total brain and total CSF volumes. The ventricle-to-brain ratio (VBR) was calculated by dividing the total ventricle volume by total brain volume and multiplying by 100. Corpus

TABLE 3. NORMATIVE VOLUMETRIC MEASURES OF INTRACRANIAL STRUCTURES OF FEMALES (NORMAL CONTROL) FOR MRI BY AGE[a]

| | | | | | *Age Decade* | |
|---|---|---|---|---|---|---|
| | *16–25* n = 20; mean = 21.30 SD = 2.47 | | *26–35* n = 24; mean = 30.71 SD = 3.07 | | *36–45* n = 22; mean = 40.68 SD = 2.98 | |
| Structure | Uncorrected | Corrected | Uncorrected | Corrected | Uncorrected | Corrected |
| Total CSF | 89.35 (36.65) | 91.54 (34.03) | 95.88 (32.73) | 103.87 (32.99) | 111.24 (31.70) | 119.01 (31.28) |
| Subarachnoid CSF | 71.99 (30.86) | 73.76 (28.88) | 79.02 (29.61) | 85.62 (30.33) | 95.83 (29.86) | 102.59 (30.13) |
| Total ventricle | 17.36 (7.15) | 17.78 (6.61) | 16.86 (7.16) | 18.25 (7.29) | 15.40 (5.00) | 16.42 (4.83) |
| Lateral ventricle | 14.73 (6.79) | 15.07 (6.37) | 14.25 (6.89) | 15.39 (7.03) | 12.79 (4.74) | 13.62 (4.64) |
| Left temporal horn | 0.25 (0.15) | — | 0.19 (0.16) | — | 0.21 (0.23) | — |
| Right temporal horn | 0.20 (0.12) | — | 0.27 (0.33) | — | 0.17 (0.14) | — |
| Third ventricle | 0.69 (0.18) | 0.71 (0.17) | 0.65 (0.24) | 0.71 (0.27) | 0.71 (0.31) | 0.76 (0.31) |
| Fourth ventricle | 1.48 (0.56) | 1.53 (0.57) | 1.49 (0.39) | 1.63 (0.45) | 1.52 (0.38) | 1.64 (0.42) |
| Gray matter | 694.78 (112.71) | 717.72 (100.01) | 647.54 (98.61) | 703.74 (75.69) | 611.71 (105.79) | 655.57 (102.85) |
| White matter | 616.21 (109.00) | 639.43 (116.34) | 586.26 (79.34) | 641.07 (90.98) | 628.68 (108.83) | 674.10 (105.67) |
| Total brain volume | 1310.99 (79.74) | 1357.15 (34.03) | 1233.79 (102.43) | 1344.81 (32.99) | 1240.39 (102.96) | 1329.67 (31.28) |
| TICV[b] | 1400.33 (93.09) | — | 1329.67 (112.37) | — | 1351.62 (110.79) | — |
| VBR[c] | 1.32 (0.53) | — | 1.36 (0.56) | — | 1.24 (0.37) | — |
| Corpus callosum | 658 (107) | 694 (103) | 636 (76) | 695 (81) | 672 (103) | 714 (104) |
| Left hippocampus | 2.55 (0.24) | 2.61 (0.24) | 2.37 (0.21) | 2.63 (0.35) | 2.49 (0.22) | 2.73 (0.21) |
| Right hippocampus | 2.64 (0.22) | 2.7 (0.21) | 2.47 (0.16) | 2.75 (0.3) | 2.44 (0.29) | 2.67 (0.25) |
| Total hippocampus | 5.19 (0.45) | 5.32 (0.44) | 4.84 (0.34) | 5.38 (0.63) | 4.93 (0.5) | 5.4 (0.43) |

[a]Measurements are given in cubic centimeters with standard deviation in parentheses.
[b]Total intracranial volume.
[c]Ventricle-to-brain ratio.

callosum surface area was traced with a semiautomated threshold technique (see Figure 10, Chapter 10, Volume II of this series) fully discussed elsewhere (Johnson, Farnsworth, Pinkston, Bigler, & Blatter, 1994; Johnson, Bigler, Burr, & Blatter, 1994), and the hippocampus was also separately segmented (Blatter *et al.*, 1995).

An expert rating system was utilized where raters would judge protocols in comparison to the expert (a board-certified neuroradiologist). All inter-rater reliabilities exceeded 0.9 with the exception of those for gray and white matter, which average approximately 0.6. This low inter-rater reliability reflects some of the problems with signal inhomogeneity and partial volume effect at the white–gray matter interface. Additional reliability data are provided by Blatter *et al.* (1995).

TABLE 3. (*Continued*)

| *46–55* n = 24; mean = 50.38 SD = 2.73 | | *56–65* n = 15; mean = 59.80 SD = 2.18 | | F score | | p value | |
|---|---|---|---|---|---|---|---|
| Uncorrected | Corrected | Uncorrected | Corrected | U | C | U | C |
| 116.30 (30.31) | 125.38 (29.28) | 147.35 (34.93) | 160.80 (40.84) | 8.04 | 10.75 | 0.0001 | 0.0001 |
| 100.79 (28.88) | 108.63 (28.36) | 123.12 (30.91) | 134.25 (35.34) | 8.02 | 10.37 | 0.0001 | 0.0001 |
| 15.52 (3.76) | 16.75 (3.58) | 24.23 (12.15) | 26.55 (13.65) | 4.34 | 5.28 | 0.0028 | 0.0007 |
| 13.16 (3.66) | 14.22 (3.62) | 21.11 (11.31) | 23.13 (12.63) | 4.17 | 5.03 | 0.0036 | 0.001 |
| 0.15 (0.11) | — | 0.25 (0.25) | — | 1.1 | — | .3603 | — |
| 0.18 (0.11) | — | 0.25 (0.21) | — | 1.13 | — | .345 | — |
| 0.75 (0.24) | 0.81 (0.24) | 1.19 (0.60) | 1.30 (0.67) | 7.88 | 8.63 | 0.0001 | 0.0001 |
| 1.27 (0.49) | 1.37 (0.50) | 1.42 (0.49) | 1.57 (0.61) | 1.04 | 1.12 | 0.388 | 0.351 |
| 580.94 (81.68) | 627.65 (69.33) | 597.59 (60.71) | 649.16 (64.30) | 4.72 | 4.45 | 0.0016 | 0.0024 |
| 645.95 (103.27) | 695.65 (76.00) | 590.55 (106.22) | 638.72 (95.18) | 1.37 | 1.56 | .2506 | 0.191 |
| 1226.89 (138.06) | 1323.30 (29.28) | 1188.14 (95.75) | 1287.89 (40.84) | 3.17 | 10.75 | 0.0017 | 0.0001 |
| 1343.19 (146.41) | — | 1335.49 (82.98) | — | 1.24 | — | 0.298 | — |
| 1.27 (0.28) | — | 2.08 (1.11) | — | 5.82 | — | 0.0003 | — |
| 632 (93) | 677 (70) | 618 (93) | 672 (106) | 0.97 | 0.6 | 0.42 | 0.66 |
| 2.38 (0.23) | 2.64 (0.29) | 2.24 (0.29) | 2.43 (0.23) | 2.78 | 1.9 | 0.04 | 0.12 |
| 2.42 (0.18) | 2.7 (0.27) | 2.32 (0.28) | 2.52 (0.21) | 3.02 | 1.55 | 0.03 | 0.21 |
| 4.8 (0.38) | 5.35 (0.53) | 4.56 (0.56) | 4.95 (0.42) | 3.01 | 1.76 | 0.03 | 0.15 |

TABLE 4. NORMATIVE VOLUMETRIC MEASURES OF INTRACRANIAL STRUCTURES OF MALES (NORMAL CONTROL) FOR MRI BY AGE[a]

| | *Age Decade* | | | | | |
|---|---|---|---|---|---|---|
| | *16–25* n = 26; mean = 23.15 SD = 1.85 | | *26–35* n = 21; mean = 30.95 SD = 3.20 | | *36–45* n = 16; mean = 41.13 SD = 2.70 | |
| Structure | Uncorrected | Corrected | Uncorrected | Corrected | Uncorrected | Corrected |
| Total CSF | 91.22 (34.41) | 82.90 (31.41) | 108.13 (39.01) | 101.07 (34.56) | 138.60 (37.95) | 129.79 (33.66) |
| Subarachnoid CSF | 72.72 (33.22) | 66.22 (30.55) | 90.09 (36.86) | 84.17 (32.92) | 118.92 (35.02) | 111.37 (31.25) |
| Total ventricle | 18.50 (6.53) | 16.68 (5.63) | 18.04 (4.40) | 16.90 (3.82) | 19.67 (6.47) | 18.42 (5.76) |
| Lateral ventricle | 15.65 (6.16) | 14.10 (5.32) | 15.05 (4.33) | 14.08 (3.79) | 16.11 (6.27) | 15.07 (5.58) |
| Left temporal horn | 0.18 (0.12) | — | 0.17 (0.15) | — | 0.24 (0.17) | — |
| Right temporal horn | 0.24 (0.25) | — | 0.26 (0.16) | — | 0.23 (0.19) | — |
| Third ventricle | 0.70 (0.21) | 0.63 (0.19) | 0.71 (0.20) | 0.67 (0.18) | 1.01 (0.39) | 0.95 (0.36) |
| Fourth ventricle | 1.73 (0.57) | 1.57 (0.51) | 1.85 (0.66) | 1.75 (0.64) | 2.08 (0.57) | 1.96 (0.57) |
| Gray matter | 766.33 (104.24) | 693.44 (73.20) | 738.34 (91.46) | 693.41 (80.32) | 740.70 (112.63) | 694.40 (94.99) |
| White matter | 740.60 (88.52) | 672.35 (79.22) | 696.51 (104.93) | 654.20 (94.20) | 666.61 (120.34) | 624.49 (101.27) |
| Total brain volume | 1506.93 (102.65) | 1365.79 (31.41) | 1434.85 (82.69) | 1347.61 (34.56) | 1407.31 (99.34) | 1318.89 (33.66) |
| TICV[b] | 1598.15 (98.30) | — | 1542.98 (88.59) | — | 1545.91 (103.94) | — |
| VBR[c] | 1.22 (0.42) | — | 1.26 (0.30) | — | 1.40 (0.45) | — |
| Corpus callosum | 680 (96) | 616 (77) | 694 (102) | 642 (82) | 678 (116) | 639 (104) |
| Left hippocampus | 2.97 (0.33) | 2.65 (0.37) | 2.68 (0.39) | 2.55 (0.24) | 2.54 (0.15) | 2.38 (0.23) |
| Right hippocampus | 2.93 (0.27) | 2.62 (0.32) | 2.8 (0.33) | 2.67 (0.2) | 2.67 (0.2) | 2.49 (0.2) |
| Total hippocampus | 5.9 (0.58) | 5.27 (0.68) | 5.49 (0.71) | 5.22 (0.43) | 5.21 (0.32) | 4.87 (0.4) |

[a]Measurements are given in cubic centimeters with standard deviation in parentheses.
[b]Total intracranial volume.
[c]Ventricle-to-brain ratio.

## STATISTICAL ANALYSIS

Descriptive statistics, including mean and standard deviations were calculated for each age and gender group. Age and gender-related effects were evaluated with Pearson correlations, paired *t*-tests, and analysis of variance.

## QUANTITATIVE NEUROIMAGING FINDINGS

Descriptive statistics for volume measurements of the selected intracranial structures are given in Tables 1 and 2 by age for females and males, respectively, along with analysis of variance results for each volume measure by age for the

TABLE 4. (*Continued*)

| 46–55 n = 15; mean = 51.07 SD = 2.66 | | 56–65 n = 15; mean = 60.60 SD = 3.04 | | F score | | p value | |
|---|---|---|---|---|---|---|---|
| Uncorrected | Corrected | Uncorrected | Corrected | U | C | U | C |
| 136.38 | 129.88 | 191.07 | 177.96 | 15.46 | 17.15 | 0.0001 | 0.0001 |
| (38.75) | (39.26) | (65.49) | (56.16) | | | | |
| 115.45 | 109.99 | 162.98 | 151.81 | 14.16 | 15.38 | 0.0001 | 0.0001 |
| (36.54) | (36.49) | (62.34) | (54.44) | | | | |
| 20.93 | 19.89 | 28.08 | 26.15 | 5.21 | 6.17 | 0.0009 | 0.0002 |
| (5.13) | (5.35) | (12.29) | (10.52) | | | | |
| 17.89 | 17.00 | 24.78 | 23.02 | 5.22 | 6.1 | 0.0008 | 0.0002 |
| (4.81) | (4.95) | (12.05) | (10.29) | | | | |
| 0.16 | — | 0.20 | — | 0.75 | — | 0.558 | — |
| (0.13) | | (0.17) | | | | | |
| 0.20 | — | 0.27 | — | 0.42 | — | 0.795 | — |
| (0.09) | | (0.21) | | | | | |
| 1.13 | 1.07 | 1.30 | 1.22 | 11.77 | 12.74 | 0.0001 | 0.0001 |
| (0.40) | (0.39) | (0.42) | (0.39) | | | | |
| 1.54 | 1.47 | 1.54 | 1.46 | 2.7 | 2.53 | 0.036 | 0.047 |
| (0.54) | (0.53) | (0.42) | (0.50) | | | | |
| 712.74 | 673.99 | 691.47 | 645.68 | 1.35 | 1.16 | 0.258 | 0.334 |
| (73.27) | (62.52) | (130.51) | (102.60) | | | | |
| 683.77 | 644.81 | 666.37 | 625.04 | 1.86 | 1.14 | 0.125 | 0.344 |
| (108.43) | (83.13) | (137.92) | (124.65) | | | | |
| 1396.51 | 1318.80 | 1357.84 | 1270.72 | 6.33 | 17.15 | 0.0002 | 0.0001 |
| (108.58) | (39.26) | (100.59) | (56.16) | | | | |
| 1532.89 | — | 1548.91 | — | 1.22 | — | 0.309 | — |
| (89.29) | | (106.09) | | | | | |
| 1.52 | — | 2.07 | — | 7.56 | — | 0.0001 | — |
| (0.43) | | (0.88) | | | | | |
| 676 | 640 | 616 | 583 | 1.71 | 1.61 | 0.15 | 0.18 |
| (108) | (64) | (83) | (72) | | | | |
| 2.39 | 2.28 | 2.5 | 2.4 | 3.52 | 1.69 | 0.02 | 0.18 |
| (0.35) | (0.33) | (0.25) | (0.23) | | | | |
| 2.39 | 2.28 | 2.61 | 2.5 | 3.9 | 2.61 | 0.01 | 0.05 |
| (0.33) | (0.28) | (0.2) | (0.18) | | | | |
| 4.78 | 4.56 | — | — | 3.87 | 2.11 | 0.01 | 0.1 |
| (0.66) | (0.59) | | | | | | |

combined normal and medical control group. Tables 3 and 4 provide similar data for only the normal control subjects; the medical controls are eliminated from this table. In Table 5, Pearson correlations are given for each volume measure with age for the combined normal and medical control groups, with similar findings in the normal-only control group presented in Table 6. Age and gender-matched normal controls were compared to the 29 medical controls. There were no significant differences in any measure. Despite there being no differences, the tables that follow are based on the separate analyses of medical and normal controls. This is provided so that the clinician or researcher can use these tables, depending on the subject composition.

Since intracranial volume should not change with age, it was anticipated that an analysis of variance for total intracranial volume by age would be nonsignificant, and it was for both females ($F(4,103) = 0.95, p = 0.439$) and males ($F(4,93) = 0.69, p = 0.599$). Likewise, the Pearson correlation was also nonsignificant for females ($r = --0.045, p = 0.647$) and males ($r = -0.196, p = 0.052$). These

TABLE 5. CORRELATIONS BETWEEN BRAIN STRUCTURES AND AGE FOR MEDICAL AND NORMAL CONTROLS[a]

| | MALES (n = 98) | | | | FEMALES (n = 108) | | | | ALL SUBJECTS (n = 206) | |
|---|---|---|---|---|---|---|---|---|---|---|
| | Uncorrected | | Corrected | | Uncorrected | | Corrected | | Corrected | |
| Structure | r | p | r | p | r | p | r | p | r | p |
| CSF | .633 | .000 | .663 | .000 | .525 | .000 | .560 | .000 | .613 | .000 |
| Sub CSF | .613 | .000 | .638 | .000 | .538 | .000 | .568 | .000 | .603 | .000 |
| Total ventricles | .362 | .000 | .424 | .000 | .143 | .139 | .172 | .076 | .299 | .000 |
| Lateral ventricles | .358 | .000 | .415 | .000 | .147 | .129 | .174 | .072 | .297 | .000 |
| Left temporal horn | .105 | .304 | | | −.028 | .771 | | | | |
| Right temporal horn | .086 | .400 | | | .028 | .771 | | | | |
| Third ventricle | .540 | .000 | .586 | .000 | .340 | .000 | .358 | .000 | .479 | .000 |
| Fourth ventricle | −.215 | .033 | −.156 | .125 | −.149 | .123 | −.134 | .167 | −.143 | .040 |
| Gray matter | −.315 | .002 | −.253 | .012 | −.333 | .000 | −.373 | .000 | −.314 | .000 |
| White matter | −.203 | .045 | −.137 | .180 | .095 | .328 | .118 | .222 | −.003 | .971 |
| Total brain | −.439 | .000 | −.663 | .000 | −.225 | .091 | −.560 | .000 | −.613 | .333 |
| TICV[b] | −.196 | .052 | * | * | −.045 | .647 | * | * | * | * |
| VBR[c] | .469 | .000 | * | * | .209 | .030 | * | * | * | * |

[a]Values of r and p are uncorrected and corrected for total intracranial volume.
[b]Total intracranial volume.
[c]Ventricle-to-brain ratio.

values were based on the mixed normal and medical controls, but nearly identical findings were obtained with the normal-only group as well (female by age comparisons nonsignificant: $r = 0.147$, $p = 0.135$, $F(4,105) = 1.24$, $p = 0.29$; male by age comparison: $r = -0.190$, $p = 0.07$, $F(4,89) = 1.22$, $p = 0.31$).

A significant difference in the total intracranial volume was observed between males ($1551 cm^3 \pm 112 cm^3$) and females ($1362 cm^3 \pm 112 cm^3$) ($t(204) = 11.86$, $p < .001$) in the mixed normal and medical control groups. For both males and females there was a statistically significant correlation between total intracranial volume and each of the other volume measures, except that of the temporal horns. All of these relationships were duplicated in the normal-only group as well (males: TICV = $1558 cm^3 \pm 97 cm^3$; females: TICV = $1352 cm^3 \pm 115 cm^3$; $t(192) = 11.9$, $p < .001$).

In order to correct for population variation and gender differences in head size, a correction was made based upon the total intracranial volume for each of the other volumes, except that of the temporal horns. This correction was performed by multiplying each measured volume by the ratio of the mean total intracranial volume for the entire sample population ($1452 cm^3$ for medical control and $1446 cm^3$ for normal control) divided by the observed total intracranial volume for each individual subject. Tables 1–4 give values for both the corrected and uncorrected volumes by decade and gender for the normal and medical control and normal-only group.

In Figures 1 and 2, bar graphs illustrate age and gender differences in several uncorrected and corrected morphometric measurements by decade for the normal control group only (essentially similar results were in the medical and

TABLE 6. CORRELATIONS BETWEEN BRAIN STRUCTURES AND AGE FOR NORMAL CONTROLS[a]

| | MALES (n = 89) | | | | FEMALES (n = 105) | | | | ALL SUBJECTS (n = 194) | |
|---|---|---|---|---|---|---|---|---|---|---|
| | Uncorrected | | Corrected | | Uncorrected | | Corrected | | Corrected | |
| Structure | r | p | r | p | r | p | r | p | r | p |
| CSF | .650 | .000 | .675 | .000 | .477 | .000 | .539 | .000 | .608 | .000 |
| Sub CSF | .633 | .000 | .653 | .000 | .490 | .000 | .545 | .000 | .599 | .000 |
| Total ventricles | .412 | .000 | .458 | .000 | .171 | .081 | .220 | .024 | .324 | .000 |
| Lateral ventricles | .401 | .000 | .444 | .000 | .172 | .080 | .218 | .026 | .318 | .000 |
| Left temporal horn | .030 | .777 | | | −.043 | .667 | | | | |
| Right temporal horn | .040 | .701 | | | −.004 | .970 | | | | |
| Third ventricle | .617 | .000 | .634 | .000 | .376 | .000 | .406 | .000 | .508 | .000 |
| Fourth ventricle | −.126 | .239 | −.076 | .476 | −.125 | .204 | −.081 | .413 | −.080 | .265 |
| Gray matter | −.257 | .015 | −.200 | .059 | −.391 | .000 | −.372 | .000 | −.294 | .000 |
| White matter | −.239 | .024 | −.179 | .092 | .050 | .611 | .126 | .199 | −.014 | .850 |
| Total brain | −.485 | .000 | −.675 | .000 | −.310 | .001 | −.539 | .000 | −.608 | .000 |
| TICV[b] | −.190 | .074 | | | −.147 | .135 | | | | |
| VBR[c] | .498 | .000 | * | * | .248 | .011 | * | * | * | * |

[a]Values of r and p are uncorrected and corrected for total intracranial volume.
[b]Total intracranial volume.
[c]Ventricle-to-brain ratio.

control group as well). In uncorrected measurements for total brain volumes, males were larger than females in all age groups, and several of these measurements remained significant even after correction. For illustrative purposes, Figure 3 presents several scatter plots depicting the linear decrease in brain volume over time with corresponding increase in CSF space. These scatter plots were based on mixed normal and medical control group, however, the normal-only group scatter plot analysis yielded almost identical results.

Significant age-related changes were observed for both males and females, although in most cases the differences were usually larger and more highly significant for males. In males, significant age-related changes in the corrected volume measurements were observed (in descending order of significance) for total brain, total CSF, subarachnoid CSF, third ventricle, total ventricle, lateral ventricle, and gray matter. In females, significant age-related correlations in the corrected volume measurements were observed (in descending order of significance) for subarachnoid CSF, total CSF, total brain, gray matter, and third ventricle. Unlike males, the females showed no significant correlation of total ventricle or lateral ventricle volume with age. Similarly, the correlation for third ventricle volume with age was lower in females than in males. Conversely, the correlation of gray matter volume with age was greater in females than in males. These correlations are given in Tables 5 and 6.

In order to assess the accuracy of the database, we compared the total brain volume obtained by analysis of the MRI in the normal and medical control with autopsy reports (Ho, Roessmann, Straumfjord, & Monroe, 1980; Dekaban & Sadowsky, 1978). The comparison (see Table 7) is somewhat limited since autop-

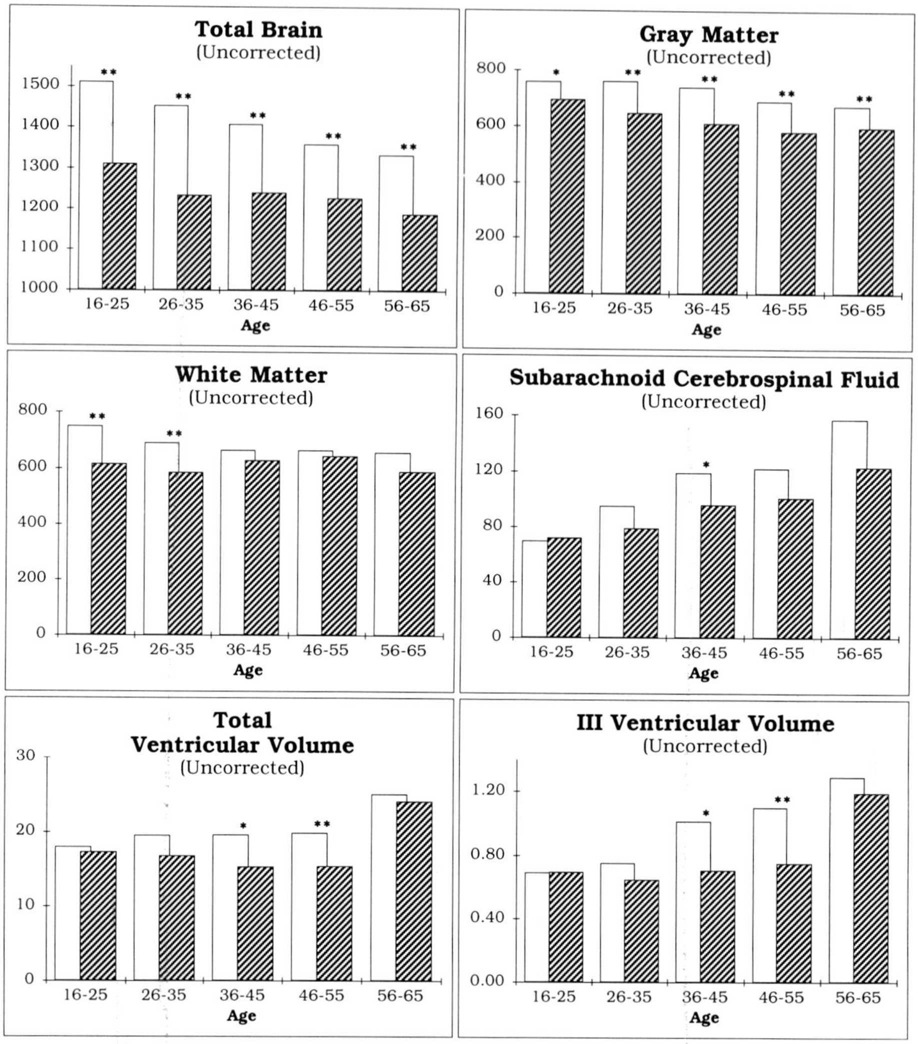


Figure 1. Bar graphs comparing selected brain structures by volume estimates according to age and gender. These are absolute values, not corrected for head size. Each bar indicates the mean volume for the decade shown below. Cross-hatched bars are female values; non-cross-hatched bars are male values. Asterisk (*) indicates a significant ($p < 0.05$) difference between males and females.

sy data report only total brain weight, and conversion to units of volume is necessary. In spite of this limitation, the comparison can provide a good indication of the overall accuracy of the technique.

In Table 7, the average brain weight, given by decade from the report of Ho *et al.* (1980) is compared with the volume measurements from the current data (average total brain volumes, combined for males and females). The weights reported in the autopsy data have been converted to volume by subtracting the reported average weight for the dura and then multiplying by 1.019, an estimate of the gravimetric density of brain based on CT data (Schwartz *et al.*, 1985). Two sample *t*-tests performed for each decade show no significant differences between the autopsy and MR volume estimates.

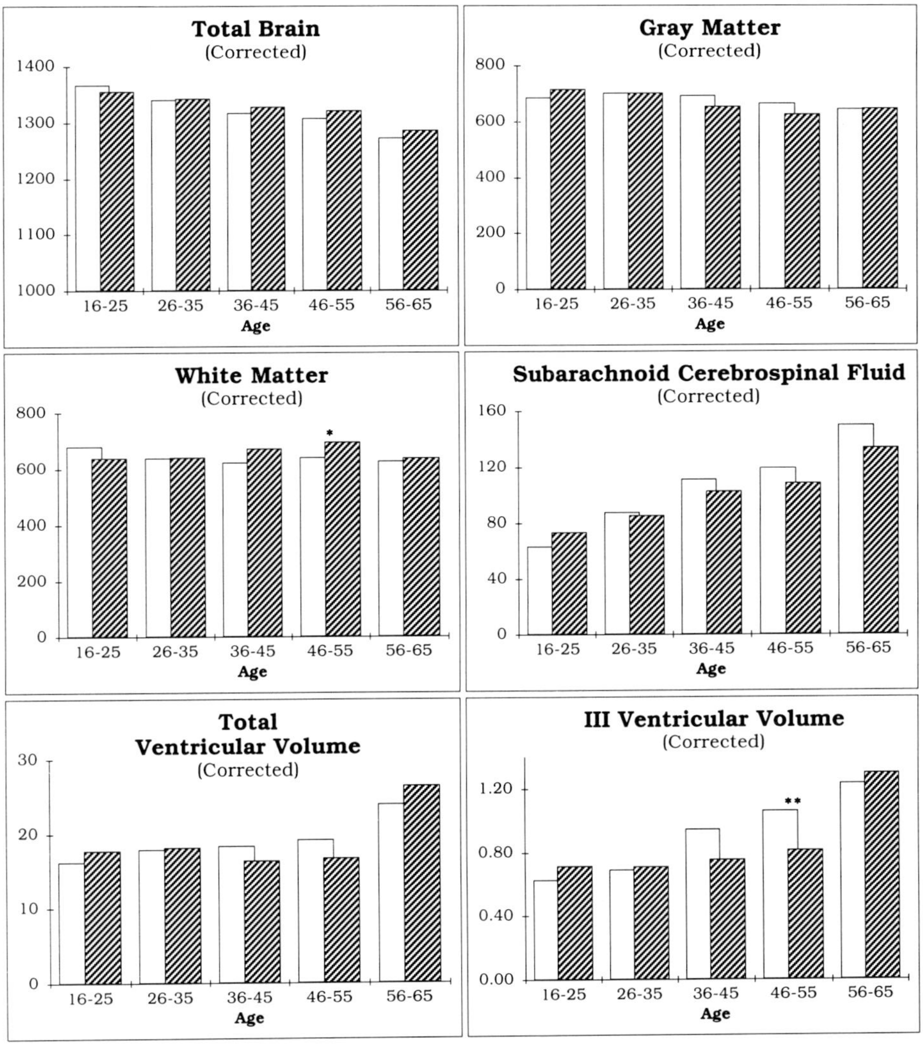


Figure 2. Bar graphs comparing selected brain structure volume estimates according to age and gender. The values are based on the absolute value corrected by head size. Cross-hatched bars are female values; non-cross-hatched bars are male values. Asterisk (*) indicates a significant ($p < 0.05$) difference between males and females.

From the autopsy reports, the difference between the average weight of female and male brains was 136 g from Ho *et al.* (1980), and 160 g from Dekaban & Sadowsky (1978) for the normal and medical control group (nearly identical findings were obtained for the normal controls as well). This is compared to 166cm$^3$ (163 g) for normal control based on the current MR data. The lower difference reported by Ho *et al.* is consistent with the relatively higher proportion of older subjects. There is also a good correlation between the autopsy and MR measures of the magnitude of brain loss with aging. Based on the regression analysis reported by Ho *et al.*, (1980), between 25 and 60 years of age there is a reduction in total brain weight of 49 g for females and 103 g for males. Averaging the first two decades of MR data shown in Table 5 (16–35, mean age 25) and comparing with the decade 56–65 (mean age of 60), there is a reduction of

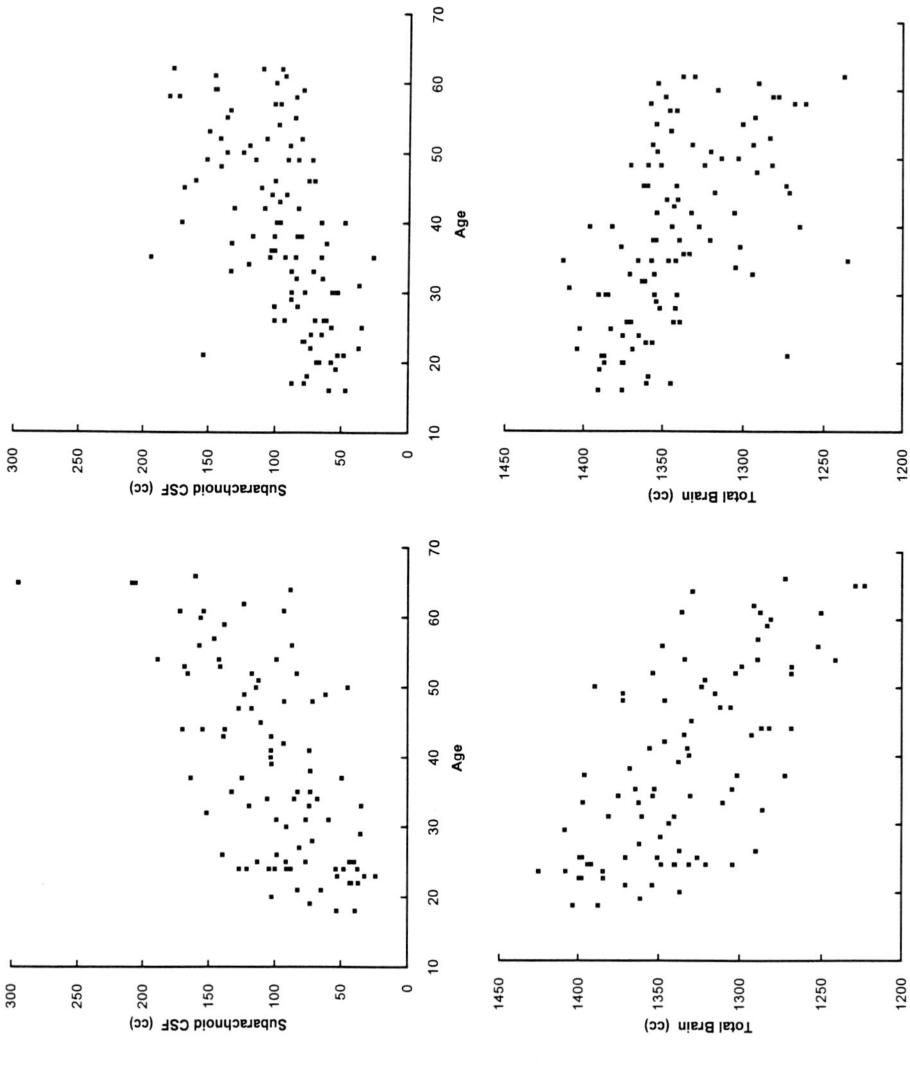


Figure 3. Scatter plots for quantitative values for total brain volume and subarachnoid CSF in a mixed medical and normal control group. Male plots on the left, female on the right.

TABLE 7. COMPARISON OF BRAIN WEIGHT BY AUTOPSY AND BRAIN VOLUME BY MRI

| AUTOPSY[a] | | | | MRI | | | | | |
|---|---|---|---|---|---|---|---|---|---|
| | | | | Medical and normal control | | | Normal control | | |
| Age | Decade | Average brain weight (g) | | Age | Decade | Measured[b] | Age | Decade | Measured[b] |
| 30 | 25–34 | 1349 ± 131 | 1342 | 31 | 26–35 | 1326 ± 143 | 31 | 26–35 | 1323 ± 137 |
| 40 | 35–44 | 1313 ± 151 | 1305 | 41 | 36–45 | 1317 ± 128 | 41 | 36–45 | 1311 ± 130 |
| 50 | 45–64 | 1290 ± 154 | 1281 | 50 | 46–55 | 1312 ± 154 | 51 | 46–55 | 1292 ± 151 |
| 60 | 55–64 | 1307 ± 159 | 1299 | 60 | 56–65 | 1287 ± 134 | 60 | 56–65 | 1273 ± 129 |

[a]Data from: Ho, Roessmann, Straumfjord, & Monroe, 1980.
[b]Weights based on gravimetric value of 1.019g/cm$^3$ and subtracting the reported weight of the dura matter; sample *t*-tests indicate no significant difference from corresponding autopsy volume estimates.

55cm$^3$ (54 g) for females and 114cm$^3$ (112 g) for males. Accordingly, the *in vivo* measurement of volume with MR techniques approximates actual tissue volumes.

An example of one possible use of this normative data set in evaluating a patient following traumatic brain injury (TBI) is provided in Chapter 10, Volume II of this series.

## QUANTITATIVE IMAGE ANALYSIS IMPLICATIONS

One of the most impressive aspects of this image analysis is how the MR volumetric data approximates postmortem data (see Table 7). This suggests that this type of quantitative image analysis has great promise in approximating real values of human brain structure in the living individual. The implications of this type of analysis are certainly obvious. A host of neurologic diseases, including essentially all of the neurodegenerative as well as cerebral vascular and traumatic diseases, result in degenerative changes in the brain. Using this neuroimaging technique, each disorder can now be plotted against a normal reference group to quantify the degree of degenerative change. This technique has the potential for yielding diagnostic classification. Similarly, the technique can be used to plot the clinical course of a variety of neurologic disorders. Tables 1–6 can be used as reference tables wherein a given patient's neuroimaging data can be compared to the databases composed entirely of normal controls (Tables 3 and 4) or those with a combination of normal and medical controls.

For example, a normative comparison such as this database would be very useful in either monitoring degenerative changes or determining clinical significance of degenerative changes in such disorders as Alzheimer's disease and other related dementias. Similar statements could be made about any of the nonstatic neurologic disorders such as vascular disease and neoplastic disease. It may be that actual treatment criteria may evolve from such quantitative image analyses.

The current data indicate that "medical controls," that is, those individuals who present a potential neurologic symptom (i.e., headache) but in whom no objective neurodiagnostic test is abnormal, compare favorably to "normal" controls. One contrary note should be made, however. Several studies (see Raz, 1989; Raz & Raz, 1990) have indicated that psychiatric controls may not be as comparable to normals as the medical controls in this study.

In terms of current MR acquisition technology, there is still a resolution problem where certain types of injury may be below the threshold of detection. For example, with any neurologic disease or disorder, subtle neuronal derangement or the earliest stages of degeneration may not be detected by MR technology because the aggregate change may not reach a threshold observable by visual inspection. However, quantitative assessment methods, particularly with multiple comparisons, may provide some means of detection of even subtle change. The potential of this technology is extremely exciting.

*Acknowledgments*

This study is based, in part, on the studies of Blatter *et al.* (1995; submitted), and the contributions of Bruce M. Burnett, Natalie Parker, Shanna Kurth, Susan Horn, and Tracy Abildskov are greatly appreciated. Financial support was provided, in part, by the Deseret Foundation and Brigham Young University, College of Social Science.

## References

Bigler, E. D., Blatter, D. D., Anderson, C. V., Johnson, S. C., Gale, S. D., Hopkins, R. O., & Burnett, B. (1996). Hippocampal volume in normal aging (16–65 years) and traumatic brain injury. *American Journal of Neuroradiology*, in press.

Blatter, D. D., Bigler, E. D., Gale, S. D., Johnson, S., Anderson, C., Burnett, B. M., Parker, N., Kurth, S., & Horn, S. (1995). Quantitative volumetric analysis of brain MR: Normative database spanning five decades (16–65). *American Journal of Neuroradiology, 16,* 241–251.

Blatter, D. D., Bigler, E. D., Gale, S. D., Johnson, S. C., Anderson, C. V., Burnett, B. M., Ryser, D., Macnamara, S. E., & Bailey, B. J. (1996). MRI-based brain and CSF quantification following traumatic brain injury: Correlation with neuropsychological outcome. *American Journal of Neuroradiology*, in press.

Clark, L. P., Velthuizen, R. P., Phuphanich, S., Schellenberg, J. D., Arrington, J. A., & Silberger, M. (1993). MRI: Stability of three supervised segmentation techniques. *Magnetic Resonance Imaging, 11,* 95–106.

Cline, H. E., Lorensen, W. E., Kikinis, R., & Jolesz, F. (1990). Three-dimensional segmentation of MR images of the head using probability and connectivity. *Journal of Computer Assisted Tomography, 14,* 1037–1045.

Coffey, C. E., Wilkinson, W. E., Parashos, L. A., Soady, S. A. R., Sullivan, R. J., Patterson, L. J., Figiel, G. S., Webb, M. C., Spitzer, C. E., & Djang, W. T. (1992). Quantitative cerebral anatomy of the aging human brain: A cross-sectional study using magnetic resonance imaging. *Neurology, 42,* 527–536.

Dekaban, A. S., & Sadowsky, D. (1978). Changes in brain weight during the span of human life: Relation of brain weights to body heights and body weights. *Annals of Neurology, 4,* 345–356.

Dudu, R. O., & Hart, P. E. (1973). *Pattern classification and scene analysis.* New York: John Wiley & Sons, Inc.

Filipek, P. A., Kennedy, D. N., Caviness, V. S., Rossnick, S. L., Spraggins, T. A., & Starewicz, P. M. (1989). Magnetic resonance imaging–based brain morphometry: Development and application to normal subjects. *Annals of Neurology, 25,* 61–67.

Gur, R. E., Mozley, P. D., Resnick, S. M., Shtasel, D., Kohn, M., Zimmerman, R. A., Herman, G. T., Atlas, S., Grossman, R., Erwin, R., & Gur, R. C. (1991). Magnetic resonance imaging in schizophrenia: I. Volumetric analysis of brain and cerebral spinal fluid. *Archives of General Psychiatry, 48,* 407–412.

Ho, K., Roessmann, U., Straumfjord, J. V., & Monroe, G. (1980). Analysis of brain weight; Adult brain weight in relation to sex, race and age. *Archives of Pathology and Laboratory Medicine, 104,* 635–639.

Jack, C. R. (1991). Brain and cerebrospinal fluid volume: Measurement with MR imaging. *Radiology, 178,* 22–24

Jernigan, T. L., Press, G. A., & Hesselink, J. R. (1990). Methods for measuring brain morphologic features on magnetic resonance images: Validation and normal aging. *Archives of Neurology, 47,* 27–32.

Johnson, S. C., Bigler, E. D., Burr, R. B., & Blatter, D. D. (1994). White matter atrophy, ventricular dilation and intellectual functioning following traumatic brain injury. *Neuropsychology, 8,* 307–315.

Johnson, S. C., Farnworth, T., Pinkston, J. B., Bigler, E. D., & Blatter, D. D. (1994). Corpus callosum surface area across the human adult life span: Effect of age and gender. *Brain Research Bulletin, 35,* 373–377.

Kesole, J. R., Cadet, J. L., Pickar, D., & Weinberger, D. R. (1988). Quantitative neuroanatomy in schizophrenia. A controlled magnetic resonance imaging study. *Archives of General Psychiatry, 45,* 533–541.

Kesslak, J. P., Nalcioglu, O., & Cotman, C. W. (1991). Quantification of magnetic resonance scans for hippocampal and parahippocampal atrophy in Alzheimer's disease. *Neurology, 41,* 51–54.

Kohn, M. I., Tanna, N. K., Herman, G. T., Resnick, S. M., Mozley, P. D., Gur, R. E., Alavi, A., Zimmerman, R. A., & Gur, R. C. (1991). Analysis of brain and cerebrospinal fluid volumes with MR imaging. *Radiology, 178,* 115–122.

Lim, K. O., Zipursky, R. B., Murphy, G. M., & Pfefferbaum, A. (1990). In vivo quantification of the limbic system using MRI: Effects of normal aging. *Psychiatry Research: Neuroimaging, 35,* 15–26.

Osborn, A. (1994). *Diagnostic neuroimaging.* St. Louis, MO: Mosby.

Pearlson, G. D., Harris, G. J., Powers, R. E., Barta, P. E., Camargo, E. E., Chase, G. A., Noga, J. T., & Tune, L. E. (1992). Quantitative changes in the mesial temporal volume, regional cerebral blood flow, and cognition in Alzheimer's disease. *Archives of General Psychiatry, 49,* 402–408.

Pfefferbaum, A., Mathalon, D. H., Sullivan, E. V., Rawles, J. M., Zpursky, R. B., & Lim, K. O. (1994). A quantitative MRI study of changes in brain morphology from infancy to late adulthood. *Archives of Neurology, 51,* 874–887.

Raz, S. (1989). Structural brain abnormalities in the major psychoses. In E. D. Bigler, R. A. Yeo, & E. Turkheimer (Eds.), *Neuropsychological Function and Brain Imaging.* New York: Plenum Press.

Raz, S., & Raz, N. (1990). Structural abnormalities in the major psychoses: A quantitative review of the evidence from computerized imaging. *Psychological Bulletin, 108,* 93–108.

Robb, R. A. (1990). A software system for interactive and quantitative analysis for biomedical images. In K. H. Hohne, H. Fuchs, and S. M. Pizer (Eds.), *3-D imaging in medicine NATO ASI Series,* F60, 333–361.

Seab, J. P., Jagust, W. J., Wong, S. T. S., Roos, M. S., Reed, B. R., & Budinger, T. F. (1988). Quantitative NMR measurements of hippocampal atrophy in Alzheimer's disease. *Magnetic Resonance Medicine, 8,* 200–208.

Watson, C., Andermann, F., Gloor, P., Jones-Gotman, M., Peters, T., Evans, A., Olivier, A., Melanson, D., & Leroux, G. (1992). Anatomic basis for amygdaloid and hippocampal volume measurement by magnetic resonance imaging. *Neurology, 42,* 1743–1750.

# 5

# Positron Emission Tomography Studies of Cognition

SUSAN Y. BOOKHEIMER

## INTRODUCTION

One of the most exciting applications of the new techniques in neuroimaging is the study of how cognitive processes are organized in the normal, functioning brain. Prior to the technological revolution in functional brain imaging, studies of human cognitive psychology on the one hand, and functional neuroanatomy on the other, were largely separate, both by the preferred methodologies used and by the professional backgrounds of scientists. On one side were the cognitive psychologists employing laboratory-based studies of normal volunteers, using paradigms such as reaction time measures in semantic priming, or dichotic and hemiretinal presentations. On the other side were neuropsychologists and neurologists who studied patients with cortical lesions, often in single case reports. The remarkable emergence of cognitive neuroscience, a *de facto* marriage of two disparate approaches to understanding human cognition, has resulted in an explosion of new scientific opportunities, reflected in a dramatic rise in training programs in cognitive neuroscience, new journals, and many publications. This merging of disciplines is particularly apparent in activation imaging studies using $^{15}O$ positron emission tomography (PET) and more recently, functional magnetic resonance imaging (fMRI). By measuring the

SUSAN Y. BOOKHEIMER Division of Brain Mapping, UCLA School of Medicine, Los Angeles, California 90024.

*Neuroimaging I: Basic Science*, edited by Erin D. Bigler. Plenum Press, New York, 1996.

amount of regional cerebral blood flow (rCBF) in the brain during performance of a cognitive task in comparison to some control condition, activation PET attempts to isolate cortical regions which are essential to the performance of a cognitive skill. These $^{15}O$ PET studies have already produced a multitude of experiments which attempt to describe, for the first time, the neuroanatomy underlying cognitive performance in the normal brain. Previously this knowledge was achieved only through studies of neurologically impaired patients.

With the tremendous increase in manpower and dollars dedicated to these admittedly costly technologies, we might wonder whether the information gleaned from imaging studies has added new information to the "brain map", or whether the studies have merely replicated results already apparent from a review of nineteenth century neurology. Indeed, many of the imaging studies to date have attempted to demonstrate well-known anatomical architecture such as the organization of the motor homunculus or of the primary visual cortex. While confirmatory demonstrations of imaging methodology were clearly necessary in the infancy of functional neuroimaging, the "new wave" of functional imaging studies should, and has begun to, emphasize the more controversial subjects in cognitive psychology. In essence, one would hope that new tools would help us answer difficult questions about the organization of cognitive skills, such as, are there separate lexicons for auditory and visual words? Is memory systems-based or process-based? Does word presentation automatically invoke semantic or articulatory codes? Ultimately, the usefulness of functional imaging techniques will be judged by their ability to answer this more difficult set of questions. With this in mind, a selective review of functional imaging studies using $^{15}O$ PET in language, memory, and higher-order visual processing follows. It is hoped that this summary will demonstrate not only the wealth of information that has been revealed with neuroimaging studies, but also how much is yet to be learned.

## Activation Imaging with Positron Emission Tomography: Methodological Considerations

Functional brain images obtained from $^{18}$Fluorodeoxyglucose (FDG) and single photon omission computed tomography (SPECT) have typically been used to make resting images. This is primarily due to the long uptake period of the radiopharmaceuticals. To obtain activation images, subjects should be in a steady state of resumptive brain activity during the uptake and scanning period. For this reason, FDG and SPECT are rarely used for activation imaging: it is difficult to argue that subjects are in a constant, homogeneous cognitive state for the long duration of time necessary to obtain functional images from these methods (often 45 minutes or greater). With the arrival of $H_2{}^{15}O$ and $C^{15}O_2$, which involve rapid uptake and scanning (on the order of 45 seconds to several minutes) it is possible to place subjects in a reasonably constant cognitive state and perform multiple scans in a single session, allowing for comparison conditions (Fox, Mintun, Raichle, & Herscovitch, 1984). Another activation technique, xenon inhalation, also measures rCBF in a reasonable time frame; however, the spatial resolution of these images is quite poor in comparison to oxygen methods and will not be discussed here.

Many PET activation studies of higher-level cognitive processing rely upon "cognitive subtraction" or "subtractive logic" to isolate purported elements of

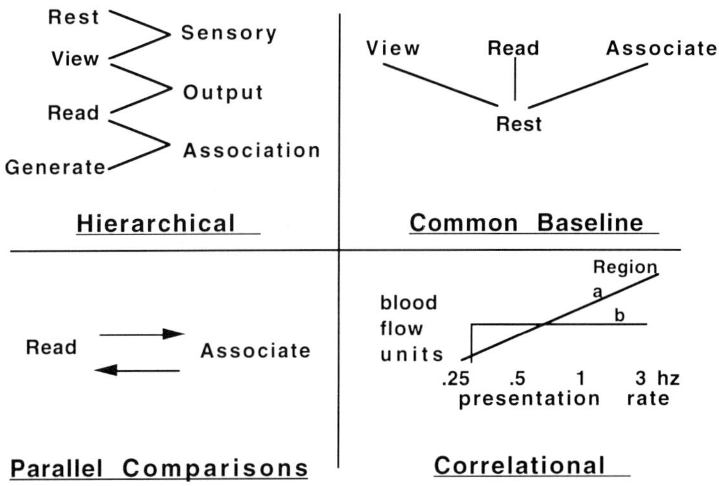


Figure 1. Four approaches to paradigm design. The *hierarchical* model presented in the upper left is taken from the study of single-word processing by Peterson *et al.* (1988). Using that example, three other design and analysis approaches are presented. The *common baseline* method tests each activation condition against the same control, regardless of its position in the presumed cognitive hierarchy. This method continues to make a hierarchical assumption, but only at one level, and is able to detect activation patterns that differ with output demands. The *parallel comparisons* approach detects differences between two cognitive tasks without relying on any hierarchical assumptions but cannot detect regions important for both tasks. Without concomitant use of a common baseline comparison, this method relies excessively on *a priori* assumptions of the cognitive components of the tasks. *Correlational* methods determine the change in blood flow in relation to some other variable of interest (in this example, presentation frequency, though many others are possible) and is a relatively new, yet promising, paradigm design.

more complex cognitive processes. In the most simplified formulation of this model, progressively hierarchical skills are subtracted from each other, under the assumption that each additional task in the cognitive hierarchy subsumes all functions in the preceding level, plus one more of interest. For example, in the widely cited study from the Washington University group (Petersen, Fox, Posner, Mintun, & Raichle, 1988), staring at a crosshatch was subtracted from passive viewing of words to reveal regions responsible for visual stimulation, word forms, and automatic semantic processing (see Figure 1). Reading the same words aloud presumably activated all the same regions, plus the motor output and articulatory centers, and subtracting the former condition from the latter then should reveal these functional areas alone. Likewise, having the subject generate a list of words and subtracting repetition of the same words should reveal the semantic association regions, at least as these authors have argued (Petersen *et al.*, 1988; Posner, Petersen, Fox, & Raichle, 1988; Raichle, 1991).

Cognitive subtraction models are attractive in that they pose a simple way to analyze and interpret PET data. Many studies since have utilized the hierarchical model; the experimental designs reflecting this approach tend to have several, highly similar task parameters. In the ideal subtraction design, input and output are kept constant, and instructions to the subject to attend to one or another aspect of the stimulus input appears sufficient to generate strikingly different and telltale cognitive maps. An extraordinary example of this approach is the visual study of Corbetta, Miezin, Dobmeyer, Shulman, & Petersen (1990). Here, moving visual stimuli differing in form and color were presented in all conditions, and subjects were instructed to make a decision based on one of these parameters. Results revealed visual-cortical regions thought to represent motion, form, and color.

Despite the wide use of subtraction methodology in PET activation experiments, few believe that cognitive skills are truly organized in an additive, hierarchical fashion. Evidence to the contrary from PET studies has been reported from several groups; for example, in our own study of silent versus oral reading (Bookheimer *et al.*, 1995), differences in the two conditions were seen not just in motor cortex, but also in angular gyrus (silent reading only), superior temporal gyrus (oral reading only), inferior temporal gyrus (silent reading), area 47 (silent reading), prefrontal cortex (silent reading), and supramarginal gyrus (oral reading). It appeared that separate cortical systems were active in oral versus silent reading, in a pattern consistent with a dual-coding model of single-word processing, but clearly inconsistent with the additive assumption. Another example is in the suppression of activity (relative decrease in rCBF) in Wernicke's area in a word generation task in comparison to a silent language task (Frith, Friston, Liddle, & Frackowiak, 1991b). Indeed, even the motor output systems engaged can be remarkably dependent upon the context in which the (same) motor behavior is performed (Raichle *et al.*, 1994). It would appear that any study in which parallel pathways are anticipated or even plausible spells trouble for the hierarchical model.

A second problem with subtraction designs, especially when comparing highly similar tasks, is that processing regions may be activated automatically, even when such processing is not required for the task. For example, presenting subjects with words may automatically activate semantic processing regions even when the subjects do not have to read or make semantic judgments. This was convincingly argued by Sergent and Signoret (1992), who demonstrated that the choice of control condition critically determined the resulting activation map. For example, a "form recognition" area in the inferior temporal lobe revealed by Corbetta *et al* (1990) was found by Sergent and Signoret (1992) to be important only in tasks where naming or labeling the visual form was required. This area has been recognized as a naming center for decades, based on lesion research (Burnstine *et al.*, 1990; Luders, 1991; Nielson, 1946). It appeared that presentation of nameable forms was sufficient to activate this object-naming area, although no language processing of the stimuli was required of the subjects. Therefore, while it is tempting to design perfectly identical tasks with slightly different processing demands in order to delineate isolated processing regions, such an approach may produce erroneous findings.

Several alternatives to strict hierarchical approaches have been proposed. The common baseline method (Bookheimer *et al.*, 1995; Demonet *et al.*, 1992) proposes using a single baseline to which all activation conditions are compared. Though this too is a subtractive method, it only necessitates a single hierarchical comparison, preventing a pileup of erroneous assumptions.

Another approach is to make parallel comparisons of activation tasks, either within or across the presumed level in a cognitive hierarchy. For example, in the Demonet *et al.* (1992) study, two tasks presumably in the same hierarchical plane (i.e., involving the same or highly similar input and output) were directly contrasted. Sergent, Zuck, Terriah, & MacDonald (1992) used three parallel conditions which differed strikingly in presumed processing demands; those regions that were active consistently across all comparisons were hypothesized to better define critical regions for the target task (in this case, face recognition). These methods, however, rely heavily on the assumption that other task parameters, such as task difficulty, attentional demands, *etc.*, are equivalent across tasks, an assumption which is in practice exceedingly difficult to demonstrate. Further, it will not reveal cortical regions important for *all* experimental tasks. One way to increase interpretability of parallel comparisons is to supplement them with a common baseline comparison, which will add to the picture cortical regions that are shared by all tasks.

A fourth alternative is a correlational approach in which the blood flow responses are matched in some statistical way to an external variable. One example is the Price *et al.* (1992) study, in which words were presented auditorily at varying rates. They found a linear relationship between rCBF and presentation rate in auditory cortex but not in Wernicke's area. Thus, the language component of the task was not identified by whether the region was activated above threshold, buy rather by how the signal was transformed. Unfortunately, this approach has been underutilized, and virtually nothing is currently known about the actual or expected hemodynamic response functions in any "higher" brain regions.

In each of these logical approaches to designing activation experiments, scientists must make assumptions about which cognitive skills are involved in each experimental condition. Not surprisingly, these assumptions are subjective and prone to dispute. For example, a task in which subjects were required to state whether a given noun was a "Canadian animal" or a "foreign animal" was stated to involve access to semantic systems as opposed to making an animal size judgment (Bub, 1991), although most would agree there is scant evidence for a cortical organization underlying animal nationality. Even outside this extreme example, it is striking to note the confidence with which the dimensions underlying complex cognitive tasks are stated in many PET studies. This is a particularly difficult problem when one considers that processing regions which are not specifically required for a task may nonetheless be engaged automatically (Posner *et al.*, 1988; Sergent, Ohta, & MacDonald, 1992). Since activation maps tell us only which brain regions are working during that experiment but neglect to mention what precisely it is they are up to, our inferences about which cognitive skills are important for a given task are not only critical, they are untestable. We ought not thereby assume that our deductions are incontrovertible.

With these cautions in mind, a review of some recent PET activation studies on complex cognitive skills follows.

## Higher-Order Visual Processing

The very earliest $^{15}O$ PET activation studies were conducted on the visual system (Fox *et al.*, 1984). Subjects wore goggles which flashed a high-contrast checkerboard, producing blood flow increases, or "activation", in primary visual cortex. Visual activation was an obvious choice in developing the technique for studies of normal physiology, since at least the primary visual pathways are relatively well understood. This was indeed confirmed in later studies of the topography of visual cortex (Fox *et al.*, 1986).

A more challenging exercise has been in understanding secondary and tertiary visual processing regions. The mechanism of higher visual processing is controversial, though it has become less so in humans with recent imaging experiments. For instance, Zeki (1973) has argued on the basis of work with nonhuman primates that there is a high degree of modularity within the visual system, such that different stimulus properties are processed in distinct cortical regions. For instance, he has identified a color processing region of the brain–area V4 located in prestriate cortex in monkeys. This modularity was confirmed in humans using PET (Zeki *et al.,* 1991). Here, subjects viewed either an abstract colored visual display or an equiluminant gray scale version of the same presentation, in comparison to a resting control. An area in lingual-fusiform gyri was active uniquely for the colored display. Likewise, the concept that there is a very specific visual motion center was tested in another study by Watson *et al.* (1993), confirming a motion-specific center, presumably human MT or V5 in the lateral occipital–temporal junction. Similar findings were reported by Corbetta *et al.* (1990). Using a selective attention paradigm, they presented an elegant study of color, form, and motion processing. Here, subjects viewed moving shapes differing in form (e.g., triangles, squares) and color. In different scans, subjects were required to report on one of these three dimensions (in a divided attention condition, they monitored all three). Although in each case subjects were presented identical stimuli and made identical motor responses, the visual characteristics upon which a decision was made differed. Accordingly, subjects showed rCBF increases in fusiform gyrus when processing form, a lateral occipital area when processing motion, and a medial occipital area when processing color. All tasks produced primary visual cortex activation. This study provided strong evidence for specialization in separate visual cortical regions on the basis of stimulus features and demonstrated that utilizing these cortical regions was under some attentional control and not merely reflexive. It also provided the basis for selective paradigms as an effective way to identify unique brain regions without the contaminating effects of different input and output skills.

Beyond stimulus processing, the visual system has been described, on the basis of nonhuman primate studies, to have separate pathways involved in further analysis of this input. Ungerleider and Mishkin (1982) demonstrated that two separate pathways emanating from striate cortex, a "where" pathway and a "what" pathway, provided parallel and independent analysis of incoming visual information. The dorsal route included the superior occipital and parietal regions and was involved in spatial aspects of the stimuli, including speed and direction of motion. The ventral pathway involving fusiform and inferior temporal (IT) gyri were associated with identifying or recognizing a visual stimulus and was sensitive to features like color and shape. In a fascinating demonstration of the similarities in humans and nonhuman primates, Haxby *et al.* (1991) at the National Institute of Health used PET to demonstrate separate dorsal and ventral systems in visual identification and spatial localization in humans. Subjects performed several match-to-sample tasks, either comparing a series of human faces (identification task) or matching the location of dot patterns (localization task). Both tasks produced increased blood flow in the lateral occipital cortex. In addition, the face matching task produced unique activation in occipito-temporal cortex. Conversely, the location matching task produced activation in lateral superior–parietal cortex. The separate activation patterns were consistent with the model developed by Ungerleider with some subtle anatomical changes.

In further studies using the same stimuli, McIntosh *et al.* (1994) used a correlational technique to uncover interactive networks important for identification versus localization. Here the analysis was conducted to determine not which regions exceeded a threshold of blood flow change, but rather which regions were correlated together in their blood flow response patterns across the range of conditions. This approach was designed to identify systems, rather than isolated regions, which worked in concert to perform complex visual tasks. They found evidence for the same dorsal and ventral visual systems as the Ungerleider and Mishkin model had demonstrated in monkeys, adding a connection of area 46 to the spatial pathway. This finding was extremely important, both for its direct hypothesis testing based on an animal model and for the use of correlational approaches to add a new dimension to activation imaging research: localization at the systems level. It is most unfortunate that few other groups have utilized this approach.

Within the ventral visual identification ("what") system, regions specialized by stimulus content have been suggested from patients with isolated agnosias. Cases of both object agnosia and prosopagnosia (face recognition) have been reported. Sergent, Ohta, & MacDonald (1992) studied localization of object and face recognition using a subtraction paradigm. Here, subjects determined the gender of a face, determined whether a famous face was an actor, or determined whether an object was living or nonliving. The comparison conditions were rest with fixation, making orientation judgments to sine wave gratings, or passively viewing faces. Using a combined method of hierarchical subtractions and parallel comparisons, Sergent identified several regions common to object and face processing—fusiform, middle temporal, and orbital gyri—and also areas unique to each task: right lingual gyrus for faces, left basal occipito–temporal for objects. Object categorization showed greater left basal temporal activation, while the right hemisphere homologue was more important for faces, but only when contrasted against gender identification and sine wave gratings. The instability of results across the various subtraction conditions in this study accents the difficulty in drawing clear conclusions in activation control tasks. While the lesion data do suggest that object and face processing are dissociated, typically it is the fusiform and not the lingual gyrus that is impaired in prosopagnosia (Hecaen, 1981). Nonetheless, this study provides evidence for further delineation of systems in the ventral visual system, one involved in categorization of a visual stimulus, that is, a "what" pathway, and a more precise discrimination of a stimulus within category that might be called a "which" pathway.

In summary, the study of visual information processing using PET has progressed from confirmatory studies of visual topography to complex visual figure analysis and to identification of separate systems involved in complex visual processing skills. The progress in this area is exemplary; as yet, no other complex cognitive system has been studied more successfully. One can hope that PET activation studies of other systems can also be studied to this extent.

## Language Processing

### Analysis of Single Words

When simply comparing the number of language functions postulated by early, lesion-based, neuroanatomical (e.g.,Penfield & Roberts, 1959) models to cognitive neurolinguistic models (Caramazza & Berndt, 1978; Coltheart, 1987), one quickly appreciates a disparity between the approaches. A "simplified" model of word reading, for example (Coltheart, 1987), posited no fewer than 12 distinct compartments, including the auditory input lexicon, phonological output lexicon, orthographic output lexicon, and an undefined "cognitive system". In contrast, Penfield's (Penfield & Roberts, 1959) summary of the known language areas of the brain include one for comprehension, one for reception, and one each for naming, reading, and spelling. What can PET studies tell us about this disparity?

To understand one basic element of language, the processing of single words, Petersen *et al.* (1988) and Posner *et al.* (1988) compared auditorily versus visually presented words in a hierarchical design (see Figure 1). In this study, normal volunteers stared at a fixation point, sat passively while viewing printed words, heard words, repeated words, or generated words. Comparisons effectively demonstrated that 1) visually presented words activate visual cortices and auditorily presented words activate auditory cortices, 2) there is a final common pathway for motor output, and 3) semantic association regions are located in the frontal lobe, around area 47. The authors also argued that there was a word form region in the medial prestriate visual cortex, but with no visual activation comparison conditions, they could not logically draw that conclusion. This study generated considerable interest and more than a little controversy, since ample lesion data suggested that language association cortex is more a posterior temporal–inferior parietal lobe function, as suggested by patients with Wernicke's aphasia (Geschwind, 1970; Hecaen & Albert, 1978). Subsequent investigators have argued, for example, that the inferior frontal activations were due to the "retrieval" process itself and not at all to the activation of stored semantic memories (Frith, Friston, Liddle, & Frackowiak, 1991a).

Several studies were subsequently conducted to test several of the findings from this study. Let us first address the "word form question", that is, the hypothesis that there is a word form region of the brain located in the medial prestriate region. The same group reported a second study in which subjects processed real words, wordlike forms constructed from false fronts, pronounceable nonwords (e.g., gorp) and unpronounceable nonwords (e.g., bggp) (Petersen, Fox, Snyder, & Raichle, 1990). Only words and pronounceable nonwords produced increased blood flow in the medial prestriate region, which they interpreted as evidence for either a loosely fitting word form center or for phonological processing in visual prestriate cortex.

There are alternative explanations. One is that the medial prestriate region is active in response to any visually familiar form. Pronounceable nonwords are typically constructed by changing a single letter of a real word. There is evidence from the lesion data that nonwords with more real word similarities are treated more like real words than nonwords with few real word similarities (Glosser & Friedman, 1990). We tested this hypothesis by comparing visually presented words and objects using PET (Bookheimer *et al.*, 1995) and found that the medial prestriate region was shared by object presentation and that the activation was greater for objects than for words. One study failed to replicate the region at all (Howard *et al.*, 1992). A difference in the latter study was that the words presented were not all concrete nouns; some were low on imagery scores. This would tend to suggest that the medial prestriate region has less to do with visual words forms *per se* than to the familiar visual images they evoked. No studies have yet tested this hypothesis directly for example, by comparing imageable with non-imageable word presentation, or by varying nonsense words on a dimension of similarity to real word forms.

Another potential word form center, the angular gyrus, was identified in a similar study of single-word processing (Howard *et al.*, 1992). Here, subjects read words, repeated words, or repeated a single word, "crime", following visual or auditory presentations. This is called a "see and say" condition. Paired comparisons between reading and "see-and-say" revealed an area in the posterior middle temporal–angular gyrus region, while a slightly more anterior region was found for auditory input only. It is unfortunate that no other study involving reading has shown activation here, and given the small number of subjects it is difficult to say with confidence whether this is a visual word center or a visual word lexicon, as the authors suggest. It was nonetheless intriguing to suggest that auditory and visual input may selectively activate separate semantic processing centers. An important and unanswered question in language research is whether and to what extent semantic processing is modality-specific. Modality specificity at this level was not found in the Petersen *et al.* (1988) study, which also involved both visual and auditory word input. Possibly, the task of reading words and auditory repetition do not sufficiently activate semantic stores to reliably assess the issue of separate lexicons. More complex semantic tasks involving cross-modality comparisons have not yet been performed.

### Phonological Processing

One controversial issue in the neuroanatomy of language concerns the role of area 44—the inferior frontal gyrus or Broca's area. When lesioned this region results in expressive language deficits, but at what level? Stimulation studies (Ojemann, Ojemann, Lettich, & Berger, 1989) have noted that this region will produce not only expression deficits but also difficulties in phoneme monitoring. This would tend to suggest that Broca's area is critically important for the production, planning, or sequencing of speech-related sounds. Phoneme processing, in contrast, has typically been attributed to the perisylvian region of the dominant hemisphere.

The role of Broca's area in phoneme discrimination was examined in a study by Zatorre and his colleagues (Zatorre, Evans, Meyer, & Gjedde, 1992). Here, subjects listened to noise bursts or a rising tone, passively listened to speech, or listened to speech and made a phoneme judgment. While primary auditory cortex was active in all sound tasks, secondary auditory cortex responded only to speech presentation, including a left posterior superior temporal gyrus (STG) region. Broca's area, however, was engaged only when subjects performed a phoneme discrimination task, even in comparison to the passive speech control. The authors attribute this finding to the role of Broca's area in "articulatory recoding in phonetic perception", while offering no direct evidence for this interpretation. For example, it is equally possible that Broca's area is a phoneme decoding region, a sequential processing or "parsing" region, or even that subjects consulted their internal lexicons, perhaps performing a semantic or word matching task as a part of the phoneme identification process. The latter hypothesis would suggest that mere presentation of language elements can activate "word" or "semantic" processing regions. Another region these authors reported in this comparison is a superior parietal region, which is close to one we recently found in a word matching task (Bookheimer *et al.*, 1993). Activation here in the Zatorre study suggests that the phoneme task may indeed have engaged "higher" language processing regions. Thus, while activation in Broca's area during phoneme monitoring is interesting and in agreement with previous stimulation studies, PET research thus far has not elucidated its precise function.

### Semantic Processing

To differentiate language processes involved in phoneme analysis from semantic analysis, Demonet *et al.* (1992) performed a series of PET experiments using auditory stimuli. In a control condition, subjects listened to tones, monitoring for a target change in pitch. In the experimental tasks, they either monitored a phoneme sequence or monitored words for a set of semantic characteristics (small animals with positive traits). They used the tone condition as a common baseline for the experimental tasks and then performed parallel comparisons of phonemes versus words and the converse. Several regions in temporal lobe, including superior middle and inferior temporal gyri, were important for both phonemes and words. Phoneme monitoring versus tones selectively activated a region in the superior portion of inferior frontal gyrus (IFG) at Brodmann's area 44–45, which the authors refer to as Broca's area. This very interesting finding replicated that of Zatorre *et al.* (1992) as well as previous stimulation data. The finding might be attributed either to phonological processing or possible subvocal articulation. The authors argue against this hypothesis, however, on the basis that previous studies suggest the supplementary motor area (SMA) has been associated with speech initiation, while the SMA was not active in the present study. Instead they suggest that this portion of the IFG might be involved in sequencing and rehearsing phoneme strings, although it is unclear how to dissociate this view from the former. No regions active for phonemes were also active for word monitoring. In contrast, the semantic task produced unique activations in four cortical regions: the left inferior parietal lobe (BA 39/40), left prefrontal cortex (BA 8,9), left inferior temporal gyrus, and the left precuneus. It is interesting that the semantic task did not selectively activate classic Wernicke's area, which certainly would have been predicted. Nonetheless, inferior parietal cortex is sometimes included in the large territory described as the receptive language area (Penfield & Roberts, 1959) and may have some unique semantic processing properties. The prefrontal region (BA 8, 9) is not traditionally thought of as a semantic processing region. Although the authors point out that this region has close anatomical connections to other language processing regions, the reason for this activation is still unclear. The inferior temporal region identified by this study is less well advertised in the language literature, but nonetheless has been well characterized as important, particularly in naming, and is variously referred to as Nielson's area (Nielson, 1946), Mill's naming center, and more recently, the basal temporal language area (Burnstine *et al.*, 1990; Luders, 1991). The small area of activation in area 47, which was posited by Petersen *et al.* (1988) as a semantic region, was found in both phoneme and semantic tasks, making this interpretation less likely.

The results of the Demonet *et al.* (1992) study are difficult to reconcile with the rather simple, traditional notion of auditory comprehension that there are primary auditory areas for phoneme analysis in the midanterior superior temporal lobe and a more posterior region called Wernicke's area responsible for language comprehension. No fewer than eight separate regions specific to language stimuli were identified here, and the stimuli were relatively simple. These results strongly suggest that more regions were active in these tasks than were probably essential to the specific characteristics of interest, that is, phoneme and semantic processing.

Wise *et al.* (1991) attempted to further delineate neural networks involved in phonological versus semantic aspects of language processing using a series of auditory word comprehension tasks. Here, subjects rested, listened to nonwords, made semantic category judgments for superordinate and subordinate nouns, made object-action semantic judgments, or generated verbs that appropriately matched a specified noun. The authors formulated their tasks based on the model of single-word processing offered by Ellis and Young (1988), which posits both a hierarchical structure to semantics and a determined sequence of events in analyzing words. Using a common baseline (rest), they found that the three tasks which involved listening to words (hearing nonwords and making the noun and verb matching judgments) all produced extensive activation in bilateral temporal cortices, as was seen in the previous studies involving word presentation. They interpreted this finding as evidence of the auditory input lexicon as proposed by Ellis and Young (1988). Parallel comparison of the two word matching tasks revealed no significant differences; further, when the category tasks and the generation task were compared to nonsense words, there were no significant increases in activity in Wernicke's area. However, a post hoc correlation analysis that used a subset of the conditions (primarily generation), in which words were presented at a slower rate, found no rate-dependent correlation in Wernicke's area, while this correlation was demonstrated in more anterior auditory and right hemisphere areas. This finding was elaborated in the study (Price *et al.*, 1992) described below. The failure to achieve clear effects in Wernicke's area in comparison to the nonsense word task was troublesome nonetheless. The authors suggested that nonword processing may have involved some activation of the semantic system. This possibility would conflict with a strictly hierarchical and modular model of auditory versus semantic processing but is in agreement with parallel, interactive, or cascade models of cognitive processing such as those proposed by McClelland and Rumelhart (1986). Finally, this study demonstrated more circumscribed activation of Wernicke's area, relative to rest only, during the generation task; this task also produced an activation of presumed Broca's area (superior area 44), the dorsolateral prefrontal cortex (DLPFC), and SMA.

A different approach to identifying critical language areas was taken by Price *et al.* (1992) in a parametric study of auditory word processing. Here, subjects listened to words presented at different rates. To analyze the data, Price performed a correlational analysis in which regions showing a linear trend were identified; they found that the bilateral primary auditory cortices and right Wernicke's homologue were responsive to input frequency; in contrast, left posterior temporal cortex, presumed Wernicke's area, was active equally when real (but not nonsense) words were presented and was insensitive to input frequency. While the study did not attempt to find a parameter to which Wernicke's area responded parametrically (e.g., semantic complexity) the study nonetheless demonstrated that cortical regions may differ meaningfully in signal transduction, suggesting a new approach to understanding the role of activated regions. The parametric approach may have the potential to differentiate *critical* structures from those involved in performance of a complex task, and it is to be hoped that others will follow this example.

### Word Generation and Retrieval

The word generation paradigm has been exploited in several studies as a means to assess semantic systems in the frontal lobe. In the single-word study by Petersen *et al.* (1988), generating a verb, or action, in response to a noun (e.g., ball-throw) produced activation in two frontal lobe regions, area 47 of the IFG and the DLPFC. The authors interpreted this finding as an indication that semantic associations are stored in the frontal lobe. This suggestion conflicts substantially with existing literature from lesion studies. Typically, semantic systems are described as a posterior temporal or inferior parietal function, or as diffusely represented in the brain. Frith *et al.* (1991a) attempted to replicate the DLPFC findings with an interesting twist: hypothesizing that this region was important not in the semantic aspects of the task but rather in the intrinsic generation component, he performed a study in which both language and non-language tasks involved intrinsic generation of a response—"willed behavior". In the language conditions, subjects repeated a word, stated the opposite of a word, or generated a response from a presented word. A set of motor tasks involving the same essential components—repetition of a response, performance of a response which was apparent by the task (moving a finger to the right of a touched finger), or intrinsic generation of a finger movement. Here, subjects moved a finger in a self-appointed and tandem direction. The intrinsic generation tasks produced activation in DLPFC for both language and motor performance. These data strongly suggested that the reason for activation in DLPFC was not because semantic systems were involved as Petersen *et al.* (1988) had claimed, but rather because intrinsic generation, that is, willed behavior, was required. It should be noted, nonetheless, that the pixel maxima in the verbal and motor tasks differed by about 7 mm in the $x$-plane and 10 mm in the $y$-plane; other papers from that group have argued that this distance is significant (Howard *et al.*, 1992). It is intriguing to hypothesize that intrinsic generation may itself have an organizational structure—in a sense, a cognitive topography within frontal cortex—but this requires further confirmation.

The language findings to date have been disappointing in one sense: even the fundamental assumptions about how the major language areas respond to simple tasks have been difficult to confirm. Broca's area has been identified most consistently not in speech but rather in phoneme tasks; during speech it would appear that several frontal regions, from area 47 up through the BA 6/8, can be engaged. Surely these do not represent the same functional zones but must be further characterized and dissociated. Wernicke's area has been most troubling to identify even in seemingly straightforward semantic tasks. One possibility is that Wernicke's area is easily engaged, perhaps automatically, to presentation of any speechlike stimuli, even when semantic judgments are clearly not required for the task. Another possibility is that the location is so variable cross subjects that group averages cannot bring changes here to threshold (see Steinmetz & Seitz, 1991, for an excellent discussion). Enticing findings by Wise *et al.* (1991) and Price *et al.* (1992) suggest that the correlational approach is better for identifying Wernicke's area than traditional subtraction threshold methods. If so, future advances in understanding language through activation imaging will rely on a very different set of approaches to paradigm design.

## Memory

In comparison to the language realm, surprisingly few memory studies have been conducted with activation imaging. Most work with patients with amnesia implicate the hippocampal formation as the essential structure in encoding and/or consolidation of new memories. As this is a rather small subcortical region and less perfused with blood than neocortex, it has consequently been rather difficult to study. Further, the size of the region, especially with respect to the spatial resolution of PET, makes precise localization of blood flow changes in hippocampus speculative at best. Nonetheless, several studies have attempted to delineate the role of the hippocampus and related structures in memory using PET.

Grasby and colleagues (Grasby *et al.*, 1993) performed a two-group design to study auditory-verbal memory function in normal volunteers. One group of subjects performed two memory tasks—a subspan and a supraspan memory task involving free recall of a word list. A second group of subjects was studied in the resting state. In the subspan task, subjects heard a series of five-word lists and immediately recalled the list. In the supraspan task, subjects heard a 15-word list which they immediately recalled; this was repeated three times. The total number of words presented and repeated was therefore equal across the scans, although the scans differed both in number of items per trial and in total number of novel items recalled. the authors presumed that the tasks differed only in that the latter uniquely invoked a long-term memory system. They did not counterbalance order of condition and thus performed additional scans, also not counterbalanced, on a subgroup of subjects. Here, six of the original group members performed the experimental tasks four more times in an ABAB design where the initial condition was always the subspan test. The authors first performed parallel comparisons of supraspan and subspan. In both groups, the supraspan task produced increases in prefrontal cortex (BA 9/10 and 46), posterior cingulate, and precuneus. The subspan task showed relative increases in superior and middle temporal gyri and the inferior precentral gyrus. In comparison to the group of subjects at rest, the subjects performing subspan task had increases in blood flow in bilateral superior temporal gyrus and thalamus, right parahippocampal gyrus, and left anterior cingulate. The same comparison with the supraspan task found similar increases in the thalamus and cingulate, cerebellum, parahippocampal gyrus, and left STG, though less than in the subspan task. Thus, the regions involved in the two memory tasks were differentiated only by the change in the superior temporal gyri—a relative decrease for the supraspan task. The authors argued that use of the ABAB design in the second study precludes order effects as explaining these results, though this conclusion is erroneous, since a decreasing trend in auditory cortex regions would have produced the same effect. While their behavioral data indicated that the tasks were indeed different (subjects performed more poorly on the supraspan task and showed primacy and recency effects), it is not at all clear that the CBF data support dissociated systems. The change in ST might reflect order effects or increased processing of auditory information in the subspan tasks merely because there were more novel items. Repetition of items has been demonstrated to reduce rCBF (Blaxton, Bookheimer, Zeffiro, Figlozzi, Gaillard, & Theodore, in press), as will be described in the memory section below. Increased blood flow in DLPFC was attributed to variability in baseline prefrontal activity, while the authors noted that this region has been associated with several aspects of memory function. Posterior cingulate activity (retrosplenial region) was interpreted broadly as important in the long-term memory system. This region receives input from prefrontal cortex and medial temporal structures in some nonhuman primates and has been implicated in rare cases in anterograde amnesia (Valenstein, Bauers, Verfaellie, Heilman, & Watson, 1987). Nonetheless, it remains unclear why this region should be active when more likely memory structures (e.g., hippocampus) were not.

The distinction between short-term and long-term memory, on which the previous study was based (Atkinson & Shiffrin, 1968), has been emphasized among memory researchers far less in recent years. More recently an emphasis on separate systems has been based on the notion that implicit memory, frequently measured by priming paradigms, differs fundamentally from explicit or declarative memory, which is impaired in patients with amnesia (Squire, 1987; Tulving & Schacter, 1990). Several studies have taken quite different approaches to addressing these questions using PET. The first was a study by Squire with collaborators from Washington University (Squire *et al.*, 1992). Prior to each PET scan, subjects studied a list of 15 words to which they made a desirability judgment. After three minutes they were scanned while performing a word stem completion task. This task, a well-studied paradigm in memory literature, presents three letters to a subject for which they must generate a word containing that stem. This task can be presented either with implicit instructions (generate the first word you can think of beginning with these letters) or explicit instructions (think back to the initial study period and generate the word previously studied). In either case, the probability of generating a previously studied word is greater than would be expected without prior exposure. The authors presented this task with implicit instructions (priming condition), explicit instructions (memory condition), in a control task in which no stems were previously studied (baseline), and in a passive condition in which subjects saw word stems but made no response. All subjects also performed two rest tasks, and the data were analyzed in a hierarchical subtraction design. Behavioral data indicated a significant memory effect for both implicit and explicit instructions (a positive priming effect). For the explicit memory versus baseline (word stem completion) comparison, rCBF increases were observed over the right medial temporal region, covering hippocampus and parahippocampal gyrus, and right prefrontal cortex. For the implicit task, the occipital cortex produced greatest increases, though subsequent analysis showed that the right hippocampus was also active in the priming task. There was also a reduction in the right lingual gyrus for the priming versus baseline comparison. The authors concluded that the right hippocampus is indeed implicated in explicit memory during word stem completion and argue that this *should* be a right hemisphere task, since it is spatial in nature. They further argue that the CBF decrease in right visual association cortex is consistent with the notion that less neural processing is required when input is repeated.

These results are problematic in several ways: the preponderance of right hemisphere activity on a task involving word generation is most unusual; even if, indeed, the visual forms of words can involve the right hemisphere, the existing data on verbal-visual dissociations in the hippocampal memory system clearly implicate the left hemisphere (Milner, 1967). Further, the pattern of results on the few regions analyzed and reported are inconsistent with a straightforward interpretation of blood flow increases with memory processing: specifically, the relationship between blood flow in the baseline condition versus fixation control differed across the regions of interest. Thus, the authors were required to posit decreased neural processing in visual association cortex and increased neural processing in hippocampus, both attributable to the memory processing aspects of the task. In fact, the dissociation between implicit and explicit memory in the right hippocampus was tested without correction for multiple comparisons and was clearly not dependent upon the difference between those conditions but was characterized by a relative decrease in the hippocampus in the baseline condition. When comparing the three described regions to fixation, only the right prefrontal cortex showed a strong dissociation between implicit and explicit instructions.

A recent study by Blaxton *et al.* (in press) extended the basic outline of the Squire *et al.* (1992) paper in a tailored baseline design yielding a very different set of results which were more consistent with the existing memory literature. Several studies have suggested that there are not separate systems involved in explicit and implicit memory; rather, these memory systems tend to be tested using different paradigms that tap different processing skills (e.g., Blaxton, 1992, 1989). In her analysis, Blaxton has argued that explicit memory is typically tested using conceptual processing tasks, while implicit memory is typically tested using perceptual processing tests. To differentiate the relative contributions of these processing functions from implicit/explicit systems, she used an elegant design in which both processing skills were tested with both presumed memory systems. In this study, subjects studied associated word pairs, which were then recalled during the PET scan under four memory conditions. In two conditions subjects received explicit instructions. In one case, they were asked to generate the second word in the word pair (conceptual-explicit); in the other they performed a similar word completion task, this time using word fragments rather than stems while instructed to think back to the original study session for correct performance (perceptual-explicit). This test is considered less likely to invoke explicit processes that the stem completion task yet demands that the subject perform a perceptual analysis of the stimuli. In two implicit conditions, subjects either freely generated an associate from the first word in a pair (conceptual-implicit) or completed a word fragment (perceptual-implicit). Two control tasks—free association and word stem completion, where all items were novel—were used as conceptual and perceptual baselines, respectively. Blaxton's data indicated that the conceptual tasks produced *decreases* in the left hippocampus and left hemisphere language zones (Broca's area, STG) regardless of whether the instructions were implicit or explicit; there was also a slight contralateral increase in the right hippocampus. In contrast, the perceptual tasks produced decreases in parietal and occipital cortex and in right hippocampus, again regardless of whether instructions were implicit or explicit. When contrasting the two explicit and implicit tasks against each other in collapsed parallel comparisons, the only difference was a slight increase in right orbitofrontal cortex during explicit task performance, but no changes in hippocampus. This data suggested that the direction of blood flow changes with studied versus nonstudied items was a *decrease* in the relevant structures; the cortical structures that showed the decrease were process-specific (conceptual versus visual perceptual), a finding consistent with the content specificity model of the hippocampus. Finally, the results did not support dissociations on the basis of implicit versus explicit instructions. In contrast to the Squire *et al.* (1992) study, Blaxton's did not require positing qualitatively different memory effects for hippocampal versus extrahippocampal brain regions, and the results are more in keeping with well-established literature on the lateralization of visual and verbal memory functions (Milner, 1967). They do, however, provide evidence for the priming effects observed in experimental studies, which probably also occurred in studies of auditory-verbal memory by Grasby and colleagues (1993a,b).

In a recent study, Shallice *et al.* (1994) attempted to elaborate on the episodic-semantic memory distinction in a verbal memory study. They used a distraction paradigm, reasoning that a difficult distractor task would lessen the involvement of episodic memory structures, a conclusion they reached on the basis of behavioral studies in normal volunteers. In the acquisition portion of the study, subjects listened to difficult paired associates, either passively or with instructions to remember them, while performing a visual motor tracking task that was either easy or difficult. The common baseline condition was passive listening to word pairs. During the *easy* distractor task, in which subjects are presumed capable of episodic memory encoding, the cortical regions activated included anterior and posterior cingulate, superior temporal gyrus, and a left frontal cortex region erroneously referred to as dorsolateral prefrontal cortex, but which was actually rather ventral near BA 46/10 based on atlas coordinates (Talairach & Tournoux, 1988). In comparison, the *difficult* distractor task, which presumably involves no episodic memory encoding, produced smaller activation of the STG, larger activation of the anterior cingulate, and no significant changes in the inferior frontal cortex and posterior cingulate. Memory performance was predictably reduced during the distractor task. Logically, this study could not demonstrate that the blood flow changes were due to the memory components of the task rather than the differences in performance of the distractor task since there was no independent evaluation of the distractor tasks. Nonetheless, the authors did so, arguing that the retrosplenial cingulate and frontal lobe regions must be the critical structures in the episodic memory encoding system. Aside from the obvious methodological problems, the frontal lobe region they refer to as DLPFC is actually far more ventral than dorsal, farther than previously reported (20 mm above the ACPC line in the z-plane; Frith *et al.,* 1991a), and closer to the frontal area reported by Petersen (1988) in a language paradigm with no memory demands.

The authors performed a second experiment in which they scanned subjects during retrieval of the same information, in comparison to a generation task and a repetition task, which was used as the common baseline comparison. During the episodic retrieval versus repetition, increases were observed in anterior cingulate, thalamus, precuneus, the right cingulate sulcus, which is (erroneously) termed right frontal cortex at BA 10/46, and the right insula, described as area 47. There was also a relative decrease in the left superior temporal region, as has been reported elsewhere. The authors suggest that this pattern of activation is consistent with our understanding of the mechanisms underlying retrieval of verbal episodic information while admitting that the lack of effects in hippocampus is problematic. In the semantic memory task (word generation minus repetition), only the anterior cingulate and bilateral thalamus were active. Unfortunately, pixel means of the regions across conditions were not presented. Since the episodic retrieval task had overall higher $z$ values even in shared regions, one wonders if differences between these conditions were not merely due to the increased attention allocated during the retrieval task. This would be consistent with the difference in magnitude of the blood flow changes in anterior cingulate they observed, an area thought to be important for selective attention. Like previous studies, this one used an activation control in lieu of a resting study. Automatic activation of memory systems may occur with mere presentation of novel information, as seen for semantic systems in the language literature. Again it would seem that a different approach to paradigm design would improve the quality of memory studies such as this one.

A different approach was indeed used in a recent study by Grasby *et al.* (1993b). In this parametrically based correlational design, 12 PET scans were performed on each participant; each subject was presented with a word list varying in length from 2 to 13 words, which they immediately recalled. A measure of long-term memory performance was extracted from the behavioral data from the subset of scans in which the word list lengths were at least seven items long; performance was judged on items in the middle of the word list, to control for primacy and recency effects. These performance measures were then correlated with blood flow in the hippocampal plane alone, reducing the need for multiple comparison corrections. Using this approach, Grasby was able to demonstrate a significant positive correlation in the left hippocampus. In contrast, short-term memory performance was not correlated with hippocampal blood flow. Other structures involved in this task could not be commented upon since the authors' decision to select a low alpha value necessitated restricting interpretation to the single plane. Given how much data was ultimately thrown out, one can guess that several analyses were performed which did not yield the desired results in hippocampus. Nonetheless, this study successfully demonstrated that positive changes in hippocampal blood flow could be observed with the right paradigm design and analysis technique.

Given the paucity of PET activation studies on memory, it is impossible to draw clear conclusions about the relative role of structures identified as key to memory from lesion and animal literature. It does appear that repetition of stimuli over a relatively short period of time produces blood flow changes that differ from the typical response we expect in activation studies, that is, memory studies can produce relative decreases in blood flow that are apparently critical to the tasks. While blood flow decreases are easily observable, how they are to be interpreted within a set of results with both positive and negative changes is not straightforward. Since both directions of change have been reported in the hippocampus, it remains to be seen how mnemonic behavior affects blood flow response dynamics in the critical memory structures.

## Music

Perhaps one of the most "human" of traits is the ability to appreciate and compose music. Inspired by the French composer Maurice Ravel, who was unable to compose music following a stroke affecting his left hemisphere (Sergent,

1993), the late Justine Sergent, herself an accomplished musician, performed a series of PET scans identifying areas of the brain specialized for reading and playing music (Sergent, Zuck, Terriah, & MacDonald, 1992). She studied ten professional pianists as they listened to scales, played scales on a keyboard, read a musical score, read a score while listening, or stared at a random dot pattern. Using a combination of hierarchical and parallel subtractions, she determined that the translation of notation into sound patterns (reading score and listening minus reading score) involved the superior supramarginal gyrus, an area near but not identical to that involved in reading. Sight reading while playing and listening produced activation in the superior parietal lobe (area 7 bilaterally and a premotor area bordering Brodmann's areas 6 and 44). These were interpreted as reflecting the integration of spatial information with a guided sensorimotor response and motor sequencing. She concluded that while these musical skills bear much surface resemblance to language—motor sequencing, prosody, a rule-guided syntax—the neural networks underlying musical performance and language are unique. Other aspects of music, including appreciation and the emotions music can evoke, will have to await future investigations.

## Discussion

Although the results of cognitive neuroimaging studies are open to liberal interpretation and remain extremely controversial, there are some general conclusions we can reach. First, the maps provided by our nineteenth-century researchers quite accurately identified and localized cognitive processing elements; when it comes to basic skills, it would be hard to argue that activation research has led to the discovery of any new ones. But the old maps were far too simple; the number of regions important for the same cognitive skills has grown; how and when these skills are utilized appears very dependent upon rather specific and often seemingly trivial differences in task demands. Possibly, cortical regions differ not just in whether they are important for task performance, but also in how they respond dynamically in response to task parameters. This kind of analysis has not been possible in lesion and stimulation studies and opens a new window of opportunity in functional neuroimaging research. Unfortunately there has as yet been little progress in understanding how different functional regions interact dynamically during performance of complex tasks. Correlational analysis, neural network approaches, and parametric designs will probably gain momentum in the next generation of PET experiments. In the end, it is very unlikely that any one experiment can answer broad and difficult questions about the organization of brain functions. Instead, many experiments from varying institutions with diverse approaches will continue to produce results subject to local interpretation, but will also reveal a slowly emerging brain map of complex cognitive skills. Nonetheless, activation imaging has clearly contributed much in a very brief time to our understanding of cognitive neurosciences. This progress should be accelerated further when functional MRI (see Chapter 12, Volume II of this series) becomes more widely available, perhaps adding a temporal processing dimension to cognitive research.

Each PET experiment generates a list of brain regions involved in some part of performing a complex task, and the number of these regions adds up quickly; the speed by which we can acquire new data seems to have outstripped the speed by which we can understand and interpret them. Having reported a set of results, usually as $z$ values of pixels representing the maximal blood flow change in a particular set of comparisons, there is still controversy over how to express the location of these changes. Although most centers have settled on reporting PET results in a common, three-dimensional coordinate space, there is still little agreement on anatomical basics, such as the bounds of Wernicke's area. The labels of functional brain regions (like Wernicke's or Broca's area) may ultimately give way to numerical descriptions of maximum areas of blood flow change, looking something like the "− 32, + 36, 0 language area". There is a recent effort by one of the pioneers in PET research, Peter Fox, and his colleagues at the University of Texas, San Antonio, to collect and tabulate activation data from different sources into a large, brain mapping database that would be available to all scientists. This database will incorporate not only PET activation data, but also data from fMRI, electrophysiology, and neuroanatomy. Programs like this will continue to increase the communication between scientists from very different backgrounds and can only bring us closer to understanding human cognition.

## References

Atkinson, R., & Shiffrin, R. (1968). Human memory: A proposed system and its control processes. In K. Spence & J. Spence (eds.), *The psychology of human learning and motivation: Advances in theory and research* (pp. 89–195). New York: Academic Press.

Blaxton, T. (1989) Investigating dissociations among memory measures: Support for a transfer appropriate processing framework. *Journal of Experimental Psychology: Human Memory and Performance, 15*(4), 657–668.

Blaxton, T. (1992). Dissociations among memory measures in memory-impaired subjects: Evidence for a processing account of memory. *Memory and Cognition, 20*(5), 549–562.

Blaxton, T., Bookheimer, S., Zeffiro, T., Figlozzi, C., Gaillard, W., & Theodore, W. (in press). Functional mapping of human memory using PET: Comparisons of conceptual and perceptual tasks. *Canadian Journal of Experimental Psychology.*

Bookheimer, S., Zeffiro, T., Gaillard, W., & Theodore, W. (1993). Regional cerebral blood flow changes during the comprehension of syntactically varying sentences. *Society for Neurosciences Abstracts, 22.*

Bookheimer, S. Y., Zeffiro, T. A., Blaxton, T., Gaillard, W., & Theodore, W. T. (1995). *Regional cerebral blood flow during object naming and word reading.* Human Brain Mapping, *3,* 93–106.

Bub, D. (1991). Paper presented at the biannual meeting of the Society for Cerebral Blood Flow and Metabolism, Miami, FL.

Burnstine, T., Lesser, R., Hart, J. J., Uematsu, S., Zinreich, S., Krauss, G., Fisher, R., Vining, E., & Gordon, B. (1990). Characterization of the basal temporal language area in patients with left temporal lobe epilepsy. *Neurology, 40*(6), 966–970.

Caramazza, A., & Berndt, R. S. (1978). Semantic and syntactic processes in aphasia: A review of the literature. *Psychological Bulletin, 85,* 898–918.

Coltheart, M. (1987). functional architecture of the language processing system. In M. Coltheart, G. Sartori, & R. Job (eds.), *The cognitive neuropsychology of language* (pp. 1–24). Sussex: Lawrence Erlbaum Associates.

Corbetta, M., Miezin, F. M., Dobmeyer, S., Shulman, G. L., & Petersen, S. E. (1990). Attentional modulation of neural processing of shape, color, and velocity in humans. *Science, 248*(4962), 1556–1559.

Demonet, J. F., Chollet, F., Ramsay, S., Cardebat, D., Nespoulous, J.-L., Wise, R., Rascol, A., & Frackowiak, R. (1992). The anatomy of phonological and semantic processing in normal subjects. *Brain, 115,* 1753–1768.

Ellis, A., & Young, A. (1988). *Human cogitive neuropsychology.* Hillsdale, NJ: Lawrence Erlbaum.

Fox, P. T., Mintun, M. A., Raichle, M. E., & Herscovitch, P. (1984). A noninvasive approach to quantitative functional brain mapping with $H_2$ (15)O and positron emission tomography. *Journal of Cerebral Blood Flow and Metabolism, 4*(3), 329–333.

Fox, P. T., Mintun, M. A., Raichle, M. E., Miezin, F. M., Allman, J. M., & Van Essen, D. C. (1986). Mapping human visual cortex with positron emission tomography. *Nature, 323*(6091), 806–809.

Frith, C. D., Friston, K., Liddle, P. F., & Frackowiak, R. S. (1991a). Willed action and the prefrontal cortex in man: A study with PET. *Proceedings of the Royal Society of London, Series B: Biological Sciences, 244*(1311), 241–246.

Frith, C. D., Friston, K. J., Liddle, P. F., & Frackowiak, R. S. (1991b). A PET study of word finding. *Neuropsychologia, 29*(12), 1137–1148.

Geschwind, N. (1970). The organization of language and the brain. *Science, 170,* 940–944.

Glosser, G., & Friedman, R. (1990). The continuum of deep/phonological alexia. *Cortex, 26,* 343–359.

Grasby, P., Frith, C., Friston, K., Bench, C., Frackowiak, R., & Dolan, R. (1993a). Functional mapping of brain areas implicated in auditory-verbal memory function. *Brain, 116,* 1–20.

Grasby, P., Frith, C., Friston, K., Frackowiak, R., & Dolan, R. (1993b). Activation of the human hippocampal formation during auditory-verbal short term memory function. *Neuroscience Letters, 163,* 185–188.

Haxby, J., Grady, C., Horwitz, B., Ungerleider, L., Mishkin, M., Carson, R., Herskovitch, P., Schapiro, M., & Rapaport, S. (1991). Dissociations of spatial and object visual processing pathways in human extrastriate cortex. *Proceedings of the National Academy of Sciences of the United States of America, 88,* 1621–1625.

Hecaen, H. (1981). The neuropsychology of face recognition. In G. Davies, H. Ellis, & J. Shepard (Eds.), *Perceiving and remembering faces* (pp. 39–54). London: Academic Press.

Hecaen, H., & Albert, M. (1978). *Human neuropsychology.* New York: John Wiley and Sons.

Howard, D., Patterson, K., Wise, R., Brown, W. D., Friston, K., Weiller, C., & Frackowiak, R. (1992). The cortical localization of the lexicons. *Brain, 115,* 1769–1782.

Luders, H. (1991). Basal temporal language area. *Brain, 114*(2), 743–754.

McClelland, J. L., & Rumelhart, D. E. (1986). *Parallel distributed processing.* Cambridge, MA: MIT Press.

McIntosh, A., Grady, C., Ungerleider, L., Haxby, J., Rapaport, S., & Horwitz, B. (1994). Network analysis of cortical visual pathways mapped with PET. *Journal of Neuroscience, 14*(2), 655–666.

Milner, B. (1967). Brain mechanisms, suggested by studies of temporal lobes. In C. Millikan & F. Darley (Eds.), *Brain mechanisms underlying speech and language.* New York: Grune & Stratton.

Nielson, J. (1946). *Agnosia, apraxia, aphasia: Their value in cerebral localization* (2nd ed.). New York: Hoeber.

Ojemann, G., Ojemann, J., Lettich, E., & Berger, M. (1989). Cortical language localization in left, dominant hemisphere. An electrical stimulation mapping investigation in 117 patients. *Journal of Neurosurgery, 71*(3), 316–326.

Penfield, W., & Roberts, L. (1959). *Speech and brain mechanisms.* Princeton, NJ: Princeton University Press.

Petersen, S. E., Fox, P. T., Posner, M. I., Mintun, M., & Raichle, M. E. (1988). Positron emission tomographic studies of the cortical anatomy of single-word processing. *Nature, 331*(6157), 585–589.

Petersen, S. E., Fox, P. T., Snyder, A. Z., & Raichle, M. E. (1990). Activation of extrastriate and frontal cortical areas by visual words and word-like stimuli. *Science, 249*(4972), 1041–1044.

Posner, M. I., Petersen, S. E., Fox, P. T., & Raichle, M. E. (1988). Localization of cognitive operations in the human brain. *Science, 240*(4859), 1627–1631.

Price, C., Wise, R., Ramsay, S., Friston, K., Howard, D., Patterson, K., & Frackowiak, R. (1992). Regional response differences within the human auditory cortex when listening to words. *Neuroscience Letters, 146*(2), 179–182.

Raichle, M. E. (1992). Memory mechanisms in the processing of words and word-like symbols. *Ciba Foundation Symposium, 163,* 198–204.

Raichle, M., Fiez, J., Videen, T., MacLeod, A., Pardo, J., Fox, P., & Petersen, S. (1994). Practice related changes in human brain functional anatomy during nonmotor learning. *Cerebral Cortex, 4*(1), 8–26.

Sergent, J. (1993). Music, the brain, and Ravel. *Trends in Neurosciences, 16*(5), 168–172.

Sergent, J., Ohta, S., & MacDonald, B. (1992). Functional neuroanatomy of face and object processing: A positron emission tomography study. *Brain, 115,* 15–36.

Sergent, J., & Signoret, J. L. (1992). Functional and anatomical decomposition of face processing:

Evidence from prosopagnosia and PET study of normal subjects. *Philosophical Transactions of the Royal Society of London, Series B: Biological Sciences, 335*(1273), 55–61.

Sergent, J., Zuck, E., Terriah, S., & MacDonald, B. (1992). Distributed neural network underlying musical sight-reading and keyboard performance. *Science, 257*. 106–109.

Shallice, T., Fletcher, P., Frith, C., Grasby, P., Frackowiak, R., & Dolan, R. (1994). Brain regions associated with acquisition and retrieval of verbal episodic memory. *Nature, 368,* 633–635.

Squire, L. (1987). *Memory and brain.* New York: Oxford University Press.

Squire, L. R., Ojemann, J. G., Miezin, F. M., Petersen, S. E., Videen, T. O., & Raichle, M. E. (1992). Activation of the hippocampus in normal humans: A functional anatomical study of memory. *Proceedings of the National Academy of Sciences of the United States of America, 89*(5), 1837–1841.

Steinmetz, H., & Seitz, R. J. (1991). Functional anatomy of language processing: Neuroimaging and the problem of individual variability. *Neuropsychologia, 29,* 1149–1161.

Talairach, J., & Tournoux, P. (1988). *Co-planar stereotaxic atlas of the human brain.* New York: Thieme Medical Publishers.

Tulving, E., & Schacter, D. (1990). Priming and human memory systems. *Science, 247*(4940), 301–306.

Ungerleider, L., & Mishkin, M. (1982). Two cortical visual systems. In D. Ingle, M. Goodale, & R. Mansfield (Eds.), *Analysis of visual behavior.* Cambridge, MA: MIT Press.

Valenstein, E., Bowers, D., Verfaellie, M., Heilman, K., Day, A., & Watson, R. (1987). Retrosplenial amnesia. *Brain, 110,* 1631–1646.

Watson, J. D., Myers, R., Frackowiak, R. S., Hajnal, J. V., Woods, R. P., Mazziotta, J. C., Shipp, S., & Zeki, S. (1993). Area V5 of the human brain: Evidence from a combined study using positron emission tomography and magnetic resonance imaging. *Cerebral Cortex, 3,* 79–94.

Wise, R., Chollet, F., Hadar, U., Friston, K., Hoffner, E., & Frackowiak, R. S. (1991). Distribution of cortical neural networks involved in word comprehension and word retrieval. *Brain, 114,* 1803–1817.

Zatorre, R., Evans, A., Meyer, E., & Gjedde, A. (1992). Lateralization of phonetic and pitch discrimination in speech processing. *Science, 256,* 846–849.

Zeki, S. (1973). Colour coding in the rhesus monkey prestriate cortex. *Brain Research, 53,* 422–427.

Zeki, S., Watson, J. D., Lueck, C. J., Friston, K. J., Kennard, C., & Frackowiak, R. S. (1991). A direct demonstration of functional specialization in human visual cortex. *Journal of Neuroscience, 11*(3), 641–649.

# 6

# Cerebral Asymmetries and Corpus Callosum Morphology

ANDREW KERTESZ

The evolution of cerebral asymmetries and their relationship to functional lateralization in the brain are important topics in anthropology, anatomy, biology, and neuropsychology. A brief review of the phylogeny of cerebral asymmetry is appropriate in this context. Evolutionary theory has assumed that functional lateralization in the higher levels of the phylogenetic scale leads to asymmetry in structure. It was believed that the more complex an organism, the more asymmetry in brain structure. In particular, cerebral asymmetries have been assumed to be related to language development in man, since Broca (1865) proposed his notion of cerebral dominance for language. Anthropological considerations led Broca (1861), among others, to believe that the frontal lobes were important for language because of the dramatic increase in their size in humans. When he produced lesion evidence in support of this he also observed that language loss usually occurred with left-sided lesions, establishing the most consistent functional asymmetry in the human brain.

Attempts have been made to link the development of language with skull asymmetries seen in the fossils of the precursors of man. For instance, the endocasts of the skull of Peking man, who lived more than a half million years ago, suggest a forward extension of the right hemisphere and a posterior exten-

ANDREW KERTESZ St. Joseph's Health Centre, Department of Clinical Neurological Sciences, University of Western Ontario, London, Ontario N6A 4V2, Canada.

*Neuroimaging I: Basic Science*, edited by Erin D. Bigler. Plenum Press, New York, 1996.

sion of the left hemisphere. Neanderthals years ago showed a distinct left occipital–right frontal petalis (bulge) 50,000 years (Holloway & De La Coste, 1982). Neanderthals also show a higher sylvian fissure on the right similar to modern man (Boule & Anthony, 1911; LeMay & Culebras, 1972). The anatomical study of modern man's brain has yielded small but reproducible asymmetries, and their relationship to functional asymmetries will be discussed below. Recently, however, definite asymmetries have been uncovered, even in birds and rats (Glick, 1985; Nottebohm & Nottebohm, 1976), suggesting that asymmetrical development of the brain occurs at all stages of evolution for various biological reasons, not only language.

Anthropoid apes appear to have more asymmetry than monkeys (Holloway & De La Coste, 1982). For instance, the orangutan has a higher posterior termination of the sylvian fissure on the right in the majority of individuals, similar to man. In the chimpanzee the left occipital pole is wider than the right (LeMay, 1976). The right hemisphere of the gorilla is longer on the right than the left, as in humans (Lemay, 1976). In the gorilla the prerolandic cortex is larger on the right side and the postrolandic area is larger on the left (Horvath, Woodward, & De La Coste-Lareymondie, 1985), similar to the right-to-left fronto-occipital torque described in man (see below).

On casual inspection, the hemispheres of humans appear symmetrical, but studies of hemispheric weight indicated a slightly heavier right side (Braune, 1891; Broca, 1885; Wagner, 1864; Wilde, 1926) and longer right side (Hoadley, 1929; Gundara & Zivanovic, 1968; Schwartz *et al.*, 1985). However, a longer left hemisphere was found by Cunningham (1892), Hrdlicka (1907), and Hadziselimovic and Cus (1966). Connolly (1950) found a sex difference in hemispheric length; the left was longer in males and the right longer in a small number of females. Lobar asymmetries have been observed on autopsied specimens, as well as on the endocasts of skulls. The most frequently investigated and confirmed petalia is the left parieto–occipital one in the majority of human and hominid skulls (Hadziselimovic & Cus, 1966), corresponding to a larger left parieto–occipital area on cross-sections (Inglessis, 1925).

Temporal lobe asymmetries were noted by Heschl (1878), who noted the regularity of two transverse gyri in the temporal operculum on the right and usually only one on the left. Pfeiffer (1936) described a larger planum temporale, the area behind Heschl's gyrus on the left side. Geschwind and Levitsky (1968) confirmed this quantitatively and found that the planum was larger on the left side in 65% of subjects, on the right side in 11%, and equal in 24%. That this might be related to language dominance on the left side generated a great deal of interest. This has been confirmed to exist in newborns by Witelson and Pallie (1973) and Wada, Clarke, and Hamm (1975), suggesting that the brain may be "prewired" at birth for language dominance on the left. Eberstaller (1884) and Cunningham (1892) observed that the left sylvian fissure was longer and more horizontal and the right curved up sharply. This was confirmed quantitatively by Rubens, Mahowald, and Hutton (1976).

Reports of asymmetries in Broca's area were less consistent. Wada *et al.* (1975) found the right side to be larger than the left, but Nikkuni *et al.* (1981) found a larger left Broca's area. It should be pointed out that the definition of Broca's area differs from study to study. In right-handed subjects the area of the frontal operculum, visible on the surface of the brain, is larger on the right, and the surface area in the sulci of this region is greater on the left (Falzi, Perrone, &

Vignolo, 1982). The frontal lobe was found to be larger on the right side on cross-sectional measures (Weinberger, Luchins, Morihisa, & Wyatt, 1982; Weis, Haug, Holoubek, & Orun, 1989). Albanese, Merlo, Albanese, and Gomez (1989) reexamined Broca's area asymmetries along with the planum temporale. In this quantitative study, 62.5% of the brain showed larger left and 12.5% larger right frontal opercular and posterior triangularis regions of Broca's area. No asymmetry was found in the more anterior portions of the inferior frontal gyrus, which is not involved with language. Although the planum temporale was not quantified, they noted that the larger side was not always the same for the frontal and temporal language area. In another study of Broca's area, measurements of the pars triangularis and the length and area of the ascending rami appeared to be larger in the right hemisphere on three-dimensional MRI measures as well as on digitized length measures on cadaver brains (Vannier *et al.*, 1991).

Subcortical structures were examined extensively by Orthner and Seler (1975), who considered sex differences as well. The left pallidum was larger in both sexes, and the right frontal region anterior to the commissure was also prominent in both sexes. A recent study confirmed a larger pallidum on the left side (Kooistra & Heilman, 1988). The pyramidal tract crosses from the left to the right in the majority of both right- and left-handers, with a great deal of variability in the pattern of the bundles (Kertesz & Geschwind, 1971).

Cytoarchitectonic asymmetries depend on the technique used and are more difficult to carry out on an adequate sample. Most of the cytoarchitectonic asymmetries were detected in the temporal lobe. Von Economo and Horn (1930) found a larger posterior temporal lobe area (designated Ta1) on the left side. Similarly in 1940, Blinkov as cited in Adrianov, 1979 found a larger Brodmann area 22 on the left Galaburda, Sanides, and Geschwind (1978) used pigment architectonics and analyzed in the planum temporale what is cytoarchitectonically defined as "tpt". The left tpt was found to be larger in the majority of cases. Geschwind and Galaburda (1985) proposed that testosterone in the embryo may retard the development of the left hemisphere and generate a compensatory growth of the right hemisphere resulting in the two sides being more equal or the right side larger. However, a subsequent study did not seem to support this hypothesis, because the total area seemed to be less in the asymmetrical traces, and this was interpreted as a developmentally reduced area on the small side (Galaburda, Corsiglia, Rosen, & Sherman, 1987).

There are hemispheric differences in the cytoarchitectonics of Heschl's gyrus that may have important neuropsychological implications. The central portion of Heschl's gyrus is the most granular and appears to be larger in the left hemisphere (Seldon, 1981). It has been postulated that this area corresponds to the representation of the frequency range of the human voice. It was found that the center-to-center distance between the cell clusters is greater on the left side (Seldon, 1981). Studies also showed that the dendritic input region for each unit on the left side is greater than on the right, but these input regions on the left are farther apart. This implies functionally that neurons in the left hemisphere can react more selectively than those on the right, which may explain the capacity to discriminate speech sounds (categorical perception) on the left (Seldon, 1985). Dendritic branching was examined in the left and right hemispheres in right-handed and left-handed individuals in several selected areas, and the left-side cortex had more neurons with higher-order branches in right-handers (Scheibel *et al.*, 1985). The patterns were reversed in the brains of the left-handed subjects.

Parietal lobe cytoarchitectonic asymmetries indicated a larger language-related area (designated PG) architectonically on the left side located on the angular gyrus and a larger area of PEG, which is a nonlanguage area of the superior parietal lobule on the right side (Eidelberg & Galaburda, 1984).

## Neuroimaging and Asymmetry

Neuroimaging asymmetries were described on angiograms and ventriculograms before CT scans became available. The left ventricle has been shown to be larger than the right, using ventricular casts and pneumoencephalograms (Last & Thompsett, 1953; McRae, Branch, & Milner, 1968). The study by McRae *et al.* (1968) was correlated with handedness and speech dominance. The left occipital horn of the lateral ventricle was larger in right-handers, but the two sides were more equal in left-handers. The asymmetry did not correlate with speech dominance as tested by the sodium amytal test. A higher sylvian point on angiograms (the end of sylvian fissure) was found on the right side by LeMay and Culebras (1972). They also observed a narrower angle of the parietal opercular arteries on the left caused by a larger left parietal operculum in right-handers. In left-handers the posterior sylvian vascular configurations were more similar on the two sides. The differences in sylvian points were also found in fetuses, indicating a genetically determined asymmetry rather than an acquired one. A study correlated the asymmetry of the sylvian point on angiograms with intracarotid amytal study in an epileptic population and confirmed a larger left parietal operculum in 66% of subjects with left-sided speech, as determined by the Wada test, and only in 35% of the subjects with speech on the right (Ratcliff, Dila, Taylor, & Milner, 1980). Another study by Strauss, LaPointe, Wada, Gaddes, and Kosaka (1985) found no correlation between sylvian point measures and amytal testing, but did find some correlation between right ear advantage on dichotic tests and posterior temporal widths on the left (right ear advantage assumes left cerebral dominance for language).

Cerebral asymmetries on the CT scan were observed by LeMay (1976, 1977). Right-handers appeared to have greater left occipital and right frontal petalia and wider left occipital and right frontal lobes (LeMay & Kido, 1978). Left-handed subjects have more symmetrical distribution or the reversed pattern. Chui and Damasio (1980), however, found no significant differences between right- and left-handers. Pieniadz and Naeser (1984) compared the autopsy findings with CT scans in 15 right-handed patients and found significant correlation only between the occipital length asymmetry on low ventricular CT slides and the length of the planum temporale. However, CT measurements only provide an approximation of anatomical asymmetries. Some of the differences between the studies are related to the differences in methodology and the difficulty in obtaining accurate midlines on CT scans with the same angulation. Most of the studies have concluded that CT scan occipital asymmetry cannot be used to predict the hemispheric side of language dominance (Naeser & Borod, 1986; Henderson, Naeser, Weiner, Pieniadz, & Chui, 1984). When language recovery and CT scan asymmetries were correlated, patients with good recovery for single-word comprehension and naming tended to have atypical asymmetry or lack of asymmetry. The assumption was that this may reflect more bilateral language (Pieniadz, Naeser, Koff, & Levine, 1983). In our study, using somewhat different

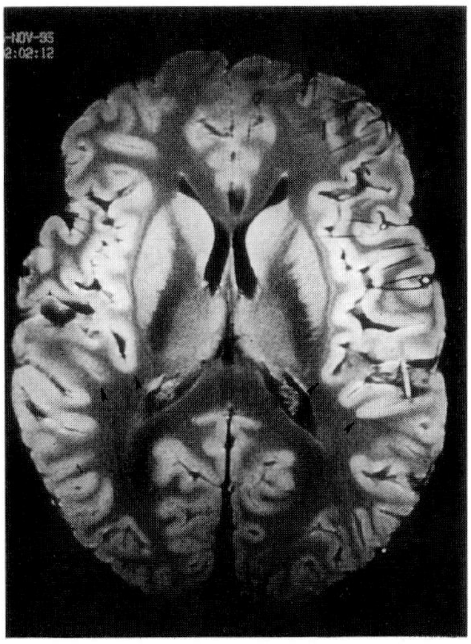


Figure 1. A postmortem MRI specimen illustrates the asymmetry of the parietal operculum. Arrows point to the angle of demarcation, sharper on the right (on the left of the figure) and extending more posteriorly on the left (equivalent to the planum temporale in the temporal region).

methodology, we could not find a difference between patients with poor and good recovery (Kertesz, 1988).

Magnetic resonance imaging (MRI) provides excellent anatomical detail and has become the method of choice for imaging brain asymmetries in the last decade. One of the asymmetries appears at the posterior opercular regions (Kertesz, Black, & Howell, 1984; Kertesz, Black, Polk, & Howell, 1986). On the axial sections across the insula, we observed a larger left planum temporale and a sharper demarcation of the opercular region from the parietal sulci on the right side in 88% of all scans in our pilot study (Figure 1) (Kertesz *et al.*, 1984).

Behavioral measures were correlated with MRI asymmetry of the parietal demarcation (planum temporale) and width measurements. A sharper parietal demarcation was found on the right side in 60% of right-handers and in only 10% of left-handers. Occipital width measurements were larger on the left in 90% of the right-handers and in only 30% of the left-handers. Anterior frontal width, parietal width, and parietal demarcation collectively discriminated between left- and right-handed groups. Hand performance appeared to correlate best with the anatomical measures (Kertesz *et al.*, 1986). Steinmetz and colleagues (1989) measured the planum temporale with a different technique. Manually traced surface lines of the temporal lobe at each sagittal MRI slice were added to obtain a total surface area which showed a significant asymmetry toward a larger left side. A similar methodology was used by Habib (1989), who showed some correlation between handedness and asymmetry coefficients.

Cross-sectional MRI area measures and linear measures were studied in a large population of 52 right-handed and 52 left-handed normal individuals divided equally between males and females (Kertesz, Polk, Black, & Howell, 1990). Areas were traced and digitized by computer from MRI scans by independent observers blind to the side of the brain and the behavioral data. Results indicated larger linear and area measurements in the right hemisphere with the exception of the anterior and posterior frontal widths and the frontal lobe area, which was larger on the left side in left-handers. A significant interaction was observed between sex and handedness in parietal asymmetries. Right-handed male and left-handed female parietal areas were larger on the left side. Further sex difference was suggested by a more symmetrical frontal length (or petalia) in females. We postulated a three-dimensional torque to explain the larger right side above the third ventricle and the slightly more prominent left side in the structures below. This is a similar torque to the right frontal and left occipital petalia measures on CT (LeMay & Kido, 1978). Temporal lobe areas were also relatively larger on the left in females. We also correlated the auditory, manual, and visual laterality measures with MRI asymmetries (Kertesz, Polk, Black, & Howell, 1992). A hand performance task of placing dots in circles (Tapley & Bryden, 1985) was the best discriminator between anatomical asymmetries. Left hemisphere dominance for this task was associated with larger right frontal and left occipital widths, confirming previous CT studies using handedness as a functional measure. Dichotic listening and hand performance interaction showed an intriguingly consistent direction: subjects who had cerebral dominance for hand performance and dichotic performance for language sounds in the same hemisphere had larger structures than those in whom dominance was crossed (Kertesz *et al.*, 1992). Visual field laterality groups showed wider left parietal width to be associated with left visual field dominance for reading, interacting with right-handedness. Auditory dominance for dichotic laterality did not, by itself, correlate with asymmetries of the temporal lobe (Kertesz *et al.*, 1992). This was replicated by Jäncke and Steinmetz (1993) using a somewhat different methodology.

The correlation of functional laterality with anatomical asymmetries is complex, and multiple factors seem to interact. Hand performance appears to be associated most consistently with cerebral asymmetries, and left hemisphere dominance on hand performance is a better predictor of the right frontal–left occipital torque than dichotic listening or visual laterality. The relationship of anatomical asymmetries to functional laterality are of great theoretical interest, but there is a practical issue of imaging certain asymmetries that may have a predictable association with language or other cognitive functions. So far, only invasive tests of cerebral dominance for language, such as the Wada test or lesion studies, could determine language dominance. More recent activation studies may provide this information as well, but they are not yet available as a practical treatment. So far, no reliable structural asymmetries were found that could reliably predict functional laterality. Nevertheless, cerebral morphometric techniques are used increasingly to interpret physiological and clinical data, including metabolic imaging. Systematic anatomical asymmetries may be important in fine localization of function, such as in stereotactic surgery, where the difference of a few millimeters may influence functional outcomes.

## Callosal Morphology and Function

The corpus callosum is a large structure connecting homologous cortical areas of the two hemispheres (for anatomical reviews see Cumming, 1970; Innocenti, 1981). Callosal connections are mostly between secondary and tertiary association areas, while primary cortical sensory or motor areas have few, if any, direct connections between homologous areas. The callosum also carries connections to heterologous cortex (Innocenti, 1981), and its main functional role may be the transfer of associative or already encoded information. Developmentally, the callosum loses fibers with maturation and evolving hemispheric specialization (Innocenti, 1981). Liepmann (as cited in Geschwind, 1965) postulated that behavioral integration of the cerebral hemispheres is accomplished by inhibitory actions of the dominant hemisphere mediated through the callosum, yet most of the reports indicated no cognitive deficit after a callosal section. By separating sensory input and response, Meyers and Sperry (1958) and Gazzaniga, Bogen, and Sperry (1965) began to clarify the functional impairment after callosal disconnection. Callosal-sectioned patients, for instance, failed to report stimuli from the left visual field, but they showed appropriate emotional responses, indicating that the stimuli were processed. They could not name objects in the left hand, but showed recognition by matching. Other examples of independent hemispheric functions, among them right hemisphere language processing (mainly single-word comprehension), were discovered (Gazzaniga & Hillyard, 1971; Zaidel, 1976).

Developmental defects of the corpus callosum, such as agenesis, are fairly frequent. This anomaly may be present without any sign of neurological dysfunction, although it is most commonly associated with seizures and varying degrees of mental retardation. The cognitive deficits are often attributed to associated cerebral defects. Agenesis by itself is considered to interfere only with integrative capacities for fine motor and visual performance. The diagnosis is often made by CT, which shows separation of the lateral ventricles, a concave shape of the medial walls, and dilatation of the caudal portion of the lateral ventricles and third ventricle. The combination of agenesis of the corpus callosum, infantile spasm, and ocular anomalies is called Aicardis syndrome (Bertoni, von Loh, & Allen, 1979).

Anatomical and anthropological studies have examined the relationship of callosal size, sex, race, and handedness. Bean (1906) measured callosal area and brain weight and concluded from scatter plots that they were related, but noted many exceptions. The ratio of callosal area to brain weight was larger in women than in men. De La Coste-Utamsing and Holloway (1982) suggested a "sexual dimorphism," as they found that the female callosum was wide at the splenium, whereas the male callosum was cylindrical and continuously narrow. In women, the callosal area was greater relative to brain weight. Witelson (1985) suggested that the corpus callosum was larger by 11% in ambidextrous subjects than in right-handers, but she found no sex difference on autopsy. No sex differences were found with MRI in several studies (Oppenheim, Lee, Nass, & Gazzaniga, 1987; Kertesz, Polk, Howell, & Black, 1987; Byne, Bleier, & Houston 1988). Nasrallah, *et al.* (1986) found that schizophrenic women had a larger callosal area on MRI than did controls.

Magnetic resonance imaging shows a clear outline of the corpus callosum on a sagittal cross-section, and this generated a great deal of interest in measuring differences between patients and normal populations to study the association of anatomical size and possible behavioral measures. We measured the MRI area of corpus callosum in 104 normal adults; 51 men and 53 women, 52 right-handers and 52 left-handers. There were no significant sex or handedness differences; sex could not be predicted from the shape of the splenium. Callosal areas did not correlate with brain size or with measures of lateralization for hand performance, dichotic listening, or visual field preference (Kertesz *et al.*, 1987). The results are contrary to the notion of increased callosal connectivity in left-handers, women, or individuals with less lateralized function. These findings differed from those of O'Kusky *et al.* (1988), which found that the area of corpus callosum varied with the direction of language lateralization.

Subregions of the corpus callosum were also studied anatomically and functionally in several laboratories. The genu, the rostrum, the body, and the splenium are more or less distinct parts of the callosum, but there is little agreement as to the appropriate, accurate, and functionally relevant way of dividing the corpus callosum. Anatomical studies on the rhesus monkey revealed that the rostral half of the corpus callosum carries the interhemispheric fibers to the various subdivisions of the frontal lobe. In the caudal half of the callosum, the parietal lobe fibers occupy a more rostral location than those of the temporal lobe. More rostral fibers tended to cross dorsally and more ventrally lying fibers in the ventral portion of the callosum both in the genu and the splenium. The precentral operculum received intrahemispheric fibers via the same portion of the posterior half of the corpus callosum as the primary somatosensory area (Pandya, Karol, & Heibronn, 1971). Such topographical organization is likely the case for human callosal fibers as well. However, the sequelae of a partial section of the corpus callosum are difficult to quantitate in humans. It requires a fairly large part of the callosum to be sectioned to effect discernible deficit. Lesions of the posterior quarter of the callosum cause some breakdown of visual transfer between the two half-brains. More anterior portions of the callosum, in addition to subserving motor function, are also able to transfer semantic information. Staged callosal section affecting first the posterior portion of the callosum allowed the transfer of semantic information about visual stimuli to the opposite hemisphere (Sidtis, Volpe, Holtzman, Wilson, & Gazzaniga, 1981). Few human studies exist concerning the topography of the corpus callosum postmortem. Vallerian degeneration was correlated with anatomical sites of lesions, and the topographical organization appeared to be similar to that of the rhesus monkey. The inferior frontal and parietal regions are connected through the rostrum and the genu of the callosum, respectively. The temporal parietal and occipital junction utilizes the splenium and the caudal portion of the body, and the superior parietal lobule and the occipital cortex use exclusively the splenium (De Lacoste, Kirkpatrick, & Ross, 1985).

Witelson (1985) divided the corpus callosum into seven equal areas and found a larger "isthmus" (the narrow segment before the splenium) in left-handers and subsequently a sex and hand interaction (Witelson, 1989). A factor structuring of the width measures of the corpus callosum, first developed for the rat, was carried out in a normal human population previously described (Denenberg, Kertesz, & Cowell, 1991). Seven factored areas were identified (Figure 2). Some of this corresponded to the traditional subregions. The only small sex difference was found in the isthmus area connecting the body with the splenium. Inconsistently right-handed males had a somewhat larger isthmus. Sex by hand interaction was found in the isthmus by Habib *et al.* (1991). Sex difference, but not a handedness difference, was found in the isthmus but not in total callosal size (Steinmetz *et al.*, 1992). More anterior portions of the body showed larger width in right-handed women and in left-handed men, indicating a sex by hand interaction (Cowell, Kertesz, & Denenberg, 1993). A similar phenomenon of laterality by sex interaction occurs in rodents, where it was found that turning bias in rats is related to corpus callosum size differently in the sexes. There was a

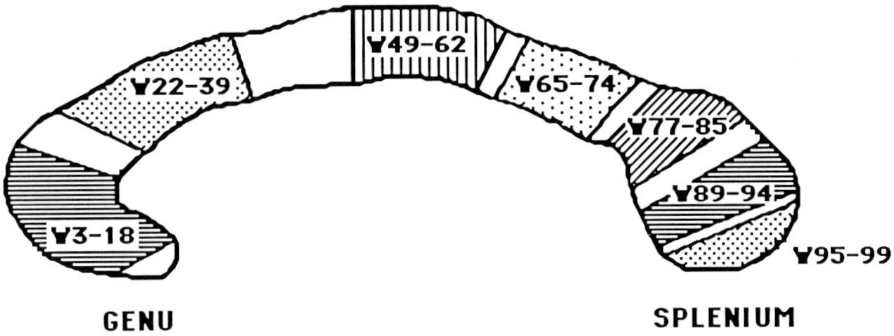


Figure 2. Factor structure of callosal width, obtained by an algorithm of best fit of equal portions on the central line of the corpus callosum. Figure reproduced with permission from Denenberg *et al.* (1991).

greater sex difference in the left-biased animals (Zimmerberg, Eddy, & Horwitz, 1988).

A recent study indicated shrinkage of the corpus callosum in aging, but only in man (Witelson, 1991). A smaller callosal area was seen in children with mental deficiency and attention deficit hyperactivity disorder (Hynd *et al.*, 1991). A larger area of the splenium of the corpus callosum was found in dyslexic adults, and female dyslexics also had larger genu and total callosal measurements when compared with males and control subjects (Duara *et al.*, 1991).

Although the left hemispheric lateralization of language and praxis has been well established, there are individual variations in the degree of laterality of these functions. This is reflected, for instance, in the relatively frequent clinical dissociations between aphasia and apraxia. Although this has been attributed to differential rates of recovery of the two functions (Kertesz & Ferro, 1984), variable cerebral laterality could also be expected to play a role. Connections in the corpus callosum could also reflect dissociations in functional laterality. In addition to the complex interaction between sex and handedness, there are individual differences in variability that contribute to the noise of the data. The variations in callosal size and shape are greater in individuals than between groups by sex and handedness or age (Bleier, Houston, & Byne, 1986). We also found it impossible to predict the gender or handedness of an individual on the basis of the size or shape of the corpus callosum (Kertesz *et al.*, 1987). The frequently dragged out "sexual dimorphism" of the corpus callosum is a misnomer. If one adds the complexity of the functions examined by some of the behavioral measures, one can easily account for the heterogeneity of behavioral profiles and callosal sizes. It is likely that callosal morphology is as multidimensional and complex as the behavior it subserves.

## References

Adrianov, O. S. (1979). Structural basis for functional interhemispheric brain asymmetry. *Human Physiology, 5,* 359–363.

Albanese, E., Merlo, A., Albanese, A., & Gomez, E. (1989). Anterior speech region: Asymmetry and weight-surface correlation. *Archives of Neurology, 46,* 307–310.

Bean, R. B. (1906). Some racial peculiarities of the negro brain. *American Journal of Anatomy, 5,* 354–432.

Bertoni, J. M., von Loh, S., & Allen, R. J. (1979). The Aicardi syndrome: Report of four cases and review of the literature. *Annals of Neurology, 5,* 475–482.

Bleier, R., Houston, L., & Byne, W. (1986). Can the corpus callosum predict gender, age, handedness, or cognitive differences? *Trends in Neuroscience, 9,* 391–394.

Boule, M., & Anthony, R. (1911). L'encéphale de l'homme fosile de la Chapelle-aux-Saints. *Anthropologie, 4,* 65–78.

Braune, C. (1891). Das Gewichtsverhältniss der rechten zur linken Hirnhälfte beim Menschen. *Archiv für Anatomie und Entwickelungsgeschichte,* 253–270.

Broca, P. (1861). Remarques sur le siège de la faculté du langage articule, suives d'une observation d'aphémie (perte de la parole). *Bulletin et Memoires de la Sociéte Ánatomique de Paris, 36,* 330–357.

Broca, P. (1865). Sur le siège de la faculté du langage articule. *Bulletin de la Société Anthropologique du Paris, 6,* 377–393.

Broca, P. (1885). Data reported by P. Topinard. In A. Delabraye & E. Delabraye (Eds.), *Éléments d'anthropologie générale* (p. 591). Paris: Lecrosnier.

Byne, W., Bleier, R., & Houston, L. (1988). Variations in human corpus callosum do not predict gender: A study using magnetic resonance imaging. *Behavioral Neuroscience, 102,* 222–227.

Chui, H. C., & Damasio, A. R. (1980). Human cerebral asymmetries evaluated by computerized tomography. *Journal of Neurology, Neurosurgery, and Psychiatry, 43,* 873–878.

Connolly, C. J. (1950). *External morphology of the primate brain.* Springfield, IL: Thomas.

Cowell, P. E., Kertesz, A., & Denenberg, V. H. (1993). Multiple dimensions of handedness and the human corpus callosum. *Neurology, 43,* 2353–2357.

Cumming, W. J. K. (1970). An anatomical review of the corpus callosum. *Cortex, 6,* 1–18.

Cunningham, D. J. (1892). *Surface anatomy of the cerebral hemispheres.* Dublin: Academy House.

De Lacoste, M. C., Kirkpatrick, J., & Ross, E. D. (1985). Topography of the human corpus callosum. *Journal of Neuropathology and Experimental Neurology, 44,* 578–591.

De La Coste-Utamsing, C., & Holloway, R. L. (1982). Sexual dimorphism in the human corpus callosum. *Science, 216,* 1431–1432.

Denenberg, V. H., Kertesz, A., & Cowell, P. E. (1991). A factor analysis of the human's corpus callosum. *Brain Research, 548,* 126–132.

Duara, R., Kushch, A., Gross-Glenn, Barker, W. W., Jallad, B., Pascal, S., Loewenstein, D. A., Sheldon, J., Rabin, M., Levin, B., & Lubs, H. (1991). Neuroanatomic differences between dyslexic and normal readers on magnetic resonance imaging scans. *Archives of Neurology, 48,* 410–416.

Eberstaller, O. (1884). Zur oberflächenanatomie der grosshirnhemisphären. *Wiener Medizinische Blätter, 7,* 479–482.

Eidelberg, D., & Galaburda, A. (1982, April). Cytoarchitectonic asymmetries in the human parietal lobe. Paper presented at the 34th annual meeting of the American Academy of Neurology, Washington, D.C.

Eidelberg, D., & Galaburda, A. M. (1984). Inferior parietal lobule. Divergent architectonic asymmetries in the human brain. *Archives of Neurology, 41,* 843–852.

Falzi, G., Perrone, P., & Vignolo, L. (1982). Right-left asymmetry in anterior speech region. *Archives of Neurology, 39,* 239–240.

Galaburda, A. M., Corsiglia, J., Rosen, D., & Sherman, G. F. (1987). Planum temporale asymmetry: Reappraisal since Geshwind and Levitsky. *Neuropsychologia, 25,* 853–868.

Galaburda, A. M., Sanides, F., & Geschwind, N. (1978). Human brain. Cytoarchitectonic left-right asymmetries in the temporal speech region. *Archives of Neurology, 35,* 812–817.

Gazzaniga, M. S., Bogen, J. E., & Sperry, R. W. (1965). Observations in visual perception after disconnection of the cerebral hemispheres in man. *Brain, 88,* 221–236.

Gazzaniga, M. S., & Hillyard, S. A. (1971). Language and speech capacity of the right hemisphere. *Neuropsychologia, 9,* 272.

Geschwind, N. (1965). Disconnection syndromes in animals and man. *Brain, 88,* 237–294.

Geschwind, N., & Galaburda, A. M. (1985). Cerebral lateralization: Biological mechanisms, associations, and pathology. *Archives of Neurology, 42,* 428–459.

Geschwind, N., & Levitsky, W. (1968). Human brain: Left-right asymmetries in temporal speech regions. *Science, 161,* 186–187.

Glick, S. D. (Ed.). (1985). *Cerebral lateralization in nonhuman species.* New York: Academic Press.

Gundara, N., & Zivanovic, S. (1968). Asymmetry in East African skulls. *American Journal of Physiology and Anthropology, 28,* 331–338.

Habib, M. (1989). Anatomical asymmetries of the human cerebral cortex. *International Journal of Neuroscience, 47,* 67–79.
Habib, M., Gayraud, D., Oliva, A., Regis, J., Salamon, G., & Khalil, R. (1991). Effects of handedness and sex on the morphology of the corpus callosum: A study with brain magnetic resonance imaging. *Brain and Cognition, 16,* 41–61.
Hadziselimovic, H., & Cus, M. (1966). The appearance of the internal structures of the brain in relation to the configuration of the human skull. *Acta Anatomica, 63,* 289–299.
Henderson, V. W., Naeser, M. A., Weiner, J. M., Pieniadz, J. M., & Chui, H. C. (1984). CT criteria of hemisphere asymmetry fails to predict language laterality. *Neurology, 34,* 1086–1089.
Heschl, R. L. (1878). *Uber die vordere quere Schlafenwindung des menschlichen Grosshirns.* Vienna: Wilhelm Braumuller.
Hoadley, N. F. (1929). On the measurement of the internal diameter of the skull in relation (i) to the prediction of its capacity (ii) to the 'pre-eminence' of the left hemisphere. *Biometrika, 21,* 85–123.
Holloway, R. L., & De La Coste, M. C. (1982). Brain endocast asymmetry in pongids and hominids: Some preliminary findings on the paleontology of cerebral dominance. *American Journal of Physiology and Anthropology, 58,* 101–110.
Horvath, D. S., Woodward, D. J., & De La Costa-Lareymondie, M. C. (1985). Cerebral asymmetries and sex differences in hominoids (abstract). *Proceedings of the Society of Neuroscience.*
Hrdlicka, A. (1907). Measurements of the cranial fossae. *Proceedings of the United States National Museum, 32,* 177–232.
Hynd, G. W., Semrud-Clikeman, M., Lorys, A. R., Novey, E. S., Eliopulos, D., & Lyytinen, H. (1991). Corpus callosum morphology in attention-deficit hyperactivity disorder (ADHD); Morphometric analysis of MRI. *Journal of Learning Disabilities, 24,* 141–146.
Inglessis, M. (1925). Untersuchungen uber Symmetrie und Asymmetrie der Menschlichen Grosshirnhemispharen. *Zeitschrift für des Gesamte Neurologie Psychiatrie, 95/96,* 464–474.
Innocenti, F. M. (1981). The development of interhemispheric connections. *Trends in Neuroscience, 4,* 142–144.
Jäncke, L., & Steinmetz, H. (1993). Auditory lateralization and planum temporale asymmetry. *Neuroreport, 5,* 169–172.
Kertesz, A. (1988). What do we learn from recovery from aphasia? In S. G. Waxman (Ed.), *Advances in neurology: Vol. 47. Functional recovery in neurological disease* (pp. 277–292). New York: Raven Press.
Kertesz, A., Black, S. E., & Howell, J. (1984). *Functional lateralization and anatomical asymmetries on nuclear magnetic resonance.* Abstract presented at Academy of Aphasia, Los Angeles, CA.
Kertesz, A., Black, S. E., Polk, M., & Howell, J. (1986). Cerebral asymmetries on magnetic resonance imaging. *Cortex, 22,* 117–127.
Kertesz, A., & Ferro, J. M. (1984). Lesion size and location in ideomotor apraxia. *Brain, 107,* 921–933.
Kertesz, A., & Geschwind, N. (1971). Patterns of pyramidal decussation and their relationship to handedness. *Archives of Neurology, 24,* 326–332.
Kertesz, A., Polk, M., Black, S. E., & Howell, J. (1990). Sex, handedness, and the morphometry of cerebral asymmetries on magnetic resonance imaging. *Brain Research, 530,* 40–48.
Kertesz, A., Polk, M., Black, S. E., & Howell, J. (1992). Anatomical asymmetries and functional laterality. *Brain, 115,* 589–605.
Kertesz, A., Polk, M., Howell, J., & Black, S. E. (1987). Cerebral dominance, sex, and callosal size in MRI. *Neurology, 37,* 1385–1388.
Kooistra, C. A., & Heilman, K. M. (1988). Motor dominance and lateral asymmetry of the globus pallidus. *Neurology, 3,* 388–390.
Last, R. J., & Thompsett, D. H. (1953). Casts of cerebral ventricles. *British Journal of Surgery, 40,* 525–542.
LeMay, M. (1976). Morphological cerebral asymmetries of modern man, fossil man, and nonhuman primate. *Annals of the New York Academy of Sciences, 280,* 349–366.
LeMay, M. (1977). Asymmetries of the skull and handedness-phrenology revisited. *Journal of Neurological Sciences, 32,* 243–253.
LeMay, M., & Culebras, A. (1972). Human brain: Morphologic differences in the hemispheres demonstrable by carotid arteriography. *New England Journal of Medicine, 287,* 168–170.
LeMay, M., & Kido, D. K. (1978). Asymmetries of the cerebral hemispheres on computed tomographs. *Journal of Computer Assisted tomography, 2,* 471–476.
McRae, D. L., Branch, C. L., & Milner, B. (1968). The occipital horns and cerebral dominance. *Neurology, 18,* 95–98.

Meyers, R. E., & Sperry, R. W. (1958). Interhemispheric communication through the corpus callosum: Mnemonic carry-over between the hemispheres. *Archives of Neurology and Psychiatry, 80,* 209–303.

Naeser, M. A., & Borod, J. C. (1986). Aphasia in left-handers: CT scan lesion site, lesion side, and hemispheric asymmetries. *Neurology, 36,* 471–489.

Nasrallah, H. A., Andreasen, N. C., Coffman, J. A., Olson, S., Dunn, V. D., Ehrhardt, J. C., & Chapman, S. M. (1986). A controlled magnetic resonance imaging study of corpus callosum thickness in schizophrenia. *Biological Psychiatry, 21,* 274–282.

Nikkuni, S., Yashima, Y., Ishige, K., Suzuki, S., Ohno, E., Kumashiro, H., Kobayashi, E., Awa, H., Mihara, T., & Asakura, T. (1981). Left-right hemispheric asymmetry of critical speech zones in Japanese brains. *Brain Nerve, 33,* 77–84.

Nottebohm, F., & Nottebohm, M. E. (1976). Left hypoglossal dominance in the control of canary and white-crowned sparrow song. *Journal of Comparative Physiology, 108,* 171–192.

O'Kusky, J., Strauss, E., Kosaka, B., Wada, J., Li, D., Druhan, M., & Petrie, J. (1988). The corpus callosum is larger with right-hemisphere cerebral speech dominance. *Annals of Neurology, 24,* 379–383.

Oppenheim, J. S., Lee, B. C. P., Nass, R., & Gazzaniga, M. S. (1987). No sex-related differences in human corpus callosum based on magnetic resonance imagery. *Annals of Neurology, 21,* 604–606.

Orthner, H., & Seler, W. (1975). Planimetrische volumetrie und menschlichen Gehirnen. *Fortschriftte der Neurologie-Psychiatrie, 43,* 191–209.

Pandya, D. N., Karol, E. A., & Heilbronn, D. (1971). The topographical distribution of interhemispheric projections in the corpus callosum of the rhesus monkey. *Brain Research, 32,* 31–43.

Pfeiffer, F. A. (1936). Pathologie der Hörstrahlung und der corticaler Hörsphare. In F. Bumke & O. Berlin (Eds.), *Handbuch der Neurologie* (pp. 523–626). Foerster: Springer-Verlag.

Pieniadz, J. M., & Naeser, M. A. (1984). Computed tomographic scan, cerebral asymmetries and morphologic brain asymmetries. *Archives of Neurology, 41,* 403–409.

Pieniadz, J. M., Naeser, M. A., Koff, E., & Levine, H. L. (1983). CT scan cerebral hemispheric asymmetry measurements in stroke cases with global aphasia: Atypical asymmetries associated with improved recovery. *Cortex, 19,* 371–391.

Ratcliff, G., Dila, C., Taylor, L., & Milner, B. (1980). The morphological asymmetry of the hemispheres and cerebral dominance for speech: A possible relationship. *Brain and Language, 11,* 87–98.

Rubens, A. B., Mahowald, M. W., & Hutton, J. T. (1976). Asymmetry of the lateral (sylvian) fissures in man. *Neurology, 26,* 620–624.

Scheibel, A. B., Paul, L. A., Fried, I., Forsythe, A. B., Tomiyasu, U., Wechsler, A., Kao, A., & Slotnick, J. (1985). Dendritic organization of the anterior speech area. *Experimental Neurology, 87,* 109–117.

Schwartz, M., Creassey, H., Grady, C. L., DeLeo, J. M., Frederickson, H. A., Cutler, N. R., & Rapoport, S. I. (1985). Computed tomographic analysis of brain morphometrics in 30 healthy men, aged 21 to 81 years. *Annals of Neurology, 17,* 146–157.

Seldon, H. L. (1981). Structure of human auditory cortex. I. Cytoarchitectonics and dendritic distribution. *Brain Research, 229,* 277–294.

Seldon, H. L. (1985). The anatomy of speech perception–human auditory cortex. In A. Peters & E. G. Jones (Eds.), *Cerebral Cortex Vol. 4.* New York and London: Plenum Press.

Sidtis, J. J., Volpe, B. T., Holtzman, J. D., Wilson, D. H., & Gazzaniga, M. S. (1981). Cognitive interaction after staged callosal section: Evidence for transfer of semantic activation. *Science, 212,* 344–346.

Steinmetz, H., Jäncke, L., Kleinschmidt, A., Schlaug, G., Volkman, J., & Huang, Y. (1992). Sex but no hand difference in the isthmus of the corpus callosum. *Neurology, 42,* 749–752.

Steinmetz, H., Rademacher, J., Huang, Y., Hefter, H., Zilles, K., Thron, A., & Freund, H-J. (1989). Cerebral asymmetry: MRI planimetry of the human planum temporale. *Journal of Computer Assisted Tomography, 13,* 996–1005.

Strauss, E., LaPointe, J. S., Wada, J. A., Gaddes, W., & Kosaka, B. (1985). Language dominance: Correlation of radiological and functional data. *Neuropsychologia, 23,* 415–420.

Tapley, S. M., & Bryden, M. P. (1985). A group test for the assessment of performance between the hands. *Neuropsychologia, 23,* 215–222.

Vannier, M. W., Brunsden, B. S., Hildebolt, C. F., Falk, D., Cheverud, J. M., Figiel, G. S., Perman, W. H., Kohn, L. A., Robb, R. A., Yoffie, R. L., & Bresina, S. J. (1991). Brain surface cortical sulcal lengths: Quantification with three-dimensional MR imaging. *Radiology, 180,* 479–484.

Von Economo, C., & Horn, L. (1930). Über Windungsrelief, Masze und Rindenarchitektonik der Supramtemporalfläche, ihre individuellen und ihre Seitenunterschiede. *Zeitschrift Gesamte Neurologie Psychiatrie, 130,* 678–757.

Wada, J. A., Clarke, R., & Hamm, A. (1975). Cerebral hemispheric asymmetry in humans: Cortical speech zones in 100 adult and 100 infant brains. *Archives of Neurology, 32,* 239–246.

Wagner, H. (1864). *Massbestimmungen der Oberfläche des grossen Gehirns.* Kassel and Göttingen: Wiland.

Weinberger, D. R., Luchins, D. J., Morihisa, J., & Wyatt, R. J. (1982). Asymmetrical volumes of the right and left frontal and occipital regions of the human brain. *Annals of Neurology, 11,* 97–100.

Weis, S., Haug, H., Holoubek, B., & Orun, H. (1989). The cerebral dominances: Quantitative morphology of the human cerebral cortex. *International Journal of Neuroscience, 47,* 165–168.

Wilde, J. (1926). Über das Gewichtsverhältnis der Hirnhälften beim Menschen. *Litvijas Universiti Raksti, 14,* 271–288.

Witelson, S. F. (1985). The brain connection: The corpus callosum is larger in left handers. *Science, 229,* 665–668

Witelson, S. F. (1989). Hand and sex differences in the isthmus and genu of the human corpus callosum: A postmortem morphological study. *Brain, 112,* 799–835.

Witelson, S. F. (1991). Sex differences in neuroanatomical changes with aging. *New England Journal of Medicine, 325,* 211–212.

Witelson, S. F., & Pallie, W. (1973). Left hemisphere specialization for language in the newborn: Anatomical evidence of asymmetry. *Brain, 96,* 641–646.

Zaidel, E. (1976). Auditory vocabulary of the right hemisphere following brain bisection or hemidecortication. *Cortex, 12,* 191–212.

Zimmerberg, B., Eddy, T., & Horwitz, N. (1988). Sex and age in the developmental patterns of behavioral laterality in rats. *Society of Neuroscience Abstracts, 14,* 1140.

# 7

# Imaging the Neurocognitive Networks of the Human Brain

ALAN GEVINS

## INTRODUCTION: EVOKED POTENTIALS AND COGNITION

Scalp-recorded neural potentials have been used for over 65 years to study brain activity associated with higher mental functions, both as continuous electroencephalography (EEG) or as event-registered evoked potentials (EPs). Even so, human cognitive neurophysiology is still young, awaiting the development of more powerful tools for measuring the minute signals of neurocognitive processes and their relationship to behavior. Such tools include recordings with arrays of many channels of EEG and magnetoencephalography (MEG), advanced signal processing, anatomical information from magnetic resonance images (MRIs), and functional metabolic information from PET, SPECT, and fMRI (functional MRI). With the most advanced techniques currently available, it is possible to obtain some indication of the location, timing, and coordination of mass neural activity associated with cognition. The EEG's advantages of subsecond temporal resolution, ability to record in a natural situation, and low cost are motivating the development of better EEG recording and analysis techniques.

---

ALAN GEVINS EEG Systems Laboratory and SAM Technology, San Francisco, California 94105.

*Neuroimaging I: Basic Science*, edited by Erin D. Bigler. Plenum Press, New York, 1996.

Evoked potentials generated during cognitive processes are usually studied by averaging a number of brief EEG segments time-registered to the stimulus or response in a simple cognitive task. Averaging enhances signals which are temporally consistent and suppresses unrelated background activity. An averaged EP waveform consists of a series of positive and negative waves. A significant change in latency, amplitude, or duration of an EP wave between two experimental conditions which differ in only one cognitive or behavioral factor is assumed to reflect the mass neural activity associated with that factor. Measurement of such changes in the amplitude and timings of peaks in the series of EP waves evoked by a cue, stimulus, response movement, or feedback allows inferences to be made about the sequence and timing of task-associated processes. These processes include prestimulus preparation, encoding of stimulus features, matching operations, comparison of stimulus codes and memory codes, evaluation of the "meaning" of the stimulus, response selection and execution, and evaluation of feedback in the context of ongoing task performance. Because of this, EPs can provide the basis for an imaging modality, used either by itself or in conjunction with MRI and fMRI when precise three-dimensional localization is needed, which can directly relate the observed neural processes to overt behavioral variables, such as response time and accuracy, in a precisely controlled manner and with subsecond temporal resolution.

## Basic Aspects of the Evoked Potential Method

The study of cogition using amplitude and timing measurements of EP waves and peaks is built on many of the same paradigms as experimental psychology. In the most general terms, it is assumed that a cognitive behavior can be viewed as a combination of sequential (and in some cases parallel) subprocesses in which information from stimuli is encoded, compared with memory and prior experience, and acted upon. The division into component subprocesses may be somewhat arbitrary and depends on the particular cognitive model used, but it is assumed that these subprocesses can be systematically manipulated and that the electrical signals generated by these processes can be distinguished at the scalp. These issues have been reviewed extensively elsewhere (Donchin, Kutas, & McCarthy, 1977; Thatcher & John, 1977; Gevins & Cutillo, 1986; Meyer, Osman, Irwin, & Yantis, 1990).

### Cognitive Factors Can Be Systematically Manipulated

Mental events cannot be measured directly, but can be inferred from measurements of the behavioral and neural activity associated with them. Psychological studies use behavioral measures such as reaction time and error to define component stages of cognitive processing and make inferences about their nature and timing. The validity of a study depends on the ability to manipulate a specific cognitive factor while holding all other factors constant. This is also the case in cognitive studies using measures of brain signals, where the issue of holding irrelevant factors constant across conditions is more difficult due to the sensitivity of neuroelectric signals to factors such as the physical properties of stimuli and motor responses (see later in this chapter).

## The Timing of Neurocognitive Processes Is Consistent



As mentioned above, the averaged EP method assumes that the component subprocesses comprising a cognitive behavior do not vary much in timing from trial to trial. This assumption should be made with the understanding that there may be a certain amount of trial-to-trial variation in the timing of neurocognitive processes which affect both the latency and amplitude of averaged EP components, since they are interdependent. For this reason, measures of the integrated amplitude (area) of an averaged waveform are sometimes used, and "latency correction" of a particular peak prior to averaging is sometimes applied (Woody, 1967; McGillem & Aunon, 1987).

## Overlapping Components Can Be Resolved

The average EP waveform is assumed to be a linear composite of spatially and temporally overlapping components generated in different neural systems. This overlap makes measurement of individual components in the composite waveform an important consideration. In some cases, relatively simple experimental or analytic methods may be effective in distinguishing components, as with the isolation of the N200 wave from the P200 by subtraction of the waveform to standard stimuli from the waveform to target stimuli. But it is not often possible to distinguish overlapping components by such simple methods. Principal components analysis (PCA) is often used as a more general solution to the problem of objective definition of components (Donchin & Heffly, 1978). The PCA method can be useful, but is limited by the fact that physiologically independent neural generators may produce neuroelectric time series which are statistically correlated. Other practical problems have been described, such as the misallocation of variance among components and arbitrariness of the rotation used (Rosler & Manzey, 1981; Wood & McCarthy, 1984). Alternatives to PCA, such as comparative factor analysis (Rogers, 1991), have been developed and may be more effective in extracting small signals in cases of low signal-to-noise ratio.

## Experimental Control Is Essential for Relating a Signal to a Particular Cognitive Factor

In any experiment, it is essential to control all factors not directly related to the intended experimental manipulation. Special attention to experimental control is required in neurocognitive studies since brain signals are sensitive to many sensory, motor, and cognitive factors. The data for testing a hypothesis should be controlled for irrelevant stimulus, physiological state, and response-related factors by experimental design and, in addition, by a *posteriori* balancing of data sets. Such factors include physical stimulus properties, frequency and ordering of stimuli, task difficulty, automatization, and the timing, force, duration, hand, and nature of response movements (Gevins & Schaffer, 1980; Gevins *et al.*, 1979a, 1979b, 1985; Gevins, Zeitlin, Doyle, Schaffer, & Calloway, 1979; Gevins, Doyle, Schaffer, Calloway, & Yeager, 1980). In addition, brain signals are known to vary with age, handedness, gender, level of autonomic arousal, alertness, fatigue, habituation, and use of caffeine, nicotine, alcohol, and drugs. These factors should be taken into account when trying to measure physiological signs of mental activity in the brain.

## Evoked Potential Events Associated with Cognition: Two Major Lines of Research

### Signals of Preparation: The Contingent Negative Variation

The first reports of EP events related not just to external (exogenous) factors, such as stimulus properties or motor activity, but to cognitive (endogenous) factors came in the 1960s. In 1964, Grey Walter reported an EP event related to expectation (Walter, Cooper, Aldridge, McCallum, & Winter, 1964). The contingent negative variation (CNV) is a low-frequency negative wave preceding an expected stimulus, often visible in the unaveraged EEG. It is best seen when two stimuli, or a cue and stimulus, are separated by a regular interval of 1 to several seconds, with a mental operation and/or response required after the second (imperative) stimulus (Figure 1). About 200 msec after the second stimulus, the CNV ends in a sharp, positive (~ 10 Hz) wave, termed the CNV resolution. The CNV was initially considered a signal of "expectancy" (Walter, 1964), but subsequent studies demonstrated that it is affected by many cognitive factors, including attention, distraction, preparation for cognitive tasks, anticipation of information, ambiguity, and task difficulty, and also by noncognitive factors such as stimulus intensity, modality and timing, reaction time and magnitude of movement, motivation, positive or negative reinforcement, noxious stimuli, arousal, age, fatigue, various types of drugs, and clinical disorders including learning disability, senility, neurosis, anhedonia, depression, and schizophrenia (reviewed in Tecce, 1972; Rockstroh, Elbert, Canavan, Lutzenberger, & Birbaumer, 1989; Brunia, 1993).

The idea that the CNV reflects preparatory processes assumes that the low-frequency, negative wave reflects a build-up of "cerebral potentiality." The scalp distribution of the CNV (with linked ear or noncephalic reference) is widespread, usually largest at the midline central electrode, but also strong at precentral and frontal sites. The CNV is thought to reflect depolarization of apical dendrites in frontal and central cortex, but the theory that this facilitates neuronal action potentials has not been proven (Rebert, 1978; Rockstroh *et al.*, 1989). The possible involvement of frontal cortex is of particular interest, since this area is thought to mediate functions such as preparation and planning (Fuster, 1989; Goldman-Rakic, 1984). The theory that the CNV reflects *task-specific* cortical activation was questioned after repeated failures to elicit localization of amplitude with different types of tasks or with attention to different sensory modalities. More recently, however, evidence for localized, task-specific cortical activation has been uncovered by recording from more scalp sites and using a spatial filter to reduce volume-conducted activity common to all electrodes (Gevins *et al.*, 1987, 1989b; Gevins & Cutillo, 1993).

The extent to which the CNV reflects cognitive association of two stimuli is still unclear. Some researchers believe that the CNV is merely a composite of two separate negative waves and does not involve a contingent association between two stimuli (Rohrbaugh, Syndulko, Sanquist, & Lindsley, 1980). But many others favor the view that the CNV does reflect a cognitive association between stimuli

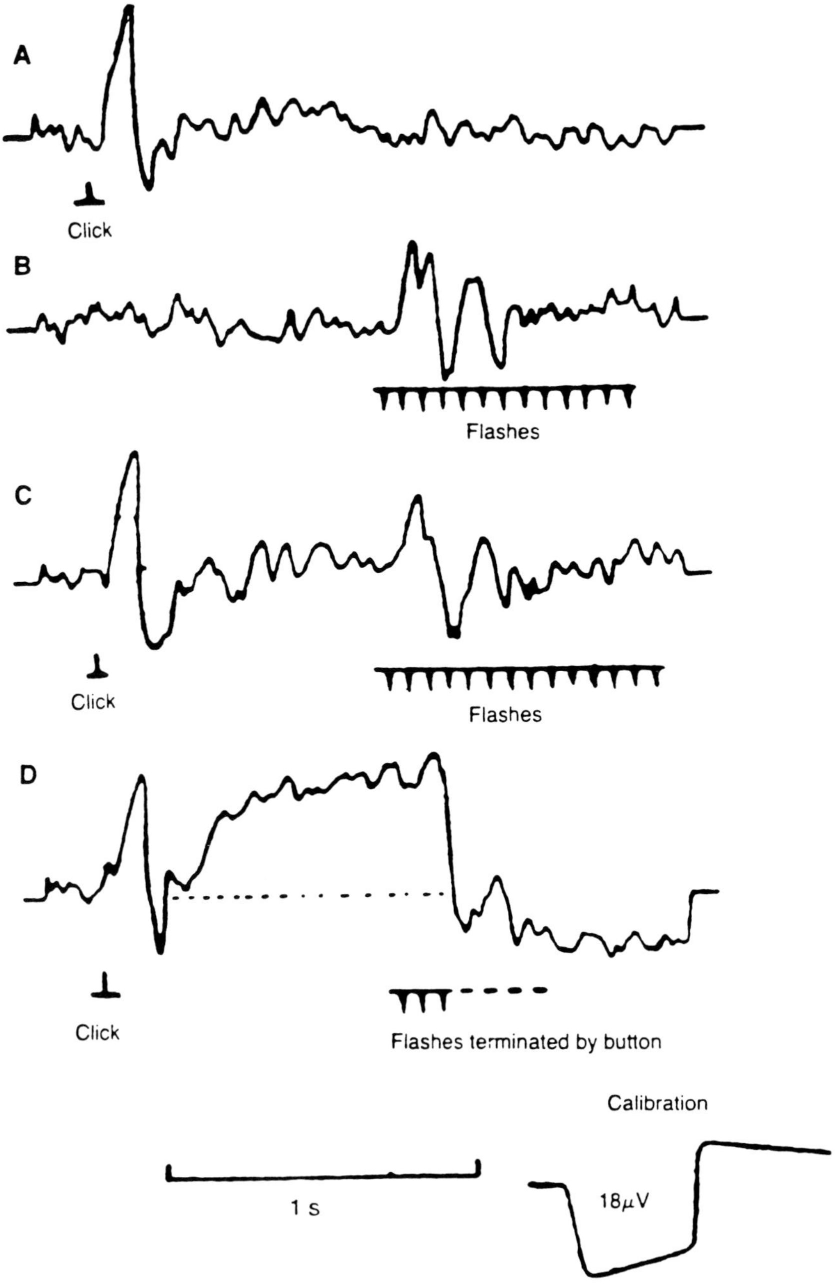


Figure 1. The contingent negative variation (CNV), reported to be related to expectation in the interval between warning and imperative stimuli, from the original report of Grey Walter (Walter *et al.*, 1964). Averaged waveforms (from vertex electrode referenced to mastoid) to unpaired click (A) and flash (B) stimuli, paired click and flash stimuli 1 second apart with no task-relevant contingency (C), and CNV in the 1-second interval between the paired stimuli when they are made contingent by requiring a response (D).

(Donchin, Kutas, & Johnson, 1974; Ritter, Kelso, Kutas, & Shiffrin, 1984). The CNV has been seen to develop when a contingency between two stimuli is learned and then decline when the contingency is removed. The CNV amplitude is reduced by distracting stimuli, instructions to change attentional focus, and equivocation of contingency through random omission of the second stimulus. Currently, the issue of "cognitive association" is an academic point. The contingent negative variation is viewed as just one type of low-frequency wave or, in traditional terms, slow potential shift (SPS). Positive and negative SPSs are frequently seen in EEG recordings, but they have been largely ignored. Researchers are beginning examine the functional significance of SPSs (Harter, Miller, Price, LaLonda, & Keyes, 1990; Rockstroh *et al.*, 1989; Rosler, 1991; Brunia, 1993). The use of many recording sites and spatial enhancement techniques make possible the detection of previously elusive variations in topography in response to task factors, as will be discussed later in this paper.

### Information and Expectation: The P300

At about the same time that the CNV was discovered, another EP wave related to cognition was found. In 1965 Samuel Sutton and colleagues reported a positive wave peaking at about 300 msec after infrequent, task-relevant stimuli (Sutton, Braren, Zubin, & John, 1965; Figure 2). This P300 or P3 peak was considered to be endogenous because its scalp topography and behavior did not depend on physical stimulus properties *per se,* but on their "meaning" in the context of the task. The P300 peak is evoked by many types of paradigms, but the strongest factors in its elicitation are stimulus frequency and task relevance. The P300 is usually accompanied by several other overlapping positive waves in the 300–1000 msec latency range. This late positive complex (LPC) includes the P3a, P3b (P300), and the slow wave (SW). An especially cogent finding in support

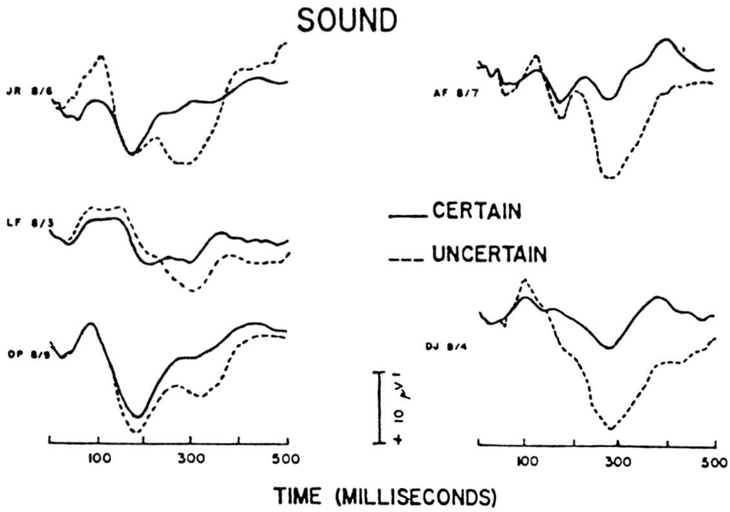


Figure 2. Illustration of P300 wave from the original report of Sutton *et al.* (1965). The P300 was observed as a large positive wave peaking about 300 msec after the stimulus in these averaged waveforms from five persons and elicited by infrequent unexpected (uncertain) and frequent expected (certain) auditory stimuli. The illustration shows one midline channel for five subjects.

of the endogenous origin of P300 was that it can be evoked by the *absence* of an expected stimulus in a regularly timed sequence of stimuli. This "missing stimulus EP," obtained by averaging brain potentials registered to the time the missing stimuli should have occurred, lacks early exogenous peaks such as N100, but contains P300 and other endogenous waves.

Although it has been extensively studied, the robust, easily elicited P300 has resisted a precise definition of its relationship to cognition and behavior. Although it is influenced by many factors in a complex manner, there are several well-established facts about its behavior (reviewed in Donchin, 1979; Duncan-Johnson & Donchin, 1982; Pritchard, 1981; Tueting, 1978; Picton, 1992). Task-relevant (attended) stimuli elicit substantial P300 waves whose amplitudes are larger for infrequent deviant stimuli, as in the popular "oddball" paradigm, where infrequent target stimuli must be distinguished from frequent non-targets. Its potential distribution (with linked ear or noncephalic reference) is maximal at midline parietal (Pz) or central (Cz) sites. The distribution is relatively unaffected by physical stimulus properties, such as modality or intensity, provided that they do not provide task-relevant information. The amplitude of P300 is inversely related to the probability of occurrence of stimuli which have different meanings in the context of the task. The behavior of P300 is usually interpreted in terms of concepts from the field of cognitive psychology. However, its behavior may fit a particular theory in one study, but not in another. There is no simple relationship between P300 and cognitive factors, and because of this, a parsimonious model of P300 and cognition has not yet been achieved. The difficulty in interpreting P300 in terms of psychological constructs has been addressed (Chase, McCarthy, Squires, & Schvaneveldt, 1984; Verlager, 1988; Picton, 1992). A few of the more popular models of P300 are discussed below.

P300 AND STIMULUS EVALUATION. One theory is that P300 is related to the process of stimulus evaluation. Increases in P300 latency and duration have been produced by increasing stimulus complexity or decreasing stimulus discriminability (Ritter, Simson, Vaughan, & Macht, 1982; Squires, Donchin, Squires, & Grossberg, 1977) and also by increasing difficulty of categorization (Ritter, Simson, & Vaughan, 1983) or the number of comparison items items in short-term memory (Ford, Roth, Mohs, Hopkins, & Koppell, 1979). Since P300 does not always precede response initiation and is not always closely correlated with reaction time, it is more likely associated with neural processes contingent on stimulus evaluation but occurring subsequent to them.

P300 AND UPDATING OF AN INTERNAL MODEL. One explanation for both the loose association of P300 latency and reaction time *and* its relation to stimulus evaluation is that P300 is associated with the updating of an internal schema or memory template *subsequent* to stimulus evaluation (McCarthy & Donchin, 1983; Tueting, 1978). Updating of internal schema cannot occur until the significance of the stimulus has been fully evaluated, while a response can be initiated as soon as the information required for performance of the current trial is extracted. Thus, P300 latency would be correlated with reaction time only indirectly. Two aspects of context updating are the amount of updating induced by the *current* stimulus and its effects on expectancy, which should be evident in the *succeeding* trial. The amount of updating aspect is supported by the larger amplitude of P300 for infrequent stimuli (global probability effect) and longer latency for disconfirming feedback (Squires, Hillyard, & Lindsay, 1973), two instances where more updating would be expected. Evidence for the expectancy aspect is the influence of local probability (i.e., the order of the immediately preceding stimuli), which can modify the relationship of global probability and P300 amplitude (Squires, Wickens, Squires, & Donchin, 1976).

P300, Controlled Processing and Capacity. The psychological concepts of controlled processing and processing capacity have been applied to the behavior of P300. Controlled processing denotes a limited-capacity central process involving conscious effort and is distinguished from automatic processing, which does not make demands on processing capacity. Examples of automatic processes are initial encoding of simple sensory features and the execution of well-practiced motor actions. Controlled processes include memory rehearsal and search, target detection, and various types of decision or judgment. It has been suggested that P300 amplitude is related to the amount of controlled processing required. In fact, in dual-task paradigms, when two tasks that require controlled processing are performed simultaneously, their interference is evident not only in measures of reaction time and accuracy, but also in P300 amplitude. The P300 to infrequent targets in an oddball task is reduced when a second task is performed concurrently, and its amplitude varies inversely with the difficulty of the second task only if that task requires controlled processing (Israel, Chesney, Wickens, & Donchin, 1980; Wickens, Kramer, Vanasse, & Donchin, 1983). Further, P300 amplitudes in two controlled processing tasks vary reciprocally with task difficulty, suggesting that P300 does indeed index the allocation of processing capacity.

The long-duration slow wave (SW), which is frequently seen as part of the late positive complex, lasts about 500 to 700 msec. The amplitude of the SW is larger for all task-relevant stimuli, whether or not they are targets, and can even occur without a P300 (Squires, Squires, & Hillyard, 1975). Its amplitude can be larger for nontargets if they require more processing than targets (Friedman, Vaughan, & Erlenmeyer-Kimling, 1981). The SW is thought to be associated with processes subsequent to stimulus evaluation, possibly the extraction of information necessary to maintain task performance over many trials. The SW was the first EP event whose scalp topography could be differentially manipulated by cognitive factors. An experimental effect is sometimes seen only in its frontal negative or parietal positive portions, suggesting that they are generated by different neural processes (Picton & Stuss, 1980; Friedman *et al.*, 1981; Picton, 1992).

## Neuroelectric Chronometrics

Chronometrics refers to methods for inferring the sequence and timing of component stages of cognitive processes from behavioral and/or neuroelectric measures. This approach was pioneered by Donders in the mid-1800s, brought back into popular use by Sternberg (1966), and most elegantly applied in the work of Posner (1978). The methods have also been reviewed in detail (Chase *et al.*, 1984; Meyer *et al.*, 1990). Chronometric analysis is well suited to EP studies, since the timing of onset, peak latency, and duration of the waves associated with various cognitive factors can, in principle, be used to measure directly the temporal relationships of processing stages (Renault, Ragot, Lesevre, & Remond, 1982; Ritter, Vaughan, & Simson, 1983). In the past 15 years the complexity of reported EP components has increased. Evoked-potential events first thought to be unitary phenomena have been found to consist of several waves, distinguishable by topography, experimental manipulations, and analytic methods. However, we will limit this discussion to the "canonical" peaks in a typical visual EP paradigm. [The labeling of waves in the following discussion (e.g., N100) follows the convention of designating polarity (P or N) and typical peak latency (e.g., 100).]

The short- and medium-latency peaks in the first 80 msec following a visual stimulus are highly exogenous: their latency and topography depend on the modality, intensity, and other physical properties of the stimulus. They reflect the arrival of the incoming sensory signals at subcortical and cortical structures. (The same is true of auditory and somatic stimuli, except that the initial processing is faster.) Next, the P100 and the following N100 are usually considered exogenous. Their topography changes with modality and agrees with the assumption that they are generated primarily in sensory cortex. They are thought to reflect the encoding of basic features of stimuli. Following the N100 there is a P200 peak which has a fronto-central topography mostly independent of stimulus modality. If the task requires selection of stimuli on the basis of simple physical properties, the long-duration, low-amplitude processing negativity, thought to be a sign of early selective attention, may also occur in this latency range (Naataanen, Gaillard, & Montysalo, 1978).

If the stimulus differs from the majority of stimuli, the N2-P3a peak complex occurs in the 200–300 msec range (Squires *et al.*, 1975). Task relevance itself is not required, and N2-P3a amplitudes vary inversely with probability. The N2 peak frequently appears to consist of two successive peaks. The earlier peak (termed N2a, mismatch negativity, or NA) and the succeeding N2b are differentially affected by task factors. Since N2b onset is delayed when NA peak latency is lengthened, and N2b peak latency is highly correlated with reaction time, N2a (or NA) and N2b are thought to reflect successive, contingent stages of task processing (Ritter, Vaughan, & Simson, 1983). If the stimulus is also *task-relevant,* P3a will be followed by the P3b (P300) peak. The behavior of P3b is affected in a complex manner by numerous cognitive factors of a higher order than the preceding waves. The P3b appears to be involved in almost all types of higher cognitive functions. Some tasks elicit later peaks such as the long-duration slow wave, peaking between 500 and 700 msec, which may reflect processes related to task performance over many trials.

A typical discrimination task might consist of identifying infrequent target stimuli in a sequence of standard stimuli and making a finger movement in response to the targets. An idealized averaged EP for the target stimuli might contain a sequence of peaks like those in Figure 3: P90 (P1), N130 (N1), P190 (P2), N225 (N2a), N250 (N2b), P290 (P3a), P375 (P3b), a slow wave peaking at 575 msec, and negative movement-related readiness potential (RP) and motor potential (MP) peaking at 500 msec. The chronology of onset, peak latency, and duration of the component waves in this hypothetical waveform, as might be derived by PCA, subtraction, topographic, or other methods, is illustrated in the lower part of Figure 3. When the onset of a particular wave precedes the peak of the previous wave, the neural process indexed by the later wave appears to begin before the culmination of the preceding stage. If experimental manipulation

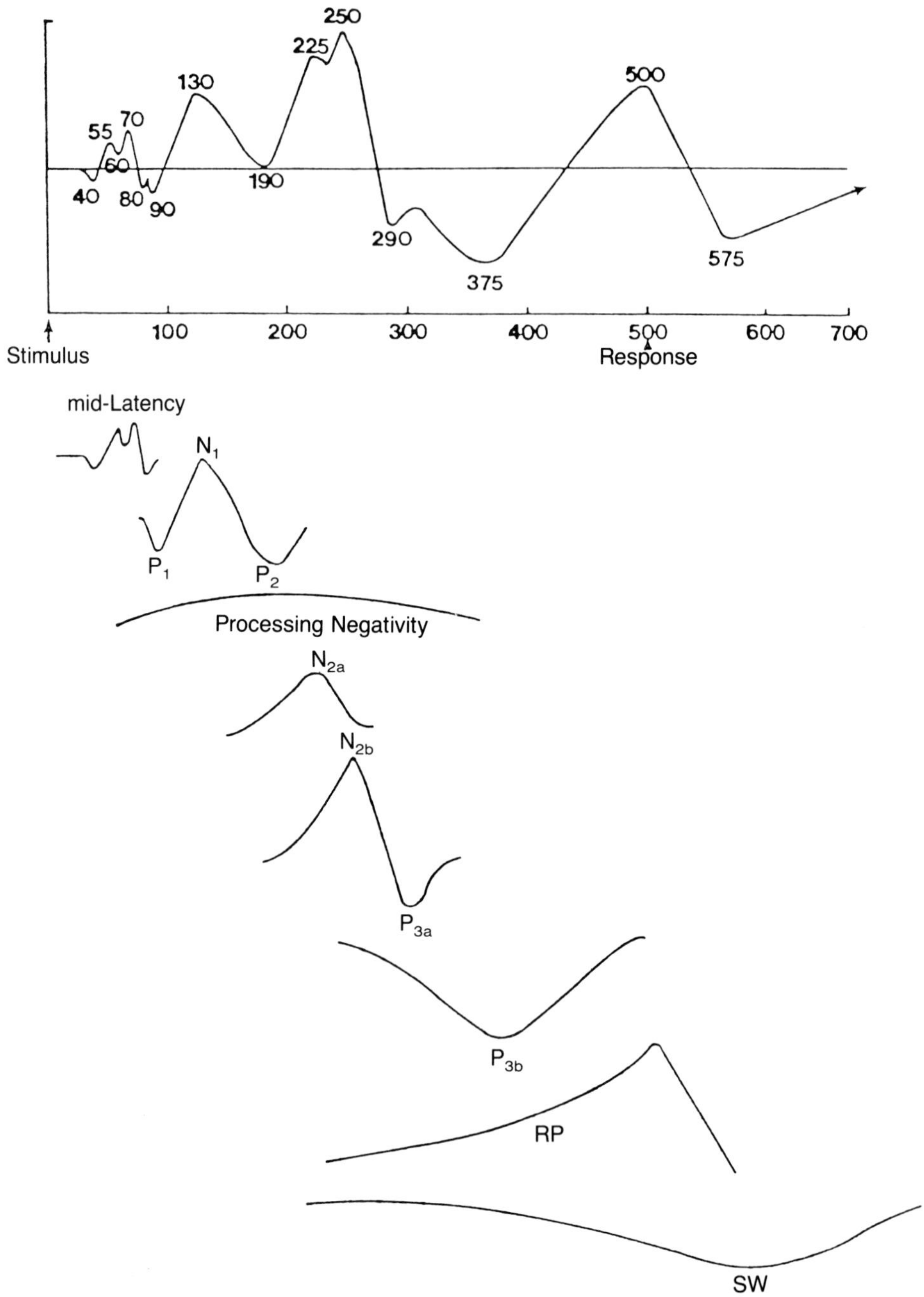


Figure 3. (top) Model averaged EP waveform for infrequent target stimuli in a hypothetical visual discrimination task, showing typical midlatency and late peaks, the processing negativity, and the movement-related readiness potential (RP) potential. (bottom) Hypothetical components of the waveform, as might be inferred by subtraction, PCA, and topographic methods, showing their onset times, peak latencies, and durations. Note the high degree of overlap of successive components. (Adapted from Gevins & Cutillo, 1986).

indicates that there is a contingency between two waves, the issue of discrete or continuous models of information transfer arises. If there appears to be no direct contingency, then the possibility of multiple parallel processing stages must be accounted for in the model (see Meyer *et al.*, 1990).

Considering the hypothetical components of the EP waveform for target stimuli of our discrimination task in terms of information flow and processing stages, we might speculate that the encoding of stimulus features begins with the mid-latency exogenous waves between 40 and 80 msec and continues through P90, N130, and P190. The feature codes necessary for identification of target stimuli must be available by the N2a peak at 225 msec since it is affected by the nature and degree of deviation from the standard stimuli. The N2b at 250 msec is sensitive to endogenous factors which do not affect N2a, and the topographies of N2b and the P3a at 290 msec are not affected by stimulus modality, suggesting that they are associated with processes involving a modality-independent common code. The increase in N2b latency caused by categorization difficulty and its high correlation with reaction time suggests that it is associated with evaluation of stimulus information required for response selection. The P3b (P300) at 375 msec is elicited only by task-relevant stimuli. Its latency is sometimes (but not always) moderately correlated with N2b latency and reaction time. Its amplitude is larger for stimuli that *match* the memorized target, but also becomes larger as the frequency of target stimuli is reduced. Considered along with its latency behavior, P3b appears to be related both to stimulus evaluation in the current trial and also to the updating of an internal schema or expectancy set subsequent to stimulus evaluation in preparation for future trials. The slow wave also seems to be related to processes subsequent to stimulus evaluation and involved in ongoing task performance, perhaps more than P3b, since it is often stronger for nontargets.

## Imaging Distributed Neural Networks

### Distributed Network Model of Cognitive Processing

Evoked potentials rarely reflect simple, unitary neural processes. This is not surprising, since neural processes are complex, as is their relationship to psychological constructs. There is little reason to expect that simple peak amplitude and latency measurements from a few electrodes could adequately characterize processes which involve the integrated activity of many widely distributed neural areas. Since each cognitive task involves the organized processing of various types of sensory, motor, temporal, and mnemonic codes, a different sequence of coordinated neural subsystems must be invoked for each particular cognitive behavior. These metastable configurations, or processing stages, typically last on the order of 100 to 200 msec and represent large-scale integrations of distributed local processing. These large-scale integrations are evident in the brief stabilities in topography of EP peaks and waves (Lehmann, 1981, 1987).

Exogenous EP components evoked by simple sensory stimuli have scalp topographies corresponding to fairly compact neuronal generators in the sensory cortex. In such instances the use of equivalent current dipole models may provide a convenient approximation of the location of the center of mass of these sources (Nunez, 1981; Fender, 1987). However, the assumption that generators are simple or compact is unrealistic in the case of higher cognitive functions. Most cognitive EP components, such as P300, have been considered to have a simple, broad, spatial distribution at the scalp (e.g., parietal maximum; bilateral symmetry), quite similar from paradigm to paradigm. This view arose from studies using only a few electrodes and employing a common reference such as linked ears. With such methods, EP events such as the P300 and contingent negative variation do indeed appear to be bilaterally symmetrical and similar in topography in different paradigms. With a few exceptions, such as the slow wave, whose anterior and posterior components can be seen to behave differently with just a few midline electrodes, most EP events reveal local between-condition differences only when recorded from many sites and spatially enhanced with signal processing methods such as Laplacian derivation (Gevins & Cutillo, 1993; Gevins *et al.*, 1989b). When this is done, many EP events exhibit local variations in topography over cortical areas appropriate to the tasks used to evoke them. In support of this view, recent intracerebral recordings of N200, P300, and slow wave components directly from prefrontal, parietal, and temporal cortices indicate that the cortical generators of these events are widely distributed and vary in amplitude and (slightly) in timing between sites just a few centimeters apart (Smith *et al.*, 1990). Also, the "mosaiclike" structure of cortical areas involved in language processes and in the storage and retrieval of linguistic memory has been demonstrated by recordings and stimulation studies in humans during surgery (Ojemann, 1978). Moreover, many neural imaging studies have demonstrated increased activation in multiple cortical areas, with a different constellation of local activations for different tasks. However, due to the long time sample required by most imaging modalities, it is not possible to determine which foci are coactive at a given time. This is not a problem for EEG (and MEG), where the temporal resolution is limited only by the rate of digitization. The resolution of the dynamic systems of focal activation with EEG, however, does require improvement in spatial sampling, head modeling, and spatial enhancement.

During the past 20 years rapid advances have been made in such neuroimaging modalities as CT, MRI, PET, and SPECT. In contrast, although the EEG is unobtrusive and has associated equipment costs an order of magnitude less than other modalities, development of the EEG as a spatiotemporal neuroimaging modality has not been fully actualized. This slow development stems to some degree from the fact that the amount of spatial information that can be recovered from the scalp-recorded EEG has been underestimated, and as a result little effort has been made to improve its spatial resolution. It is true that the ability to infer electrical sources from scalp-recorded brain electrical or magnetic activity has fundamental physical limitations, but it is also the case that to date the spatial information obtained from EEGs has been technologically limited by poor spatial sampling and underutilization of modern signal enhancement methods. Since computing expense is no longer an obstacle to extensive signal processing, major advances in EEG technology can now be expected. The remainder of this paper deals with technological developments which have taken place at the EEG Systems Laboratory and SAM Technology over the past decade. These include improved spatial sampling and anatomical registration of electrodes, realistic head modeling from MRI, and two analytic techniques: finite element deblurring, which reconstructs the potential distribution at the level of the cortex, and evoked potential covariance, which quantifies spatiotemporal relationships among signals generated by distributed "functional networks." Examples of experimental results employing these methods are presented.

INCREASED NUMBER OF ELECTRODES AND MEASUREMENT OF ELECTRODE POSITIONS. Adequate spatial sampling is a requirement for extracting detailed information about neurocognitive processes. The 19-channel "10/20" montage commonly used in clinical and research EEG recordings has an interelectrode distance of about 6 cm on a typical adult head. This spacing may be good enough for detecting signs of gross pathology, but is insufficient for resolving the finer topographical differences which are important in studying human perceptual and cognitive processes. In our more recent studies we have increased the number of scalp electrodes to 59 and 124 (Gevins, 1988; Gevins, Brickett, Costales, Le, & Reutter, 1990; Gevins, Le, Morton, & Reutter, 1993), which provide interelectrode distances of about 3.5 cm and 2.25 cm, respectively. This is within the 2.5-cm range of the 3 dB point on the cortex-to-scalp point spread function, that is, the size of the scalp representation of a small, discrete neuronal source at the level of the cortex. In addition to increasing the number of recording electrodes, accurate signal enhancement techniques require that the spatial positions of those electrodes be known. At present we measure the three-dimensional positions of scalp electrodes with a commercial magnetic digitizer with an accuracy of about 3 mm.

REALISTIC HEAD MODEL AND ALIGNMENT OF ELECTRODE POSITIONS WITH SCALP SURFACE. Most attempts to increase the three-dimensional spatial resolution of brain electrical events have relied on spherical head models that are uniform in their conductive properties. However, the head is not spherical, nor are its tissue layers uniform in thickness. These sources of error are reduced by utilizing an accurate anatomical representation of a subject's brain surface, skull, and scalp. We have developed methods to process MRIs for accurate rendering of the scalp surface, alignment of electrode positions with this scalp surface, and construction of structured finite elements (tetrahedral volumes) to model the tissue layers of scalp, skull, and cortical surface (Gevins, Brickett, Costales, Le, & Reutter, 1990; Gevins, Le, Brickett, Reutter, & Desmond, 1991; Le & Gevins, 1993).

### COMPUTING POTENTIALS AT THE CORTICAL LEVEL: FINITE ELEMENT DEBLURRING (FED)

The usefulness of an increased number of scalp electrodes is mainly limited by the distortion of neuronal potentials as they are conducted through the highly resistive skull. This distortion amounts to a spatial low-pass filtering, which causes a blurring of the potential distribution at the scalp. Thus, signs of local variations in neurocognitive processes cannot be measured at the scalp without sufficiently fine spatial sampling and some technique for spatial sharpening. There are a number of methods for reducing this distortion, from the computationally simple to computationally intensive. One currently popular method is the Laplacian derivation (Hjorth, 1975; Nunez, 1981, 1989; Nunez, Pilgreen, Westdorp, Law, & Nelson, 1991), which we have used with considerable success in this laboratory. The spatial detail can be improved by using a finite element model of the cortex, cerebrospinal fluid, skull, and scalp to estimate the potentials which would actually be recorded on the surface of the brain. We call our implementation of this method finite element deblurring (FED) (Gevins *et al.*, 1991; Le & Gevins, 1993).

Magnetic resonance images are used to construct a realistic model of the head in the form of many small tetrahedral elements representing the tissues of scalp, skull, and brain. By assigning each tissue a conductivity value, it is possible to calculate the potential at all finite element vertices by using Poisson's equations. The actual conductivity value of each type of tissue is not definitively known, so textbook values are currently being used. However, the procedure is well behaved with respect to this source of uncertainty (see below). It is important to stress that FED is not a source localization method, but rather a "downward continuation" method for reconstructing the distribution of neural potentials as might be recorded at the superficial cortical surface. Further development of FED is under way, including better estimation of conductivity values, better models of the skull, and development of methods to remove ambiguities due to tangentially oriented cortical and subcortical sources.

Deblurring of Somatosensory EPs. Exploratory studies of the FED technique focused on the spatial enhancement of sensory EPs (Gevins *et al.*, 1994). A steady-state stimulation protocol was designed to activate localized areas of visual, auditory, and somatosensory cortex. The somatosensory stimulation consisted of 0.2-msec electrical pulses generated by a Grass S11 stimulator at a rate of 14.93 Hz and applied with a Grass bipolar stimulation electrode affixed to the left hand middle finger and the right hand index finger. Artifact-free epochs of 0.523 seconds were averaged and digitally re-referenced to compute a linked mastoid reference. After application of a Hanning window, the discrete Fourier transform was used to compute the real and imaginary spectral components, and the waveforms were reconstructed from frequency components harmonically related to the stimulation frequency. Figure 4 shows the scalp topography of 122-channel somatosensory steady-state EPs with linked mastoid reference and the computed cortical-level potentials after deblurring. The deblurred somatosensory response clearly contains two foci of activity near the central sulcus and the presumed sensory cortices of the stimulated fingers.

Deblurring of Language-Related EPs. The FED method was used to increase the spatial resolution of EPs related to component processes of reading. A simple cued matching task required four types of judgment: graphic comparison of unfamiliar nonletter character strings, phonemic comparison of pronounceable letter strings (pseudowords), semantic judgment of antonymy, and grammatic judgment of noun-verb agreement (Gevins, Cutillo, & Smith, 1995). Each trial began with a cue indicating the condition. One second later the first stimulus was presented, followed after one second by the comparison stimulus. The conditions were presented in random order, as were the occurrence of match and mismatch trials. We analyzed these component processes of reading on 43 Laplacian derivation time series computed from 59 scalp-recorded channels in nine subjects. In addition, the FED method was performed for two subjects for whom MRI-based head models had been constructed. Several striking between-condition differences were evident in highly localized EP patterns. For example, larger amplitude EP peaks occurred in the grammatic condition (relative to the other language conditions) at 422 msec after the first stimulus and about 300 msec after the second stimulus. These potentials were localized to a region over

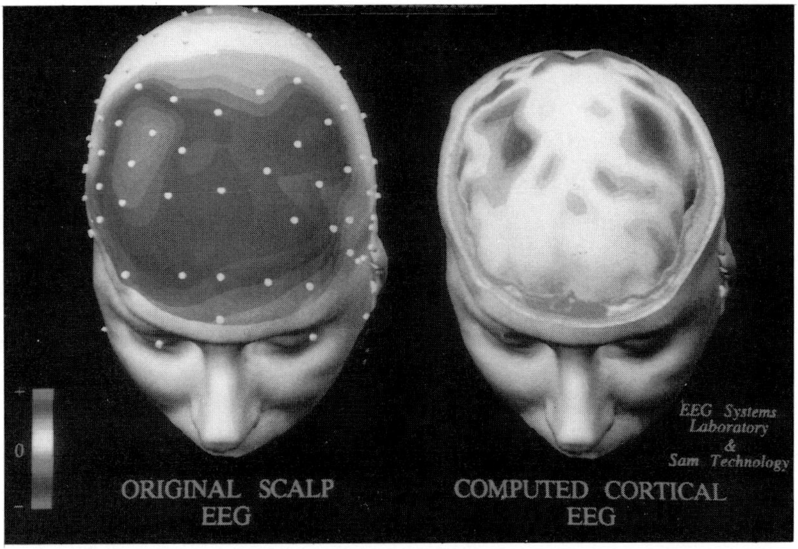


Figure 4. Spatial sharpening of somatosensory EPs with the deblurring method. Spatial detail is greatly improved by the finite-element deblurring (FED) procedure. The 124-channel topographic map of the deblurred steady-state evoked potentials elicited by repetitive electrical stimulation of the left-hand middle finger and right-hand index finger (right) contains more useful localizing information than the original potential recorded at the scalp surface with a linked-mastoids reference (left). The high-resolution map clearly shows two separate foci of activity corresponding to stimulation of the two fingers. For a color representation of this figure, see the color insert facing page 178.

what is presumably Broca's language area of the left prefrontal cortex (Figure 5), which is consistent with what is known about the functional neuroanatomy of this region.

VALIDATION WITH SIMULATIONS AND SUBDURAL GRID RECORDING. The validity of the deblurring procedure was assessed in several ways. Simulations using a spherical conducting model were performed to compare the obtainable exact solutions with the results of the deblurring method (Le & Gevins, 1993). The scalp, skull, and brain surfaces were modeled with concentric (upper) hemispheres of radii 9, 8.5, and 8 cm, each containing 127 points which defined the vertices of 432 prism-shaped finite elements. The conductivity ratio of scalp/skull and brain/skull tissues was taken to be 80. Six current dipoles which varied in depth, orientation, and horizontal location were selected. For each dipole the potentials at the vertices of the cortical surface and the outer surface of the scalp were calculated as the forward solution for a three-sphere model. Then these scalp surface potentials were submitted to the deblurring procedure in order to compute the corresponding potentials over the cortical surface. The resulting error variance ranged from 0.2 to 1.2% for the six cases, which shows good agreement between the deblurred potentials and the exact solutions. In another simulation study, the effect of varying the conductivity ratio was determined. It was found that changing the conductivity ratio by ± 50% from the standard ratio of 80 had small effects on the deblurred results. Different conductivity ratios produced slight differences in the exact size of active regions, but did not substantially change the pattern of focal activations (Le & Gevins, 1993).

A more direct verification of the FED method was performed by comparing

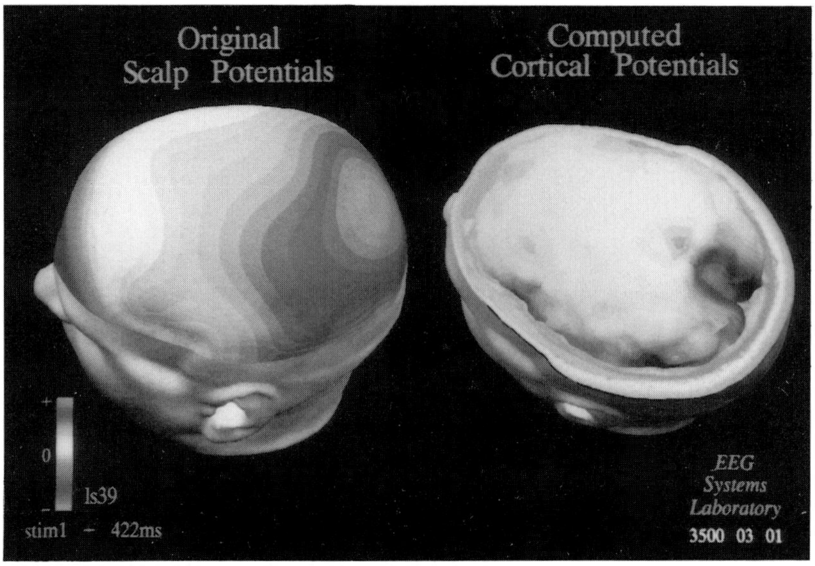


Figure 5. High-resolution EPs during linguistic encoding. Data corresponding to an EP peak in an individual subject 422 msec after the first visual stimulus in a grammatic condition of a linguistic judgment task. In this condition the stimulus was a closed-class word (i.e., syntactic function word such as a pronoun), which is known to involve processing in Broca's language area. The distinguishing feature of activation of the grammatic condition is evident as a positive potential over a circumscribed area of left prefrontal cortex (right), a feature which is not apparent in the conventional linked-mastoid EP map (left). The focus of activation over the posterior visual cortex in response to the visually presented stimuli occurred for both closed- and open-class words. For a color representation of this figure, see the color insert facing page 178.

cortical EPs recorded from a 64-channel intracerebral grid over the somatosensory cortex of an epileptic patient with the deblurred potentials computed from scalp-recorded EPs obtained prior to surgery. Stimulation consisted of 14.93-Hz electrical pulses delivered to the fingers of the hand contralateral to the hemisphere where the grid was to be placed. The cortical grid was aligned with the MRI-based head model using a postimplant radiograph. The deblurred potentials were quite similar to the actual recorded cortical potentials (Figure 6). Additional validation studies with patient data are under way.

## Integration of Spatial and Temporal Information: Evoked Potential Covariance

Topographical maps of brain electrical events typically display the voltage over the scalp measured at a single time point, usually corresponding to an EP peak. However, the peak latency of an EP component often varies substantially among electrode sites, or, as is the case with slowly varying potentials such as the contingent negative variation, may not exist as a discrete time point. In such cases the selection of a time point for mapping becomes somewhat arbitrary. We have been developing methods for characterizing EP components in terms of their spatial topography and temporal relationships between electrode sites. We refer to this method as evoked potential covariance (EPC). It evolved from the theory that neural processes underlying higher cognitive functions involve the coordination of activity in widely distributed cortical and subcortical areas (Liv-

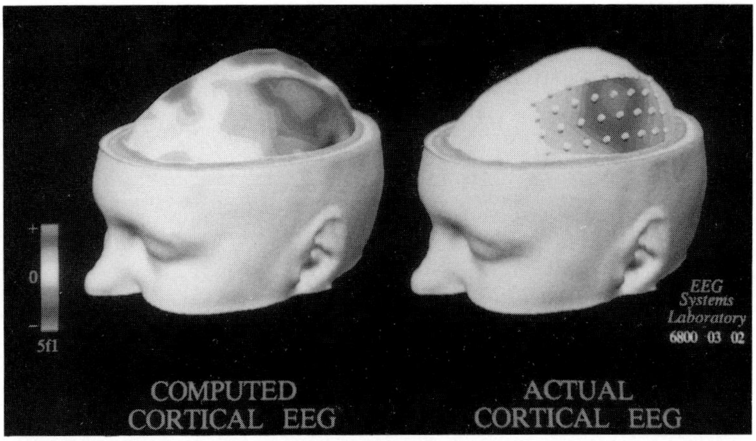


Figure 6. Validation of the deblurring method by intracerebral recording. Response to steady-state electrical stimulation to the fingers of the right hand. Deblurred (left) and actual (right) cortical EPs from an epilepsy patient who had a 64-channel electrode grid implanted for diagnostic reasons. The left focus of cortical activation in the deblurred potentials is very similar in topography to the actual recorded cortical EP. (The scalp EP data were recorded prior to implantation of the grid.) For a color representation of this figure, see the color insert facing page 178.

anov, 1977; Gevins, 1987). Since the EP waveform delineates the time course of event-related mass activity of a neural population, the coordination of two or more local populations should be demarcated by a relationship between the morphology of their EP waveforms with consistent time delay (Gevins & Bressler, 1988). If the relationships are linear, as they often appear to be, this coordinated activity might be measured by the lagged covariance between EP waves recorded from different sites. Nonlinear statistical association measures have been used to characterize seizure onset but have not yet been applied to EPs (Lopes da Silva, Kamphuis, van Neerven, & Pjin, 1990). Of course, a significant covariance of this type does not prove direct corticocortical interactions, and the interpretation of functional relationships between sites is ambiguous for recordings made at the scalp. We merely note here that statistical interdependence between channels is a necessary, but not sufficient, feature of the signals of functional neural networks. With regard to the degree to which it is possible to measure the activity of cortical networks immediately underlying each electrode, there is no general solution to the question of the origin of particular scalp-recorded signals, but the ambiguity can be mitigated by careful experimental design and by the use of spatial filters such as the Laplacian derivation which selectively emphasize superficial (i.e., cortical) sources (Nunez, 1981). Ongoing work on clarifying tangentially oriented cortical sources will also facilitate the interpretation of EPCs. While any interpretation of EPCs in terms of the underlying neural processes is tentative, our results to date have been highly consistent with the known functional neuroanatomy of the cerebral cortex. We have found the EPC method to be useful for characterizing the changes in neurocognitive processes associated with simple cognitive tasks. Our current EPC procedure is given below, and an illustrative example of the covariance between two channels is depicted in Figure 7.

1. A sufficient data set is recorded using a minimum of about 30 electrodes.
2. Artifact contamination is eliminated by automated methods with subsequent visual inspection.

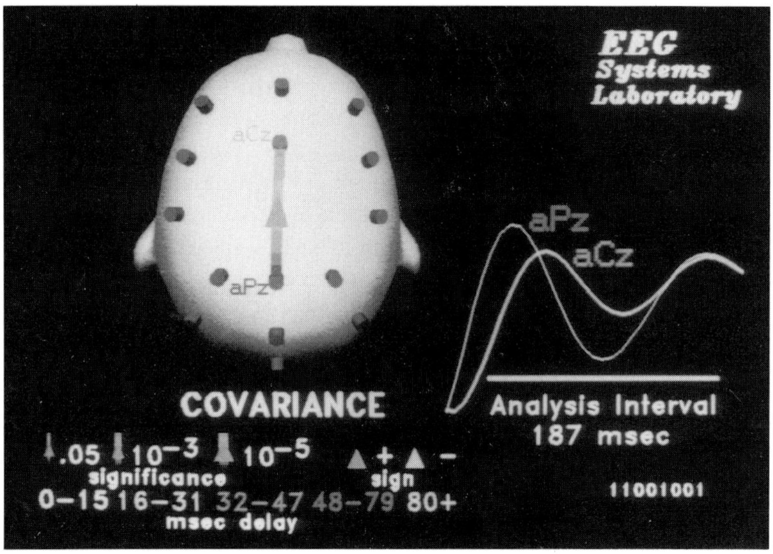


Figure 7. Illustration of the evoked potential covariance (EPC) method. The EPC line on the top view of a model head is computed from the evoked potential segments recorded from scalp positions aPz and aCz. The EPs were spatially sharpened with a Laplacian derivation and then filtered to extract activity in a 4–7 Hz band; the EPC analysis interval was 187 msec wide. The width of an EPC line indicates the relative significance of the covariance between the two EP segments, with the scale appearing above the word "significance." The color of the arrow indicates the sign of the covariance (same color as line = positive; skin color = negative), and the arrow points from the leading to the lagging electrode, unless there is no delay. In this case, the EPC between aPz and aCz has a $p$ value $< 10^{-5}$, the aPz waveform leads the aCz waveform by 16–31 msec, and the covariance is positive. For a color representation of this figure, see the color insert facing page 178.

3. Pairs of data sets are formed which differ significantly only in factors related to the hypothesis in order to avoid false results due to spurious differences in stimulus, motor, or behavioral factors (Gevins *et al.*, 1985). These data sets are averaged by registration to stimulus or response.
4. A Laplacian derivation or deblurring procedure is applied to spatially sharpen the evoked potentials.
5. Analysis intervals and digital filter characteristics are determined from the EP waveforms. Since the components of the EP waveform serve to indicate the onset and duration of the stages of neural processing (Gevins & Cutillo, 1986), the location and width of EPC analysis intervals are determined by the timing and duration of EP peaks and waves. EPC analysis intervals typically range from 100 msec in the case of early visual sensory peaks to 375 msec in the case of the contingent negative variation or the late positive complex.
6. Following application of a suitable digital filter tuned to the duration of each analysis interval, multilag cross-covariance functions are computed between all pairwise channel combinations of the averaged EPs in that interval. The features used to characterize the EPC are the maximum value of the cross-covariance function and the lag time of this maximum.
7. For descriptive purposes, the significance of each EPC magnitude is determined by reference to the standard deviation of the "noise EPC". A distribution of noise EPCs is computed by first averaging random inter-

vals selected from each trial in the data set and then performing EPC analysis on these noise averages. The threshold for significance is increased using Duncan's correction procedure, with the number of channels used as an estimate of the number of independent dimensions.

8. ANOVA and post hoc *t*-tests are used to compare EPC magnitudes between conditions. The similarity of EPC patterns between conditions is measured by estimating the correlations between them. The estimate comes form a distribution-independent "bootstrap" Monte Carlo procedure, which also yields a confidence interval for the estimates. The correspondence between conditions of the sites of highest covariance is determined using the Spearman rank order correlation test.

EPCs of Preparatory Processes in a Visuomotor Task. Even in the highly controlled conditions of the laboratory, a stimulus does not fall upon a blank brain state. There is always some degree of expectation based on what has occurred during preceding trials, because the perceptual interaction between oneself and one's environment always occurs in the context of prior experience. One way to study the effects of expectation in the laboratory is to deliberately manipulate expectation by means of a get-ready cue. Our EPC analysis has been particularly fruitful in this area. A clear example of the modulation of "functional networks" associated with preparatory processes was evident in the cued prestimulus interval of a skilled motor response task (Gevins *et al.*, 1989a, 1989b). The task required the production of precisely graded pressure responses of an index finger proportional to visually presented numbers from 0 to 9. The responding hand was indicated by a cue presented one second before the stimulus. Feedback was presented one second after the response as a two-digit number indicating the exact response pressure made.

Sixteen Laplacian derivation (LD) EP waveforms were computed from 26 channels of scalp-recorded data. EPC patterns were computed from all pairwise combinations of the 16 LD channels in an interval from 500 to 875 msec following the cue, which spanned the late CNV wave. Figure 8 shows this prestimulus interval for preparing right and left hands for those trials for which the response was accurate (i.e., trials where the response error was less than the mean error over the session). The "functional networks" for trials with subsequently accurate responses have covariances of the left prefrontal electrode with electrodes overlying the same motor, somatosensory, and parietal areas involved in the actual response execution of either right or left hand. The preparatory pattern preceding inaccurate responses differed markedly from that preceding accurate responses (Gevins *et al.*, 1987, 1989b). In the case of inaccurate right hand trials, the prestimulus EPCs between the left frontal site and posterior left hemisphere sites were fewer and weaker than for accurate trials. On the other hand, inaccurate left hand trials were associated with a complex pattern involving EPCs between the left frontal site and posterior sites over both hemispheres. In effect, the inaccurate left hand pattern appeared to be a mixture of "accurate" right and left hand EPC patterns. Thus, it is possible that inaccurate performance of this task is due in part to a failure of prefrontal regions to properly prepare and coordinate the activity of cortical areas used in response execution, which agrees with the view of the CNV as a sign of mobilization of cerebral potentiality in preparation for motor or mental processes.

Coordination of task-related functional networks also appears to play an

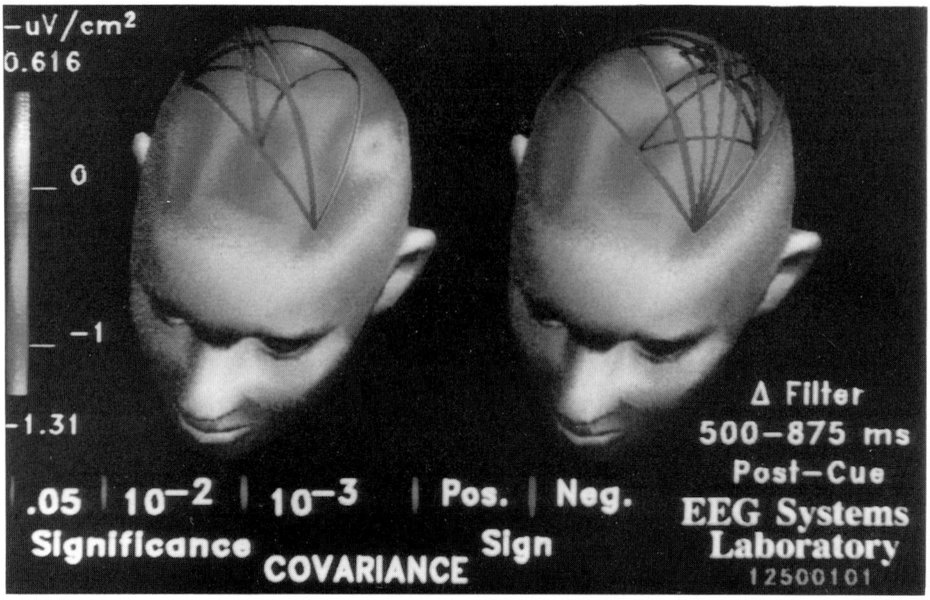


Figure 8. Preparatory evoked potential covariance (EPC) patterns preceding accurate responses. EPC patterns were computed in an interval from 500 to 875 msec after seven subjects had been cued to prepare to make a response with the index finger of either the right or left hand when the next numeric stimulus was presented. Data are shown only for those trials for which the response was subsequently accurate. The preparatory pattern preceding accurate responses have covariances of the left prefrontal electrode with electrodes overlying the same motor, somatosensory, and parietal areas that were involved in actual response execution. For a color representation of this figure, see the color insert facing page 178.

important role in tasks involving working memory, that is, the process of holding a small amount of information in awareness while it is used in some mental operation. We have studied EPCs during a working memory task using a modified version of the finger pressure task described above (Gevins & Cutillo, 1993). In this version of the task, a graded pressure response was made with the index finger of the right hand according to the magnitude of the visual stimulus number (from 1 to 9) presented two trials previously. The original right hand motor response task was also included as the control task. The working memory (WM) and control tasks both had the same stimulus, response, and feedback, but the WM task differed in the use of working memory to maintain two digits and compare the stimulus number two trials back with the current stimulus number. Twenty-seven EEG channels were used to determine 18 nonperipheral Laplacian derivation channels, and EPC patterns were computed between all pairwise combinations of the 18 channels. During the period before onset of the number stimulus, in the interval spanning the late CNV, EPC patterns in both tasks were focused on electrodes over midline and bilateral premotor and right posterior areas (Figure 9). However, EPC differences were observed in the WM task, with the WM condition being distinguished by stronger EPCs overall and strong left-sided EPCs between precentral and left anterior parietal and left anterior occipital sites. This result suggests that the functional network associated with the remembering of two numbers can be measured without confounding by sensory or motor processes during the prestimulus preparatory interval. Further, like the above-mentioned result for the bimanual task, these results suggest that the

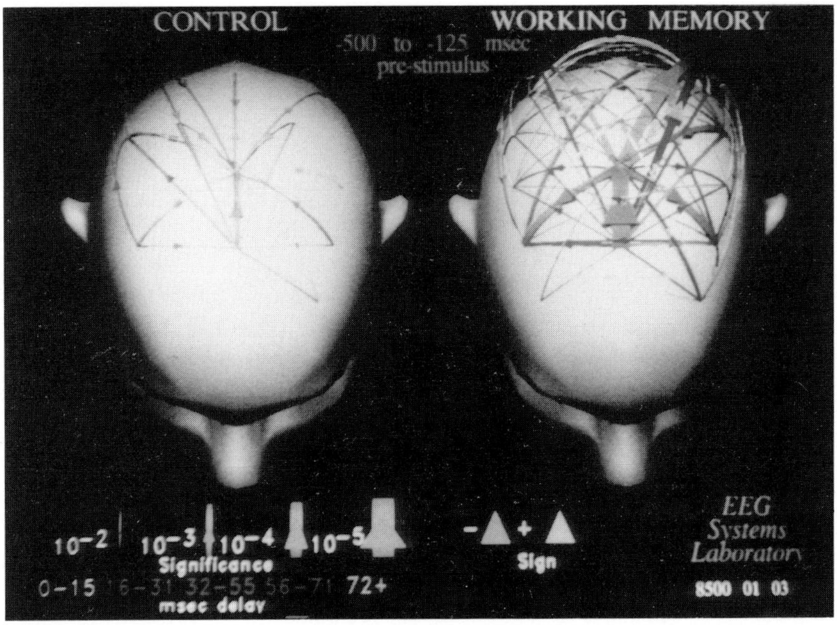


Figure 9. Preparatory evoked potential covariance patterns during working memory and control tasks. EPCs during a 375-msec interval when five subjects were staring at a blank computer screen awaiting the next stimulus number. In the working memory task (right), subjects had to remember the two stimulus numbers from the preceding two trials, while in the control task (left), they did not. The working memory task is distinguished by stronger EPCs overall and strong left-sided EPCs among anterior occipital, anterior parietal, and anterior central sites, which overlie cortical areas that may be involved in the maintenance of the numbers in working memory. (From Gevins & Cutillo, 1993.) For a color representation of this figure, see the color insert facing page 179.

higher-order motor and left parietal cortical areas which appear to be activated during this preparatory period are the same areas that participate in subsequent stimulus processing.

Preparatory functional networks appear to be altered by prolonged task performance, an observation which may be important for understanding and predicting the performance failures that tend to occur in fatigued individuals. The working memory task described above was used to search for predictive, leading indicator neuroelectric signs of operational fatigue accompanying extended sessions of task performance (Gevins *et al.,* 1990). This experiment was conducted with the participation of five highly skilled Air Force test pilots who practiced the tasks extensively prior to the EEG recording session. Changes in the EPCs during the prestimulus interval (when the pilots were staring at a blank screen, remembering the last two numbers, and awaiting the next stimulus number) were observed over the course of a recording session that lasted from 10 to 14 hours (Figure 10). The changes appeared after 7–8 hours of performance, but before behavioral signs of fatigue became significant. The overall magnitude of the covariances was reduced, and the topographic pattern changed from one strongly focused on midline central and precentral and left posterior sites to one focused primarily on right-sided precentral and parietal electrodes. It appeared that sustained task performance affected the "neural strategy" for remembering the two numbers several hours before statistically significant behavioral signs of fatigue were evident.

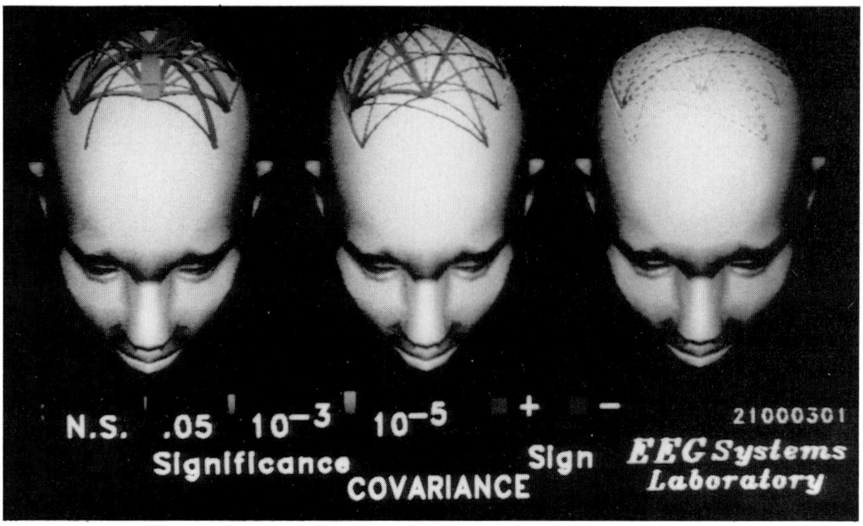


Figure 10. Preparatory evoked potential covariance patterns during sustained performance of a working memory task. The three heads in this illustration show changes in EPC patterns in the prestimulus interval of the working memory task as the five subjects performed the working memory and control tasks for 10 to 14 hours. [Same subjects and task as Figure 9; the pattern on the left (Alert) is the same as that on the right side of Figure 9.] The fatigued pattern at the right occurred late in the session after a significant decline in task performance. The middle pattern (becoming fatigued) occurred just prior to the appearance of significant signs of performance decrement after 7–8 hours of work. Not only did the EPC magnitude decline greatly from the alert to the becoming fatigued condition, but the EPC pattern also changed, the emphasis shifting from midline central and precentral and left parietal sites to the right anterocentral and parietal and left central sites. For a color representation of this figure, see the color insert facing page 179.

All the findings discussed above deal with EPCs in the cue-to-stimulus interval during the late CNV wave. Similarly detailed findings were also obtained in these studies in intervals spanning the P300 wave in stimulus and feedback evoked EPs. It is clear that a great deal more useful information is available in both the CNV and P300 than is obtainable by simple amplitude and latency measures.

## Conclusion: Neuroimaging and Behavior

Recent advances in EEG recording and analysis technology have significantly improved measurement of the neurophysiological basis of perception, cognition, and behavior. These include increasing the number of recording electrodes, increasing topographic detail with spatial signal enhancement techniques, and developing methods to characterize distributed neurocognitive processing. The development of these technologies is by no means complete. In improving spatial deblurring, ambiguities must be resolved due to cortical generators which are not directly under a recording electrode, as is the case with signals from primary auditory areas in temporal cortex which project upward to the vertex. This will require development of better methods of "multiple/distributed source analysis", which is under way at several laboratories. Also, concerns about the validity of the evoked potential covariance measure in characterizing functional networks must be addressed (Bressler, Coppola, & Nakamura, 1993). Once this has been accomplished, high-resolution evoked potentials, both by themselves and in combination with other functional neuroimaging modalities, will likely play an increasingly important role both in cognitive neuroscience and in clinical neurology.

The ability of high-resolution neuroelectric imaging to accurately characterize the subsecond spatiotemporal dynamics of neurocognitive processes and relate them directly to behavioral variables in carefully controlled paradigms should promote the development of more accurate models of mental functioning. The simple stimulus-response format of most experimental designs has given rise to models of cognitive functioning with a predominantly passive tone. The brain is assumed to react to a stimulus, and the succeeding processing stages leading to response are inferred from measures of reaction time, evoked potential peak latencies, and so on. However, experience, observation, and inference all suggest that cognitive processes are highly interactive and proactive. Perception is a synthesis of sensation, current brain state, and past cognitive experience. This synthesis relies on a continuously updated, dynamic, internal representation of one's self and environment. Sensory and effector systems are actively used to probe the environment for information relevant to maintaining and updating the self/world model. Each perception, each action, is incorporated into the internal model, and new perceptions and actions are in turn influenced through the model's role in directing attentional and conceptual processes. Although the frontal cortex appears to play a pivotal role in these high-level cognitive processes, it would be simplistic to consider the frontal lobes as exclusively executive in nature. Rather, it is likely that many areas are involved in a constellation of rapidly changing functional networks that provide the delicate balance between stimulus-locked behavior and purely imaginary ideation. It is these evanescent patterns of neural activity which are the focus of all functional imaging modalities, but neuroelectric imaging is especially suited for dealing with their dynamic nature.

The impressive images of the various types of modern imaging modalities should always be tempered with an assessment of their ability to relate the observed neural activity to behavior. Even in clinical settings, the ability to image *functional* aspects of brain activity, such as perception, language, working memory, and so on, is a valuable adjunct to the information obtained from the basic imaging of structural lesions and clearly pathological neural (e.g., epileptiform) activity. Because each individual's brain differs far more in functional neuroanatomy than in physical structure, it is very valuable to be able to locate the areas involved in performance of specific cognitive functions, rather than inferring them on the basis of population norms or through more invasive methods. Further, the fine temporal resolution afforded by neuroelectric imaging provides a unique degree of precision in correlating neural activity with behavior and thus with the stages of cognitive processes. It is challenging, but fruitful, to design experimental situations that emphasize this dynamic and interactive nature of cognition. With even further advances in brain imaging technology, we can hope to achieve increasingly detailed and direct measurements of the organization and interrelationships of sensory and higher cognitive processes.

*Acknowledgments*

Our research is sponsored by competitive grants and contracts from the following agencies of the United States government: the National Institute of

Mental Health, Air Force Office of Scientific Research, National Institute of Neurological and Communicative Diseases and Strokes, Office of Naval Research, National Science Foundation, and National Institute of Alcoholism and Alcohol Abuse. Thanks to Brian Cutillo for major help in writing this paper and to all my colleagues and friends at EEG Systems Laboratory and SAM Technology whose efforts produced the original methods and results described here.

## References

Bressler, S., Coppola, R., & Nakamura, R. (1993). Episodic multi-regional cortical binding at multiple frequencies during visual task performance by the macaque monkey. *Nature, 366,* 153–155.

Brunia, C. (1993, July). Waiting in readiness: Gating in attention and motor preparation. *Psychophysiology.*

Chase, W., McCarthy, G., Squires, K., & Schvaneveldt, R. (1984). Mental chronometry. In E. Donchin (Ed.), *Cognitive psychophysiology: Event-related potentials and the study of cognition* (pp. 249–301). Hillsdale, NJ: Erlbaum.

Donchin, E. (1979). Event-related brain potentials: A tool in the study of human information processing. In H. Begleiter (Ed.), *Evoked potentials and behavior* (pp. 13–75). New York: Plenum.

Donchin, E., & Heffley, E. F. (1978). Multivariate analysis of event-related potential data: A tutorial review. In D. Otto (Ed.), *Multidisciplinary perspectives in ERP research* (EPA-600/9-77-043, pp. 555–572.) Washington, DC: U.S. Government Printing Office.

Donchin, E., Kutas, M., & Johnson, R. (1974). The CNV does not behave like a 'motor potential.' *Electroencephalography and Clinical Neurophysiology, 37,* 484–496.

Donchin, E., Kutas, M., & McCarthy, G. (1977). Electrocortical indices of hemispheric utilization. In S. Harnard, R. W. Doty, L. Goldstein, J. Jaynes, & A. Krauthamer (Eds.), *Lateralization in the nervous system* (pp. 339–384). New York: Academic Press.

Duncan-Johnson, C., & Donchin, E. (1982). The P300 component of the event-related potential as an index of information processing. *Biological Psychology, 14,* 1–52.

Fender, D. (1987). Source localization of brain electrical activity. In A. Gevins & A. Remond (Eds.), *Handbook of electroencephalography and clinical neurophysiology: Vol. 1* (Rev. series, pp. 355–403). New York: Elsevier.

Ford, J., Roth, W., Mohs, R., Hopkins, W., & Koppell, B. (1979). Event-related potentials recorded from young and old adults during a memory retrieval task. *Electroencephalography and Clinical Neurophysiology, 47,* 450–459.

Friedman, D., Vaughan Jr., H. G., & Erlenmeyer-Kimling, L. (1981). Multiple late positive potentials in two visual discrimination tasks. *Psychophysiology, 18,* 635–649.

Fuster, J. (1989). *Prefrontal cortex: Anatomy, physiology and neuropsychology of the frontal lobe.* New York: Raven Press.

Gevins, A. S. (1987). Correlation analysis. In A. S. Gevins & A. Remond (Eds.), *Handbook of electroencephalography and clinical neurophysiology: Vol. 1. Methods of analysis of brain electrical and magnetic signals* (pp. 171–193). Amsterdam: Elsevier.

Gevins, A. S. (1988). Recent advances in neurocognitive pattern analysis. In E. Basar (Ed.), *Dynamics of sensory and cognitive processing of the brain* (pp. 88–102). Heidelberg: Springer-Verlag.

Gevins, A. S., & Bressler, S. L. (1988). Functional topography of the human brain. In G. Pfurtscheller (Ed.), *Functional brainimaging* (pp. 99–116). Bern: Hans Huber.

Gevins, A. S., Bressler, S. L., Cutillo, B. A., Illes, J., Miller, J., Stern, J., & Jex, H. (1990). Effects of sustained mental work on functional brain topography. *Electroencephalography and Clinical Neurophysiology, 76,* 339–350.

Gevins, S. A., Bressler, S. L., Morgan, N. H., Cutillo, B. A., White, R. M., Greer, D., & Illes, J. (1989a). Event-related covariances during a bimanual visuomotor task, Part I: Methods and analysis of stimulus- and response-locked data. *Electroencephalography and Clinical Neurophysiology, 74,* 58–75.

Gevins, A. S., Brickett, P., Costales, B., Le, J., & Reutter, B. (1990). Beyond topographic mapping: Towards functional-anatomical imaging with 124-channel EEGs and MRIs. *Brain Topography, 3*(1), 51–64.

Gevins, A. S., & Cutillo, B. A. (1986). Signals of cognition. In F. H. Lopez da Silva, W. Storm Van Leeuwen, & A. Remond, (Eds.), *Handbook of electroencephalography and clinical neurophysiology: Vol. 2* (pp. 335–381). Amsterdam: Elsevier.

Gevins, A. S., & Cutillo, B. C. (1993). Spatiotemporal dynamics of component processes in human working memory. *Electrocephalography and Clinical Neurophysiology, 87,* 128–143.

Gevins, A. S., Cutillo, B. A., Bressler, S. L., Morgan, N. H., White, R. M., Illes, J., & Greer, D. (1989b). Event-related covariances during a bimanual visuomotor task, Part II: Preparation and feedback. *Electroencephalography and Clinical Neurophysiology, 74,* 147–160.

Gevins, A., Cutillo, B., & Smith, M. (1995). Regional modulation of high resolution evoked potentials during verbal and non-verbal matching tests. *Electroencephalography and Clinical Neurophysiology, 94,* 129–147.

Gevins, A. S., Doyle, J. C., Cutillo, B. A., Schaffer, R. E., Tannehill, R. S., & Bressler, S. L. (1985). Neurocognitive pattern analysis of a visuospatial task: Rapidly shifting foci of evoked correlations between electrodes. *Psychophysiology, 22*(1), 32–43.

Gevins, A. S., Doyle, J. C., Schaffer, R. E., Callaway, E., & Yaeger, C. (1980). Lateralized cognitive processes and the electroencephalogram. *Science, 207,* 1005–1008.

Gevins, A. S., Le, J., Brickett, P., Reutter, B., & Desmond, J. (1991). Seeing through the skull: Advanced EEGs use MIRs to accurately measure cortical activity from the scalp. *Brain Topography, 4* (2), 125–131.

Gevins, A., Le, J., Martin, N., & Reutter, B. (1993). High resolution EEG: 124-channel recording, spatial enhancement and MRI integration methods. *Electroencephalography and Clinical Neurophysiology, 90,* 337–358.

Gevins, A. S., Morgan, N. H., Bressler, S. L., Cutillo, B. A., White, R. M., Illes, J., Greer, D. S., Doyle, J. C., & Zeitlin, G. M. (1987). Human neuroelectric patterns predict performance accuracy. *Science, 235,* 580–585.

Gevins, A., & Schaffer, R. (1980, October). A critical review of electroencephalographic correlates of higher cortical functions. *CRC Critical Reviews in Bioengineering.*

Gevins, A. S., Zeitlin, G. M., Doyle, J. C., Schaffer, R. E., & Callaway, E. (1979). EEG patterns during "cognitive" tasks. II. Analysis of controlled tasks. *Electroencephalography and Clinical Neurophysiology, 47,* 704–710.

Gevins, A. S., Zeitlin, G. M., Doyle, J. C., Yingling, C. D., Schaffer, R. E., Callaway, E., & Yaeger, C. L. (1979a). Electroencephalogram correlates of higher cortical functions. *Science, 203,* 665–668.

Gevins, A. S., Zeitlin, G. M., Yingling, C. D., Doyle, J. C., Dedon, M. F., Schaffer, R. E., Roumasset, J. T., & Yaeger, C. L. (1979b). EEG patterns during "cognitive" tasks. I. Methodology and analysis of complex behaviors. *Electroencephalography and Clinical Neurophysiology, 47,* 693–703.

Goldman-Rakic, P. (1984). The frontal lobes: Uncharted provinces of the brain. *Trends in Neuroscience, 7,* 425–429.

Harter, M., Miller, S., Price, N., LaLonda, M., & Keyes, A. (1990). Neural processes involved in directing attention. *Journal of Cognitive Neurophysiology, 1*(3), 223–237.

Hjorth, B. (1975). On-line transformation of EEG scalp potentials into orthogonal source derivations. *Electroencephalography and Clinical Neurophysiology, 39,* 111–118.

Israel, J., Chesney, G., Wickens, C., & Donchin, E. (1980). P300 and tracking difficulty: Evidence for multiple resources in dual-task performance, *Psychophysiology, 17,* 259–273.

Le, J., & Gevins, A. (1993). Method to reduce blur distortion from EEGs using a realistic head model. *IEEE Transactions on Biomedical Engineering, HO,* 517–528.

Lehmann, D. (1981). Spatial analysis of evoked and spontaneous EEG potential fields. In N. Yamaguchi & K. Fujisawa (Eds.), *Recent advances in EEG and EMG data processing* (pp. 117–132). Amsterdam: Elsevier.

Lehmann, D. (1987). Principles of spatial analysis. In A. S. Gevins & A. Remond (Eds.), *Handbook of electroencephalography and clinical neurophysiology: Vol. 1. Methods of analysis of brain electrical and magnetic signals* (pp. 309–354). Amsterdam: Elsevier.

Livanov, M. N. (1977). *Spatial organization of cerebral processes.* New York: Wiley.

Lopes da Silva, F., Kamphuis, W., van Neerven, J., & Pjin, J. (1990). Interdependence of EEG signals: Linear vs. nonlinear associations and the significance of time delays and phase shifts. *Brain Topography, 1*(2), 9–18.

McCarthy, G., & Donchin, E. (1983). Chronometric analysis of human information processing. In A. Gaillard & W. Ritter (Eds.), *Advances in psychology: Vol. 10. Tutorials in event-related potential research: Endogenous components* (pp. 251–268). Amsterdam: Elsevier.

McGillem, C. D., & Aunon, J. I. (1987). Analysis of event-related potentials. In A. S. Gevins & A.

Remond (Eds.), *Handbook of electroencephalography and clinical neurophysiology: Vol. 1. Methods of analysis of brain electrical and magnetic signals* (pp. 131–169). Amsterdam: Elsevier.

Meyer, D., Osman, A., Irwin, D., & Yantis, S. (1990). Modern mental chronometry. *Biological Psychology, 26,* 3–67.

Naataanen, R., Gaillard, A., & Montysalo, S. (1978). The N1 effect of selective attention reinterpreted. *Acta Psychologica, 42,* 173–187.

Nunez, P. L. (1981). *Electric fields in the brain: The neurophysics of EEG.* New York: Oxford University Press.

Nunez, P. (1989). Estimation of large-scale neocortical source activity with EEG surface Laplacians. *Brain Topography, 2,* 141–154.

Nunez, P., Pilgreen, K., Westdorp, A., Law, S., & Nelson, A. (1991). A visual study of surface potentials and Laplacians due to distributed neocortical sources: Computer simulations and evoked potentials. *Brain Topography, 4*(2), 151–168.

Ojemann, G. (1978). Organization of short-term verbal memory in language areas of human cortex evidence from electrical stimulation. *Brain and Language, 5,* 331–340.

Picton, T. (1992). The P300 wave of the human event-related potential. *Journal of Clinical Neurophysiology, 9,* 456–479.

Picton, T. W., & Stuss, D. T. (1980). The component structure of the human event-related potentials. In H. Kornhuber & L. Deecke (Eds.), *Progress in brain research: Vol. 54. Motivation, motor and sensory processes of the brain: Electrical potentials, behavior and clinical use* (pp. 17–49). Amsterdam: Elsevier.

Posner, M. (1978). *Chronometric explorations of mind.* Hillsdale, NJ: Erlbaum.

Pritchard, W. (1981). Psychophysiology of the P300. *Psychological Bulletin, 90,* 506–540.

Rebert, C. (1978). Electrogenesis of slow potential changes in the central nervous system: A summary of issues. In D. Otto (Ed.), *Multidisciplinary perspectives in event-related brain potential research* (EPA-600/9-77-043, pp. 3–11). Washington, DC: U.S. Government Printing Office.

Renault, B., Ragot, R., Lesevre, N., & Remond, A. (1982). Onset and offset of brain events as indices of mental chronometry. *Science, 215,* 1413–1415.

Ritter, W., Kelso, S., Kutas, M., & Shiffrin, R. (1984). Preparatory processes. In E. Donchin (Ed.), *Cognitive psychophysiology: Event-related potentials and the study of cognition,* (pp. 179–219). Hillsdale, NJ: Erlbaum.

Ritter, W., Simson, R., & Vaughan, H. (1983). Event-related correlates of two stages of information processing in physical and semantic discrimination tasks. *Psychophysiology, 20,* 168–179.

Ritter, W., Simson, R., Vaughan, H., & Macht, M. (1982). Manipulation of event-related potential manifestations of information processing stages. *Science, 218,* 909–911.

Ritter, W., Vaughan, H., & Simson, R. (1983). On relating event-related potential components and stages of information processing. In A. Gaillard & W. Ritter (Eds.), *Advances in psychology: Vol. 10. Tutorials in event-related potential research: Endogenous components* (pp. 143–158). Amsterdam: Elsevier.

Rockstroh, B., Elbert, T., Canavan, A., Lutzenberger, W., & Birbaumer, N. (1989). *Slow cortical potentials and behavior.* Baltimore: Urban and Schwartzberger.

Rogers, L. (1991). Determination of the number and waveshapes of event related potential components using comparative factor analysis. *International Journal of Neuroscience, 56,* 219–246.

Rohrbaugh, J., Syndulko, K., Sanquist, T., & Lindsley, D. (1980). Synthesis of the contingent variation brain potential from noncontingent stimulus and motor elements. *Science, 208,* 1165–1168.

Rosler, F. (1991). Perception or action: Some comments on preparatory negative potentials. In C. Brunia, G. Mulder, & M. Verbaten (Eds.), *Event-related brain research, Electroencephalography and Clinical Neurophysiology, (Suppl. 42),* 116–129.

Rosler, F., & Manzey, D. (1981). Principal components and Varimax-rotated components in event-related potential research: Some remarks on their interpretation. *Biological Psychology, 13,* 3–26.

Smith, M., Halgren, E., Sokolik, M., Baudena, P., Musolino, A., Liegeois-Chauvel, C., & Chauvel, P. (1990). Intracranial topography of the P3 event-related potential during auditory oddball. *Electroencephalography and Clinical Neurophysiology, 76,* 235–248.

Squires, K., Hillyard, S., & Lindsay, P. (1973). Cortical potentials evoked by feedback confirming and disconfirming an auditory decision. *Perception and Psychophysics, 13,* 25–31.

Squires, K., Wickens, C., Squires, N., & Donchin, E. (1976). The effect of stimulus sequence on the waveform of the cortical event-related potential. *Science, 193,* 1142–1146.

Squires, N., Donchin, E., Squires, K., & Grossberg, S. (1977). Bisensory stimulation: Inferring decision-related processes from the P300 components. *Journal of Experimental Psychology: Human Perception and Performance, 3,* 299–315.

Squires, N., Squires, K., & Hillyard, S. (1975). Two varieties of long latency positive waves evoked by unpredictable auditory stimuli in man. *Electroencephalography and Clinical Neurophysiology, 38,* 387–401.

Sternberg, S. (1966). High speed scanning in human memory. *Science, 153,* 652–654.

Sutton, S., Braren, M., Zubin, J., & John, E. (1965). Evoked potential correlates of stimulus uncertainty. *Science, 150,* 1187–1188.

Tecce, J. (1972). Contingent negative variation (CNV) and psychological processes in man. *Psychological Bulletin, 77,* 73–108.

Thatcher, R., & John, E. (1977). *Functional neuroscience: Vol. 1. Foundations of cognitive processes.* Hillsdale, NJ: Erlbaum.

Tueting, P. (1978). Event related potentials, cognitive events and information processing: A summary of issues and discussion. In D. Otto (Ed.), *Multidisciplinary perspectives in event related brain potential research* (EPA-60019-77-043, pp. 159–169. Washington, DC: U.S. Government Printing Office.

Verlager, R. (1988). Event-related potentials and cognition: A critique of the context updating hypothesis and an alternative interpretation of P3. *Behavioral and Brain sciences, 11*(3), 343–356.

Walter, W. G. (1964). The contingent negative variation: An electrocortical sign of significant association in the human brain. *Science, 146,* 434–437.

Walter, W. G., Cooper, R., Aldridge, V., McCallum, W., & Winter, A. (1964). Contingent negative variation: An electrical sign of sensorimotor association and expectancy in the human brain. *Nature, 203,* 380–384.

Wickens, C., Kramer, A., Vanasse, L., & Donchin, E. (1983). Performance of concurrent tasks: Physiological analysis of reciprocity of information-processing resources. *Science, 221,* 1080–1082.

Wood, C., & McCarthy, G. (1984). Principal components analysis of event-related potentials: Simulation studies demonstrate misallocation of variance across components. *Electroencephalography and Clinical Neurophysiology, 59,* 249–260.

Woody, C. D. (1967). Characterization of an adaptive filter for the analyses of variable latency neuroelectric signals. *Medical and Biological Engineering, 5,* 539–553.

8

# Magnetoencephalography

Andrew C. Papanicolaou and Ina M. Tarkka

## Introduction

Neuromagnetism is the study of magnetic signals generated by currents arising from ion movements in the nerve cells of the peripheral system or the brain. The measurement of magnetic fields produced by intracranial currents is the proper subject of magnetoencephalography (MEG), which has developed into a noninvasive functional imaging technique.

The first concrete step toward the goal of measuring biomagnetic signals and localizing their sources was taken in the early 1960s when magnetic signals produced by current sources in the human heart were recorded (Baule & McFee, 1963). This was a very important step in that it demonstrated the feasibility of such recordings in spite of the fact that heart signals are about a million-fold weaker than the magnetic field of the earth and that these measurements were undertaken with instruments that, from the present perspective, may be considered quite crude. The second significant step was taken when Cohen, Edelsack, and Zimmerman (1970) described the use of the Superconducting Quantum Interference Device (SQUID) inside a magnetically shielded room to improve the quality of the magnetic signals from the heart. The third milestone occurred two years later when Cohen (1972) made the first successful measurements of

Andrew C. Papanicolaou Department of Neurosurgery, University of Texas Health Science Center–Houston, Houston, Texas 77030. Ina M. Tarkka Department of Neurosurgery, University of Texas Health Science Center–Houston, Houston, Texas 77030.

*Neuroimaging I: Basic Science*, edited by Erin D. Bigler. Plenum Press, New York, 1996.

magnetic fields generated by intracranial currents, the same ones that give rise to the EEG, thus launching the technique of magnetoencephalography. Identification of the origin of such neuromagnetic signals was successfully attempted soon after and continues to be the central theme of most MEG investigations.

Traditionally, intracranial currents, associated with ongoing cerebral activity, specific brain responses to sensory stimuli, and activity specific to preparation and execution of movement, have been recorded noninvasively from the scalp as time-varying voltages. Electroencephalography (EEG), the study of spontaneous brain electrical activity and sensory and motor event-related potentials (ERPs) or evoked potentials (EPs), is based on the correspondence between intracranial currents and the resulting voltages on the scalp. On the basis of this correspondence, it has been possible to address a variety of questions regarding the nature of normal and pathological brain processes, since many such processes are reflected in the morphology of voltage variations on the surface of the head.

Among those issues which cannot be addressed satisfactorily using surface voltage measurements is the precise localization of the sources of the currents that give rise to specific aspects or components of the voltage waveforms, whether these are particular EP components or epileptic spikes seen in the EEG record. One factor that complicates precise localization of intracranial current sources on the basis of EEG records is volume conduction. If, for instance, one wants to localize the source of current evoked by a visual stimulus, one has to contend with the fact that this primary current, flowing in a cluster of cells in the visual cortex, has given rise to extracellular secondary currents which, spreading throughout the conducting volume of the head, surface onto the scalp and are thus recorded as voltages. This effect is unavoidable, since the cluster of cells where the primary currents originate is inevitably embedded in surrounding tissues such as other neurons and glial cells, cerebrospinal fluid, meninges, skull, and scalp. Each of these tissues have different conductivity values, and their geometric arrangement is quite irregular. For example, the skull, although it can be approximated by a sphere, owing to its apertures and differential topical thickness, is clearly not a homogeneous conductor. The irregularities of this conductor result in sufficient distortion of the volume currents that translate the primary currents to the scalp, rendering precise localization of the intracranial source extremely problematic. It is mainly this difficulty that MEG can overcome by virtue of the following specific advantages of magnetic recordings.

Currents generated in any source, or electric charge moving along any conductor, are associated with magnetic fields. In the case of neural sources, such magnetic fields represent, quite accurately, physiologically important ion movements in cells resulting in primary currents. In addition, magnetic fields, unlike the volume-conducted currents, are minimally distorted as they emerge through tissue layers of varying conductivity. Moreover, the main boundary of the brain is represented by the skull, the shape of which roughly approximates a sphere. This feature provides the advantage of rendering the component of the magnetic field perpendicular to the scalp substantially insensitive to volume currents. In other words, this component of the magnetic field, at least in a first approximation, directly reflects the primary currents generated by the intracranial source. These considerations account for the contention that MEG measurements may supplement or replace voltage measurements as a means of localizing the sources of functionally significant cerebral events.

Source localization is based on the solution of the inverse problem, that is, the evaluation of source configuration from a measured distribution of electric potentials and magnetic fields on the body surface. Unfortunately, the inverse problem has no unique solution, as an infinite number of equivalent source configurations may account for the measured distribution. The only way to overcome this difficulty is to choose a suitable model source that is sufficiently realistic from a physiological point of view yet mathematically tractable and to calculate the theoretical distribution of fields and potentials from that source, that is, to solve the forward problem. A best fit between calculated and measured values may eventually result in an acceptable answer to the source localization problem. A suitable model source is a current dipole that consists of a short current element *i* along an oriented segment of length *l*. This model can represent reasonably well a cerebral source, such as a column of cortical cells with the primary current corresponding to intracellular currents and with the volume currents standing for extracellular ionic flows. Such a model, properly referred to as an equivalent current dipole model, has been extensively used in MEG applications.

The magnetic induction *B* (often referred to as magnetic field) associated with this source is evaluated by means of the Biot-Savart law which, when the current dipole is immersed in an infinite medium with homogeneous conductivity, states:

$$B = \mu_o Q \sin\theta / 4\pi r^2$$

The formula gives *B* in units of tesla, when *r* is expressed in meters and the moment *Q* in ampere-meters ($Am^{-1}$). The expression contains some significant features. First, the strength of the field generated by the dipole is inversely proportional to the square of the distance *r* from the current source and has an axial symmetry with respect to the dipole axis, from which the angle $\theta$ is calculated. Indeed, the field has maximum strength for $\theta = 90°$, whereas it vanishes in the direction of the current ($\theta = 0°$). The quantity $\mu_o = 4 \times \pi\ 10^{-7}$ newton/ampere is the magnetic permeability of free space and has practically the same value for living tissue such as the brain, the skull, and the scalp.

It was commented earlier that an intracranial equivalent current dipole, the source of neuromagnetic signals, most likely consists of a column of cells with similar orientation acting as a unit. The feasibility of the idea that a set of cells rather than a single neuron is the source of recorded signals is readily apparent: no single neuron can produce a sufficiently strong field to be detected outside the head. It has been estimated (e.g., Williamson & Kaufman, 1987) that the strength of the dipole moment *Q* for the activity of a pyramidal neuron is on the order of $3 \times 10^{-13}$ $Am^{-1}$. On the basis of the Biot-Savart law, it is possible to evaluate the intensity of the magnetic induction at a distance of about 4 cm from the neuron. The resulting value of $B = 2 \times 10^{-17}$ tesla is too small to be detectable by any available instrument. Therefore, a large number of synchronously active cells with parallel orientation are needed to produce the magnetic fields measured outside the scalp. Williamson and Kaufman (1987) estimated that $10^4$ to $10^5$ neurons should be simultaneously involved in the generation of a magnetic evoked response.

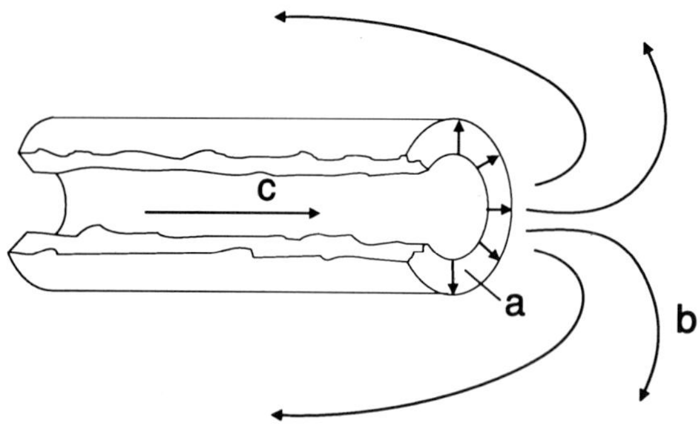


Figure 1. A schematic illustration of three types of currents in and around a segment of neural element: (a) transmembrane, (b) extracellular, and (c) intracellular currents. Intracellular currents are the most likely source for neuromagnetic signals recorded on the surface of the head.

It is worth commenting on the precise nature of activity of the cell column which gives rise to the possible currents that account for the recorded magnetic signals: transmembrane currents, extracellular currents, and intracellular currents (Figure 1).

Transmembrane currents, which are also involved in axonal depolarization, may be rejected as candidates of neuromagnetic signals because of their radial symmetry around the cell body, dendrites, and axonal trunk. As can be easily demonstrated, any pair of transmembrane current flows with opposite direction will produce fields of opposite sign, resulting in a net null field. In addition axonal conduction, in general, is an unlikely candidate because it is extremely fast, rendering the necessary synchronization of all axons in the column highly unrealistic.

Equally unlikely as sources of neuromagnetic signals are extracellular currents which propagate through the volume conductive tissue and complete the circuit of any primary dipole source. Volume currents could give rise to extracranial magnetic signals when the conductor through which they propagate is confined by nonspherical boundaries. In the case where the conducting volume is a homogeneously conducting sphere, the component of the magnetic field normal to the sphere surface sums to zero (for a more detailed exposition of these issues, see Lewine, 1990). For the present, let us simply observe that the interior surface of the cranium approximates sufficiently the requirements of sphericity and homogeneous resistivity to effectively nullify fields due to volume currents. Consequently, the remaining candidate as the main source of neuromagnetic signals is intracellular current. It has already been suggested that axonal currents, whether transmembrane or intracellular, are unlikely to contribute to extracranial fields. It remains then to consider dendritic currents. These are slower than axonal ones; therefore, the required synchronization of the dendritic activity of many cells in a column is very possible. Moreover, sufficient numbers of apical dendrites of pyramidal cells in the cortex are in fact arranged in parallel. Therefore, synchronized intracellular dendritic currents can, in principle, constitute an equivalent current dipole of the kind that would account for the recorded magnetic signals on the head surface.

## Instrumentation

The next issue to consider is how to measure these signals and localize their sources. The SQUID is sufficiently sensitive to detect neuromagnetic fields. This

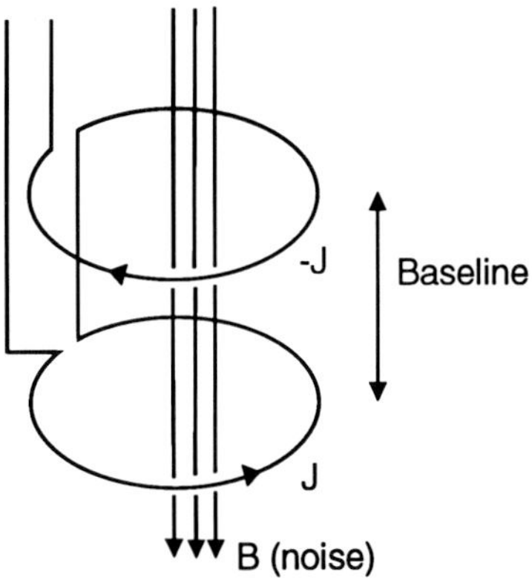


Figure 2. A diagram of a first-order gradiometer with magnetic flux of spatially uniform intensity threading through both loops. The distance between the two loops is critical for the degree of suppression of such environmental magnetic noise. A baseline of about 5 cm is optimal for recording of cerebral activity with a first-order gradiometer.

device does not directly sense the field to be measured; rather, it is coupled to the "external world," the magnetic field, by a superconducting loop of wire, and that field of interest is inductively detected by a coil with appropriate geometry, namely a magnetometer. Because the neuromagnetic signals are feeble compared to the ambient magnetic noise in the environment (due to the earth's field and to the presence of magnetic objects and instruments fed with power lines), their measurement was practically impossible before technical innovations such as magnetically shielded rooms and superconductive instrumentation were introduced. A third important innovation was the gradiometer, a device used to selectively cancel out magnetic signals emanating from environmental sources located far from the source of neuromagnetic signals of interest, thus improving the signal-to-noise ratio. An axial gradiometer, in its simplest configuration, consists of a pick-up loop and the compensation coil (Figure 2). A field $B$ coming from an external noise source with an amplitude that is uniform in space will thread through both loops of the gradiometer and will induce in each of them currents of equal strength but of opposite direction ($J$, $-J$), which will add to zero net current. The distance between these two loops (baseline) determines the achievable suppression of remote interfering signals; the smaller the baseline, the higher the suppression. When the distance between the two loops is sufficient with respect to the typical distance of a neural source from the lower coil, the signal from that source will not be significantly attenuated due to the effect of the upper compensation coil, whereas any field generated far away from the gradiometer will be canceled out if its amplitude is practically uniform in space. A short baseline of 30 mm, for example, may suppress signals from deep sources in the body. A baseline of about 80 mm is optimal for magnetocardiography, whereas a baseline of about 50 mm is optimal for magnetoencephalography (e.g., Dössel *et al.*, 1991).

This two-loop device, or first-order gradiometer, is optimal for recordings in magnetically shielded environments. The second-order gradiometer with three loops has higher noise suppression but there is a loss of signal that can be coupled to the SQUID. This device is insensitive both to uniform fields and to gradients, that is, magnetic noise that does not have uniform amplitude in space. With such second-order gradiometers, one can selectively cancel out or substantially attenuate the unwanted signals originating from the power lines needed to power instruments in the immediate vicinity of the recorded brain activity.

Magnetically shielded rooms have improved tremendously the neuromagnetic signal-to-noise ratio of MEG measurements. Yet even with these improvements, obtaining clear neuromagnetic signals would still remain problematic without superconducting instrumentation. In order to be superconducting, the entire sensing apparatus (gradiometers and the SQUID) are immersed in a liquid helium bath contained in a fiberglass superinsulated dewar at an "operating" temperature of 4.2 K. The tail of the dewar, which can contain anywhere from 1 to 122 or more gradiometers, is placed close to the subject's head in order to record evoked or spontaneous brain activity.

The very same properties of magnetic signals that facilitate localization of primary current sources also limit the range of MEG applications. Neuromagnetic signals are largely due to primary currents. However, not all active primary current sources produce fields measurable outside the scalp. For example, if it is assumed that a number of intracranial sources of equal strength are active at a given moment, only those oriented tangentially to the skull have the best chance of being detected. Dipolar sources oriented obliquely or radially with respect to the skull have little or no chance at all of being detected, as is the case with sources markedly distal to the head surface. Fortunately, a variety of brain events of interest such as auditory, visual, and somatosensory evoked responses and some pathological processes have cortical sources close to the head surface and do lend themselves to measurement. In the next section we present several experiments to elucidate the process of measuring neuromagnetic evoked responses and localizing their sources.

## Source Localization of Auditory Evoked Fields

Long-latency auditory evoked potentials consist of two well-known components, the N1 or N100 and P2 or P200. The purpose of the experiment summarized below was to identify and locate the sources of N1 and P2 (for details, see Papanicolaou *et al.* 1990; Papanicolaou, Rogers, Baumann, Saydjari, & Eisenberg, 1990).

Eight subjects were tested individually in a magnetically shielded chamber (Vacuumschmelze (GmbH). A 1-kHz tone generated by a stimulus module (Nicolet Compact Four), 50 msec in duration (10 msec rise/fall and 30 msec plateau), with an intensity of 105 dB normal hearing level (nHL), was channeled to the subject's left ear through a plastic tube terminating in an insert earphone (Nicolet TiP-300). Similarly, white masking noise of 60 dB nHL was channeled into the other ear. The stimulus delivery rate was varied at random between 0.33/second and 0.5/second around a mean of 0.4/sec. Fifty responses to the stimulus were averaged at each measurement point.

Measurements of the averaged evoked magnetic fields (EMFs) were made over the right hemisphere of each subject. A neuromagnetometer (BTI model 607) with seven second-order gradiometers was used, allowing simultaneous measurements of EMFs at seven scalp positions about 1 cm apart and covering a circular area about 6 cm in diameter. The seven gradiometers were placed in a cryogenic container (dewar), the tail end of which was placed successively at eight adjacent locations covering the recording area of interest over the scalp. These eight locations were sampled in a different random order for each subject. At each location, averaged EMFs to the tone were obtained from all seven channels before the dewar was placed over the next location. Consequently, a total of 56 averaged EMFs was obtained with each subject, resulting in an adequate magnetic field map.

The precise location of each of the 56 measurement points on the scalp was determined electronically with reference to a Cartesian coordinate system anchored on three landmarks (fiduciary points) on each subject's head: two preauricular points and the nasion. The line passing through the two preauricular points served as the *y*-axis of the system. The line perpendicular to the *y*-axis passing through the nasion served as the *x*-axis, and the line perpendicular to the *xy* plane passing through the *xy* origin served as the *z*-axis. Subsequently, three small, easily identifiable lipid markers (vitamin pills), placed over the three fiduciary points, served to define the coordinate system for MRIs. Following EMF recordings, MRIs were obtained from each subject. Serial coronal MRI sections, 0.5 cm in thickness, were taken at right angles to the plane defined by the fiduciary points (the *xy* plane). Thus, using a common coordinate system for the MEG and MRI data made it possible to visualize and define for each subject the anatomical structures onto which the calculated sources of the EMF responses were projected.

The average EMF waveforms were characterized by two major peaks very similar to those of the typical EP waveform. The first peak, corresponding to the N1 of the EPs, occurred with an average latency of about 83 msec, and the second, corresponding to the P2, had an average latency of about 172 msec.

At two points on the head surface the EMFs were of maximal amplitude. These points are referred to as maxima or extrema. As it happens, the amplitude and the polarity of the EMF components change systematically between the two extrema. The amplitude gradually reduces to zero at the measurement point between the two extrema and also reverses polarity. This point indicates the approximate location of the source that gave rise to the waveforms.

The EMFs recorded beyond the maxima also gradually decrease in amplitude. The polarity of EMFs changes around the midpoint because in magnetic recording, "polarity" simply represents the two directions of the magnetic flux, one for the flux exiting and the other for reentering the head as it circles around the intracranial source. The gradual decrease in amplitude beyond the maxima is due to the fact that the flux strength decreases as a function of the square of the distance from the source (see the Biot-Savart law above). Finally, the amplitude decrement as a function of the distance from either extreme toward the midpoint between the extrema is due to the fact that the flux lines closest to the underlying source are more tangentially oriented with respect to the measurement surface (the scalp) and thus least likely to be sensed by the gradiometer loops that are tangential to the scalp.

The amplitude of all EMF components at any given point during the temporal evolution of the waveforms can be plotted. The resulting isocontour field map or isofield map at any point in time represents the instantaneous magnetic flux distribution over the entire recording surface at that point. Figure 3 shows the flux distribution at one point in time, 85 msec post stimulus, where the N1m component reaches peak amplitude.

Given such an isofield map one can make a rough estimate of the position of the dipole by assuming that the local curvature of the surface that contains the two extrema (the maximum and the minimum) is negligible (i.e., the surface is a flat boundary). In that case the source would be located at a depth $d = D/\sqrt{2}$, where $D$ is the distance between the extrema and, due to the symmetry of the

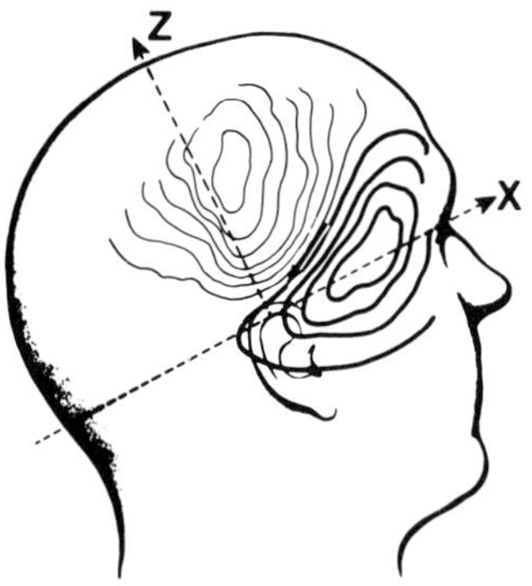


Figure 3. An isofield map for the N1m component of an auditory evoked field at 85 msec post stimulus. Thick lines show the magnetic flux out of the skull and the thin lines represent flux entering the skull.

isofield pattern, below the midpoint between the extrema on the recording surface. Thus, the farther apart the extrema, the deeper the location of the source.

Of course, this method of dipole localization is crude and rarely, if at all, used in such studies. Instead, alternative steps are usually taken for this purpose. First, we determine the best-fitting sphere, that is, the sphere that best approximates the curvature of the actual surface of the head as defined by several measurement points. Determination of the best-fitting sphere is necessary because the calculations used to estimate the position and other dipole parameters are generally based on a spherical model of the head. Essentially, two factors must be known about the geometry of the head: the radius of the best-fitting sphere and its center within the coordinate system that contains the head.

In the present case, the best-fitting sphere was defined on the basis of the local curvature of each subject's head over the auditory cortex (see, for example, Hämäläinen, Ilmonemi, Knuutila, & Reinikainen, 1987, for procedural details). The relationship between the best-fitting sphere, on the basis of which the locus of the dipolar sources is estimated, and the subject's head containing the actual sources is easily calculated. The center of the head is defined by the fiduciary points; therefore, the relation between it and the center of the best-fitting sphere that is defined by the curvature at a particular part of the head can be calculated. In this way, the location parameters of the estimated source can be translated into the actual head space.

As mentioned earlier, the dipole estimation algorithm is an attempt to solve the inverse problem through a series of iterative solutions of the "forward problem", whereby a set of dipole parameters (three of position in the Cartesian coordinate system, one of dipole strength, and one of orientation) is used to calculate a magnetic field on the surface of the best-fitting sphere. Then the calculated and the empirically obtained fields are compared using a least squares estimation or a correlation procedure. If the degree of difference between the two fields is high, the dipole parameters are changed and a new field is calculated. This iterative procedure stops when no further reduction in the differences between the two fields or no further improvement in the correlation of the two fields is obtained. At this juncture, the dipole parameters of the last solution are considered estimates of the actual location, orientation, and strength of the source that gave rise to the recorded magnetic field. Yet these estimates are not unique. An entirely different combination of values for the five dipole parameters could give rise to the same best-fitting field. This uncertainty is intrinsic to

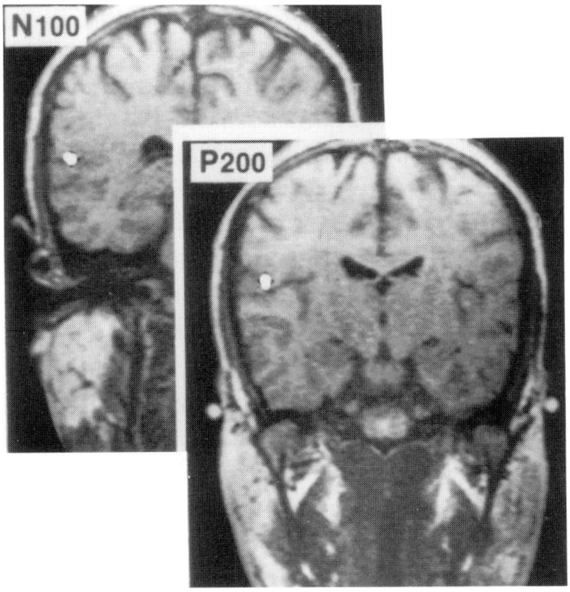


Figure 4. MRI images from one subject show the estimated locations of the N1m and the P2m sources at 85 msec and 160 msec, respectively. The N1m corresponds to the electrically recorded N100 component and the P2m to the P200 component. The N1m source was, on the average, about 1 cm posterior to the P2m source.

all inverse problem solutions and cannot be overcome theoretically. Consequently, it is always desirable to confirm independently the validity of the localization parameters. Confidence in the validity of the localization may increase if the sources are localized in cortical regions which are already suspected (or known) to generate the particular responses recorded. These regions can be visualized by projecting the estimated dipolar sources onto each individual subject's MRI. A detailed procedure for aligning the MEG and the MRI spaces is provided in Rogers *et al.* (1991). The estimated sources of the N1m and P2m components in the studies under discussion were found to be approximately on the floor of the Sylvian fissure, that is, the auditory area of the cortex. Moreover, the source of the P2m peak was always found to be approximately 1 cm anterior to that of the N1m as shown in Figure 4.

These estimates of the sources of the early cortical responses to auditory stimuli were subsequently confirmed in the context of other studies undertaken to estimate the sources of the late P3 component, which appears under certain conditions on the ERP waveforms (see below). The reliability of the N1m and P2m localization estimates has been confirmed by several investigators (Hari *et al.*, 1980, 1987; Rif, Hari, Hämäläinen, & Sams, 1991).

## Source Localization of Somatosensory Evoked Fields

The work of Tarkka, Laudahn, Braeutigam, and Bromm (1993) provides a clear example of source localization of somatosensory responses. A 19-channel biomagnetometer was used in a magnetically shielded room. In this particular MEG system, the channels are arranged in a hexagonal configuration, with an interchannel distance of 2.5 cm. The SQUIDs are connected to axial, first-order gradiometers with a 2.0-cm diameter and a 5-cm baseline. The channels are arranged in parallel. The sources of short-latency somatosensory evoked fields (SSEF) were estimated in seven healthy volunteers. The stimulation consisted of

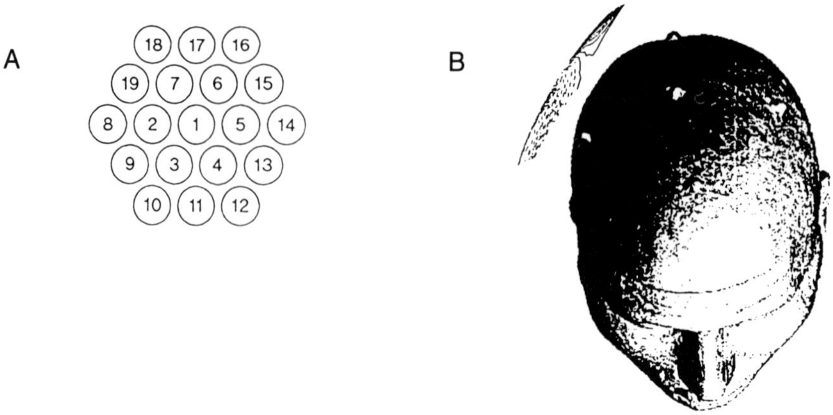


Figure 5. (A) Configuration of the 19 channels in the biomagnetometer (Philips) used for the somatosensory evoked field (SEF) studies discussed in the text. (B) Position of the dewar in relation to the head for recording the SEF.

electrical pulses of 0.2-msec duration delivered at the left median nerve at the wrist with an intensity that reliably produced a distinct motor twitch in the abductor pollicis muscle. One hundred fifty sweeps were averaged, using a recording window of 50 msec. The maxima and minima of the SSEF were found within one dewar placement, with the gradiometer array roughly centered at site P4 (according to the International 10-20 System), that is, over the primary somatosensory cortex area contralateral to the stimulated wrist. The configuration of the 19 channels and a schematic representation of the recording location in relation to the subject's head are shown in Figure 5.

The recorded magnetic waveforms of two individual subjects are shown in Figure 6, where the average waveforms of the 19 channels are superimposed to facilitate the view of the maxima and minima of the different waveform components. A large number of electrical recordings have indicated that these responses consist of two clear components, one at 20 msec and another at about 30 msec from the onset of the stimulus (e.g., Allison *et al.*, 1989; Kimura, 1989). The magnetic equivalents of these components are shown in Figure 6. At any given time point, the amplitudes of the 19 simultaneously recorded responses can be plotted to create the isofield map. Figure 6 presents the maps at the 20 msec and 30 msec time points. These maps display a dipolar pattern that justifies the application of the single current dipole model to estimate the source location and orientation generating the pattern. The single current dipole localization estimate was made, again assuming the head to be a homogeneous sphere. In Figure 6 the size of the arrow representing a dipole is relative to the strength of the dipole, which explains why the arrows have different sizes.

In order to project the calculated dipoles onto a particular head image, measurements of the head position with respect to the gradiometers were made before and after the SSEF recordings. The technique used was similar to the one described earlier. In this case three coil sets were used, each consisting of three orthogonal coils. The coil sets were placed on fiduciary head locations which were later used to align the MEG recording space and the MRI space (Fuchs & Dössel, 1992). Subsequently, the dipoles were projected onto the reconstructed

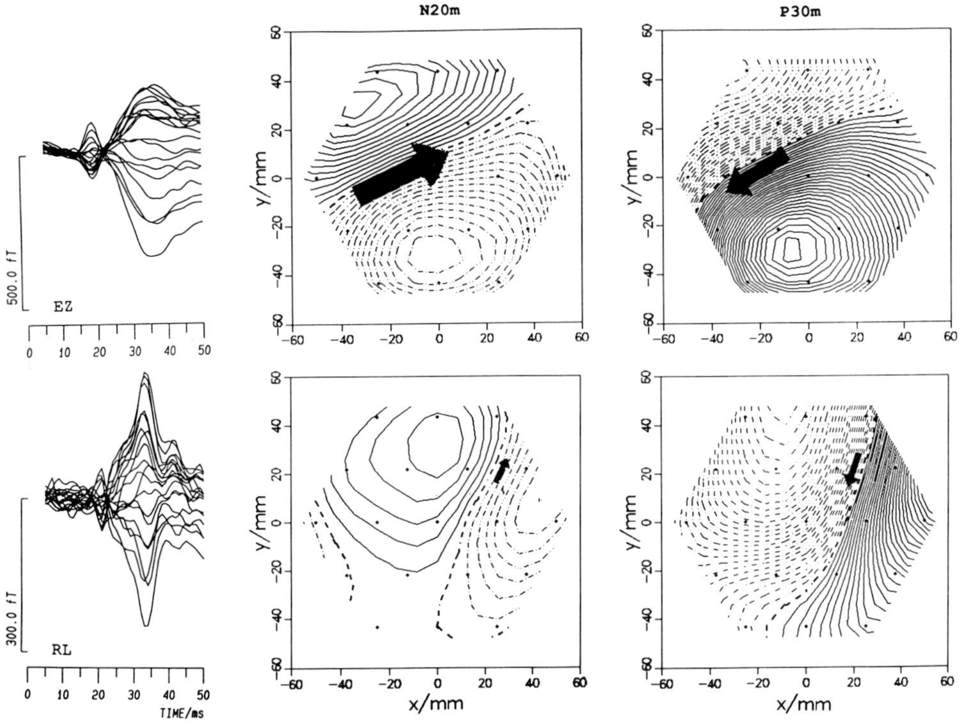


Figure 6. Short-latency somatosensory evoked fields of two healthy subjects. On the left the recordings of 19 channels are superimposed to illustrate the phase reversal at 20 msec and 30 msec points post stimulus. On the right are the contour maps created on the basis of the waveforms shown on the left. These maps represent flux amplitude at 20 msec and 30 msec post stimulus. The solid lines indicate the flux exiting and dotted lines indicate the flux entering the skull. The projections of the single dipoles are also shown. The small black dots within the contour lines indicate where the gradiometers were located.

head image. In Figure 7, the dipole generating the short-latency component at 32 msec is projected onto the head image. The contour maps of two recording locations at this time point are plotted as is the location of the single dipole generating the field. One superficial difference between this procedure and the one used in the previous studies is the designation of the axes that define the head space. In the present case, the axis passing through the preauricular points is called the *x*-axis and the one perpendicular to it passing through the nasion, the *y*-axis.

The dipole location parameters of the 20 msec and 30 msec components for each of the seven subjects are given in Table 1 in the coordinate system defined above. Previously, the source for the N20m component had been reported to occur about 10 mm lateral to the P30m source (Huttunen, Hari, & Leinonen, 1987; Tiihonen, Hari, & Hämäläinen, 1989). The depths of the SSEF generators for the N20m and the P30 were determined by several groups and varied from 12 mm to 41 mm (Huttunen *et al.*, 1987; Tiihonen *et al.*, 1989). The depths we found in individual subjects varied from 11 mm to 33 mm, showing no consistent difference in the depths of the two dipoles (Tarkka *et al.*, 1993). The strengths of the dipoles were different ($p < 0.01$), with the P30m source almost four times stronger than the N20m source. The relationship of N20m and P30m (with the

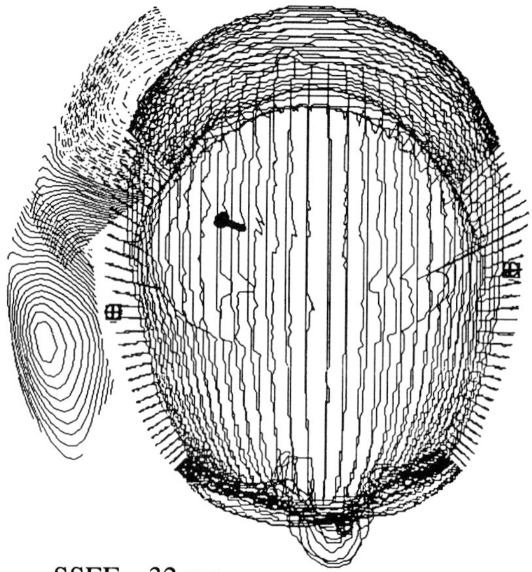


Figure 7. A single dipole for the 30 msec component of the short latency somatosensory evoked potential is projected into the head space. The head drawing has been created on the basis of the individual's MRI.

P30m source at a more medial location) was in agreement with previous reports (Huttunen *et al.*, 1987; Tiihonen *et al.*, 1989).

Long-latency somatosensory evoked fields (SEFs) have also been studied using the 19-channel biomagnetometer. For example, Bromm, Laudahn, and Tarkka (1992) used brief intracutaneous electrical stimuli to the left middle finger tip of healthy subjects. This type of electrical stimulation activates the most superficial nociceptive afferents. To accomplish this stimulation, the superficial layer of the skin is removed with a small drill and the tip of the electrode (diameter = 1 mm) is placed into the small "hole." The stimulus intensity was slightly above the threshold for pain and the stimulus repetition rate was 0.3 Hz. Recordings of 40 responses were averaged over a 256-msec window. The averaged waveforms from one subject are shown in Figure 8. Only one dewar placement was necessary to record these long-latency SEFs. The dewar was centered

TABLE 1. LOCATIONS OF THE COMPONENTS OF SHORT-LATENCY SOMATOSENSORY EVOKED FIELDS[a]

| | N20m | | | P30m | | |
|---|---|---|---|---|---|---|
| Subject | X | Y | Z | X | Y | Z |
| MI | 39.5 | 7.1 | 94.4 | 37.1 | 4.9 | 94.6 |
| HB | 36.6 | 0.5 | 99.2 | 22.6 | 6.3 | 92.1 |
| VK | 41.8 | 10.0 | 88.5 | 35.9 | 10.8 | 93.9 |
| RL | 52.5 | −7.1 | 81.5 | 37.7 | −10.7 | 76.6 |
| PC | 52.6 | −4.0 | 91.1 | 35.2 | 2.9 | 90.3 |
| MR | 42.3 | 7.1 | 99.4 | 40.5 | 9.7 | 90.7 |
| EZ | 37.3 | 1.5 | 84.2 | 32.7 | −0.9 | 81.8 |
| Mean | 43.2 | 2.2 | 91.3 | 34.5 | 3.3 | 88.6 |
| SD | 6.6 | 6.3 | 6.8 | 5.6 | 7.3 | 6.8 |

[a]Values are given in millimeters from the origin of the coordinate system located at the intersection of the x, y, and z axes. Positive x values point toward the right ear, y values toward the nasion, and z values toward the roof of the head.

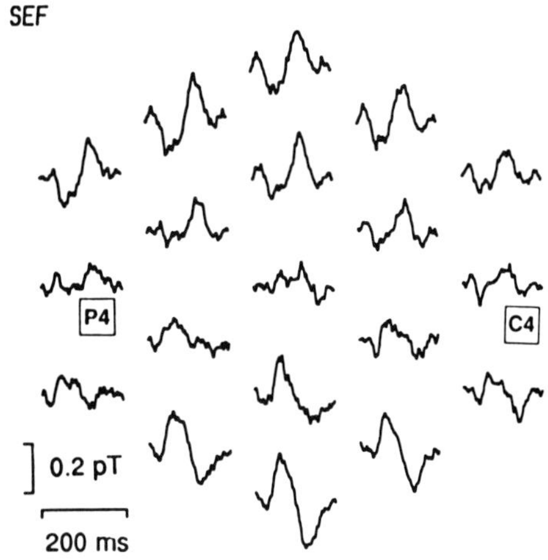


Figure 8. Long-latency somatosensory evoked fields recorded with the 19-channel biomagnetometer. A phase reversal of the 145 msec component can be readily observed. The recording location is indicated by the International 10-20 System markers.

above the parietal cortex, between C4 and P4 scalp locations. A clear phase reversal can be observed for the major long-latency SEF component at 145 msec, with the maximum appearing closer to the top of the head and the minimum toward the ear.

Electrical recordings of these long-latency pain-related somatosensory evoked potentials have revealed a major negative component at about 145 msec poststimulus over the vertex in studies of healthy subjects (Bromm, 1989). No recognizable components of long-latency SEFs were found when MEG recordings were performed at the vertex. This demonstrates effectively that at the vertex the magnetic field of this component is perpendicular to the electric field and, thus, cannot be recorded there. The field of maximum amplitude was detected when the 19-channel magnetometer recorded contralateral parietal cortex activity, as shown in Figure 8. The location of the phase reversal of the 145 msec component in the waveforms already suggests where the dipole may be found. This late pain-related SEF component may be generated in the secondary sensory area (S II). This source location has also been confirmed by positron emission tomography (PET) (for details of pain localization by PET, see Talbot *et al.*, 1991).

## Source Localization of the P3m Component

The P3 is a late positive component of evoked potentials produced in response to rare, unpredictable, or task-relevant stimuli. Because it is specific to such special stimulus events, is much less subject to change as a function of the physical features of the stimulus than earlier components, and occurs in all sensory modalities, it has often been viewed as an index of higher, cognitive processing in the brain. As such, it has been widely used to assess information processing in normal subjects and various clinical populations for diagnostic and prognostic purposes (e.g., Squires, Wickens, Squires, & Donchin, 1976; Kutas, McCarthy, & Donchin, 1977; Brown, Marsh, & LaRue, 1983; Goodin, Squires, & Star, 1978; Hansch *et al.*, 1982; Papanicolaou *et al.*, 1984).

The notion that the P3 reflects cognitive operations as well as sensory ones reflected by earlier components of the EP that are common to both frequent and rare stimuli has led to a corollary: the P3 is independent of the modality of the evoking stimulus (e.g., Pritchard, 1981; Rösler, 1983; McCarthy & Woods, 1987) and it reflects the activity of a brain region or module dedicated to cognitive processing of input from all sensory modalities, that is, a region outside the sensory modality-specific areas. Two main modules have been proposed as likely sources of the P3: the hippocampus (e.g., Halgren *et al.*, 1980) and the thalamus (e.g., Desmedt & Debecker, 1979; Yingling & Hosobuchi, 1984). However, more recent data shows that visual and auditory P3s have different scalp distributions and different sensitivities to subject variables, indicating that the P3 may be modality-specific (Johnson, 1989a). Data on P3 obtained with temporal lobectomy patients point to the same conclusion (Johnson, 1989b). Moreover, Knight, Donatello, Woods, and Clayworth (1989) have shown that lesions in the temporal-parietal junction abolish P3s to target stimuli and unexpected novel stimuli, implicating that region in the generation of the late positive component. Thus, electrophysiological studies involving both scalp and intracranial recordings, taken together, leave the issue of the P3 sources unsettled (see, for example, Regan, 1989).

Once magnetoencephalography was established as a method of localizing sources of primary intracranial currents (Cohen, 1968; Williamson & Kaufman, 1987; Romani, Williamson, Kaufman, & Brenner, 1982; Barth, Sutherling, Engle & Beatty, 1984; Hari & Lounasmaa, 1989), it became possible to identify the brain region generating the P3 through noninvasive methods. The first attempt at MEG localization of the P3 to visual stimuli, though not definite, pointed to the hippocampus as the likely source (Okada, Kaufman, & Williamson, 1983). A second attempt, involving auditory stimuli, favored the temporal cortex (Gordon *et al.*, 1987), whereas preliminary data from another study with both visual and auditory stimuli pointed, once again, to the hippocampus (Lewine *et al.*, 1990).

The MEG study summarized here represents a systematic attempt to identify the P3 sources. To maximize the chance of collecting clear and more readily interpretable data, the following procedures were employed. First, the standard oddball paradigm was used. Second, given our previous successful attempts at estimating the sources of the earlier components of auditory evoked responses, namely N1m and P2m (e.g., Papanicolaou *et al.*, 1990; Rogers, Papanicolaou, Baumann, & Eisenberg, 1990; Rogers, Papanicolaou, Baumann, Saydjari, & Eisenberg, 1990) we chose the auditory modality using the same type of stimuli and the same procedure for their delivery as before. Moreover, we recorded EPs to both tones (the frequent and the rare) from Cz (vertex in the International 10–20 system), as is usually done, to verify that the waveforms obtained under these circumstances replicated those routinely recorded in the context of the auditory oddball paradigm while magnetic evoked fields (MEFs) to the same stimuli were simultaneously recorded.

Eight normal adult volunteers participated in this study. The standard auditory oddball procedure was used involving presentation of a series of simple, frequent (80%) and rare (20%) tones at a rate of 1/second. The frequent tone was 1 kHz and the rare tone 2 kHz. The tones, each 50 msec in duration (10 msec rise/fall and 30 msec plateau) and 95 dB nHL in intensity, were generated by a stimulus generator (Nicolet Compact Four) situated outside the magnetically shielded room (Vacuumschmelze GmbH) in which the recordings were made. The tones were delivered to the subject's left ear through a 3-ft-long plastic tube of .053-in inner diameter terminating in an insert earphone (Nicolet TiP-300). Continuous white noise was similarly delivered to the right ear.

The subject's task was to count the rare tones that occurred at unpredictable intervals and to report the number at the end of each recording trial. All subjects were able to perform this simple task with ease.

Evoked potentials and MEFs were simultaneously recorded with the band-pass set between 0.3 and 50 Hz at a 200-Hz rate over a 2-sec window such that the responses to the rare tone and the frequent tone immediately preceding it were included in the same epoch. Evoked potentials were recorded at Cz with mastoid reference. Electrode impedances were below 5 kOhms. All signals were subsequently filtered digitally (0.3–20 Hz). The MEFs were recorded using the BTI neuromagnetometer (model 607). The dewar was placed successively over several scalp locations in a different random order for each subject. In each location, seven averaged MEFs were recorded. Seven such placements were necessary to sample the entire field and find the extrema for each MEF component for both tones. In one subject, channels over the parietal areas were contaminated with alpha activity and were excluded from analysis. The remaining 28 channels were sufficient to sample activity specific to each component.

In addition, some midline and occipital recordings were made, but we found no evidence of MEFs to either the rare or frequent stimulus at those sites surrounding the temporal, frontal, and parietal regions of the right hemisphere. The exact scalp location of each MEF recording point was determined electronically using a BTI probe position indicator (PPI) with reference to a Cartesian coordinate system anchored on three fiduciary points, two preauricular points and the nasion, on each subject's head (see above). The same three fiduciary points, marked with vitamin pills, were used to define the same coordinate systems for the MR images so that coronal sections perpendicular to the *xy* plane could be selected at the appropriate points along the *x*-axis, on which subsequently estimated MEF source loci could be projected.

At each measurement point, 100 MEFs were averaged. The duration of the experiment was between 4 to 5 hours per subject. This included one or two intermissions at the subject's request.

The results of the experiment are summarized in Figure 9. The EP waveforms to the frequent stimulus were characterized by the N1 and P2 peaks, and the EPs to the rare stimulus featured, in addition, the late P3 peak, lending credence to the contention that a brain module, in addition to those generating the N1 and P2 peaks common to both tones, produces the late P3 peak that is specific to the rare tone (see Figure 9). The EP waveforms presented in the figure are the grand averages of responses over all subjects and recording sessions. This accounts for the smoothness of the waveforms and the apparently small amplitude of the P3. The small amplitude is due to the fact that peak latencies varied slightly among subjects and recording sessions. The average peak amplitude of the P3 computed across the subjects was 4.7 $\mu$V, which is to be expected given the fact that the stimuli were monoaurally presented with contralateral masking noise. A regression analysis of amplitude on consecutive recording sessions showed no significant attenuation ($p > 0.73$). One of the reasons for introducing intermissions between recording sessions was to avoid P3 habituation over the entire recording period. Moreover, the waveform characteristics found on the individual averaged responses are represented in the grand average, and they are the ones typically obtained in the context of the oddball paradigm.

In contrast to the EPs, the simultaneously recorded MEFs to the two tones display a striking waveform similarity in that they each contain all three peaks,

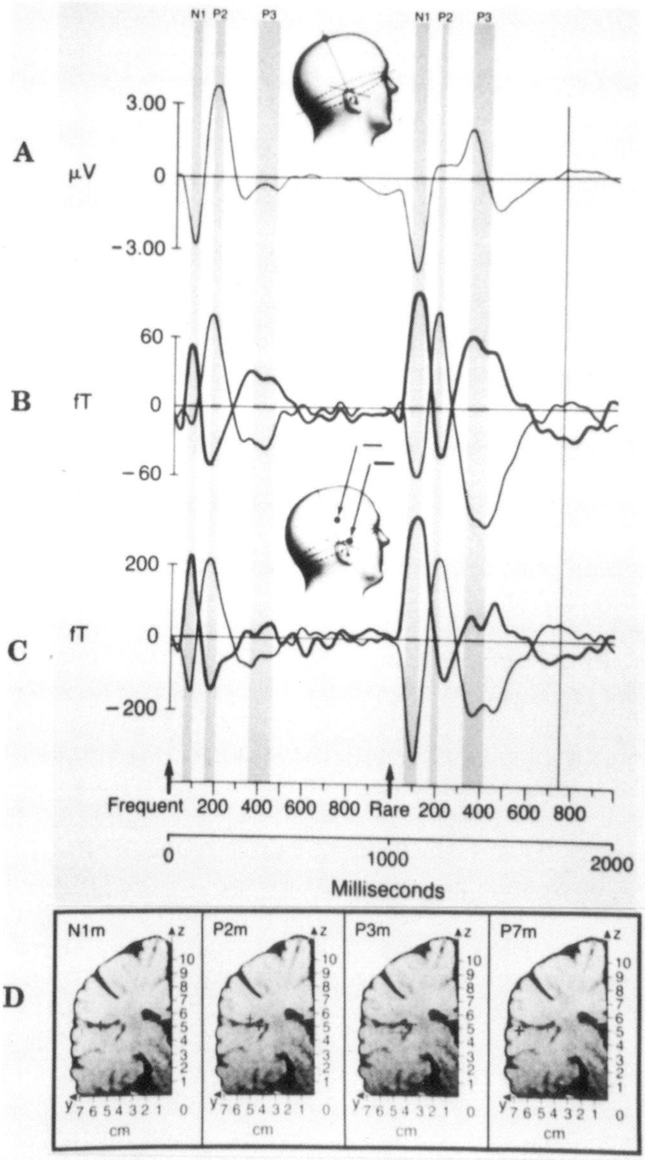


Figure 9. EP and MEF waveforms to the rare tone and to the preceding frequent tone. Shaded regions represent the latency ranges at which good-fitting dipoles were obtained for each component. (A) Grand average of EPs recorded from Cz as indicated by the inserted head model. (B) Grand average of simultaneously recorded MEFs from the extrema of the P3m component as indicated by the inserted head model. Heavy line represents waveforms at the maximum extremum. Thin line represents those at the minimum extremum. (C) MEFs of one subject recorded at the same two scalp locations. Notice clear reversal of all components of the frequent and rare tone responses. (D) Coronal section of MR image of one subject's head. The section is made at the point along the posterior-anterior (*x*) axis, which corresponds to the mean position of the equivalent dipoles of all MEF components. The center of each cross corresponds to the mean location of equivalent dipoles along the vertical (*z*) and medial-lateral (*y*) axes. The standard error of these mean locations along the *y*-axis was 2 mm for the N1m to the rare tone and for the P2m to both frequent and rare tones; 3 mm for the N1m and P3m to the frequent and the P7m to the rare tone; and 6 mm for the P3m to the rare and 8 mm for the P3 to the frequent tone. Along the *z*-axis it was 3 mm for the N1m to both tones, for the P2m to the frequent tone, and for the P3m to the rare tone; 4 mm for the P3m to the rare tone; 2 mm for the P3m to the frequent tone and the P7m to the rare tone; and 10 mm for the P7m to the frequent tone. Thin lines represent sources of MEF components to the frequent, and thick lines represent those to the rare tone.

N1m, P2m, and P3m (see Figure 9B, C), suggesting that the same processing operations, whether sensory or cognitive, occurred for both tones. The discrepancy between the EP and MEF waveforms, the former suggesting a difference between the kind of brain responses to the two stimuli and the latter merely a difference in intensity, will be deferred for later discussion.

Isofield maps for all three components of MEFs to both stimuli were constructed for each subject. In addition, isofield maps to a very late component labeled P7m, obtained in response to the frequent stimulus from three subjects and to the rare stimulus from four subjects, were also constructed. This component is seen clearly in the MEFs to the rare tone in figure 9B and C.

The isofield maps obtained were all suggestive of a separate source for each of the four components. The assumption that the source of an MEF component is an equivalent current dipole is justified when the isofield maps show only two clear extrema, that is, high and opposite values of the magnetic field at a point in time, as was the case with the present data. Though it is possible that two or more simultaneously acting dipoles can generate fields of similar shape, (see, for example, Weinberg, Stroink, & Katila, 1987), unless there are independent indications that multiple dipoles of approximately known location are active, using a single dipole model is more parsimonious.

Such isofield maps were not obtained at a single latency for each peak, but rather during a range of latencies covering the unfolding of each peak in each subject, indicating that a series of successively active single dipolar sources may be responsible for each peak (e.g., Rogers *et al.*, 1990a). Consequently, the procedures for estimating the equivalent dipole parameters on the basis of the recorded magnetic fields were applied at successive 5-msec intervals throughout the 2-sec duration of the MEFs to the frequent and rare stimuli.

Estimations of the source parameters of each MEF component were made for a best-fitting sphere approximating each subject's head, the radius of which was defined by the curvature of a 10-cm diameter circular region of the scalp over the temporal area, electronically determined using the PPI. Subsequently, the calculated and the empirically determined fields were compared. As mentioned earlier, if the degree of similarity of the two fields is low, the parameter values are changed and a new field is calculated. This iterative procedure terminates when no further increase in the correlation of the two fields is obtained.

The results of the dipole fitting routine, that is, the estimated values of a dipole's position, orientation, and strength, were considered valid in this case only if the theoretical magnetic field calculated on the basis of these values correlated better than .90 with the empirically obtained field at a particular latency reflected in the isofield map at that latency.

The average range of latencies around each peak for which such good-fitting, single dipole solutions were obtained is represented by shaded areas in Figure 9. Good fits for the P3m component to the rare tone MEFs were obtained from all eight subjects and to the frequent tone MEFs from six subjects. Good fits for the P2m component were obtained from six and seven subjects to the rare and frequent tone MEFs, respectively, and good fits for the N1m component were obtained for the MEFs to both tones in all but one subject.

The spatial arrangement of successive equivalent dipoles was consistent across subjects for the N1m component to the rare tone, indicative of a sequential progression anteriorly over an average range of about 1 cm on the floor of the Sylvian fissure, as previously observed (Papanicolaou, Rogers, Baumann, Saydjari, & Eisenberg, 1990b; Rogers *et al.*, 1990b). A wider range, again on the floor of the Sylvian fissure but extending from the medial to lateral direction, was observed for the dipoles responsible for the P3m to both the rare and frequent tone MEFs. No other consistent spatial movements of dipoles were observed. Therefore, the values of each of the three spatial parameters (i.e., for the posterior-anterior, medial-lateral, and vertical directions) of successive dipoles for each peak were averaged for each subject, thus providing the mean locus of the source of each component. Subsequently, group averages and variances for the locus, orientation, and strength of the sources of each peak were computed.

The mean locations of each component's sources for the frequent and rare tone responses are juxtaposed on the MR section of one of the subjects (Figure 9D). This particular coronal section was chosen because it corresponds to and includes the mean locations of all sources along the posterior-anterior direction, since no significant differences in source location among peaks or between tones were observed along this direction. Also, it allows visualization of the progressive activation of sources corresponding to the P3m along the medial-lateral direction on the floor of the visible Sylvian fissure. A more detailed exposition of the precise mapping of the P3 sources compared to the N1 sources onto the appropriate MRI sections for each subject (Rogers *et al.,* 1991) shows that dipolar sources that account for part of the P3m component are scattered between the insula and the thalamus, whereas the dipolar sources that account for most of the evolution of the P3m are clustered on the auditory cortex. That is, the individual MRI data verify that (a) much of the auditory P3m is generated by a series of sources on the surface of the temporal lobe and (b) the deep sources have more variable localization, thus making it difficult to assess with certainty whether they arise in the thalamus or the hippocampus.

The mean locations of all components of the responses to both tones were clearly in the vicinity of the auditory cortex. This is certainly the case for the N1m, the P2m (as it was in our previous studies), and the very late P7m component in the few subjects that it was seen. However, in the case of the P3m, whose source first appeared medially and moved laterally during a period of 74 msec and 90 msec for the rare and frequent responses, respectively, one cannot say with equal certainty that the entire component is generated in the auditory area. At its earliest appearance, the P3m source was located, on the average, 3.08 cm from the center of the head for the rare and 3.54 cm for the frequent response, a locus midway between the lateral edge of the thalamus and the medial edge of the auditory cortex. In view of the fact that with MEG localization error increases with the distance of sources from the surface of the head, the most reasonable conclusion is that while the activity associated with the onset of the P3m may originate in the thalamus or the hippocampus, most of the observed activity that accounts for the entire P3m component originates in the auditory cortex. Most certainly, the present data do not support the notion that the P3m is generated in the hippocampus or the thalamus exclusively or that it represents cognitive evaluation of only rare and unpredictable target stimuli, but they do provide some indications of possible involvement of the thalamus and/or the hippocampus in initiating the P3 response.

Comparison of the locus, orientation, and strength of the components of the MEFs to the rare and frequent stimuli also suggested a quantitative rather than qualitative difference in the brain responses to these two stimuli. The averaged parameter values for the sources of each MEF component were com-

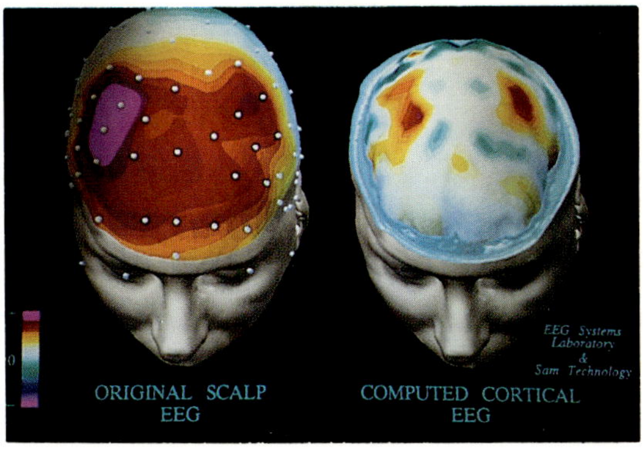


**7.4.** Spatial sharpening of somatosensory EPs with deblurring method. Spatial detail is greatly improved by the finite-element deblurring (FED) procedure. The 124-channel topographic map of the deblurred steady-state evoked potentials elicited by repetitive electrical stimulation of the left-hand middle finger and right-hand index finger (right) contains more useful localizing information than the original potential recorded at the scalp surface with a linked-mastoids reference (left). The high-resolution map clearly shows two separate foci of activity corresponding to stimulation of the two fingers.

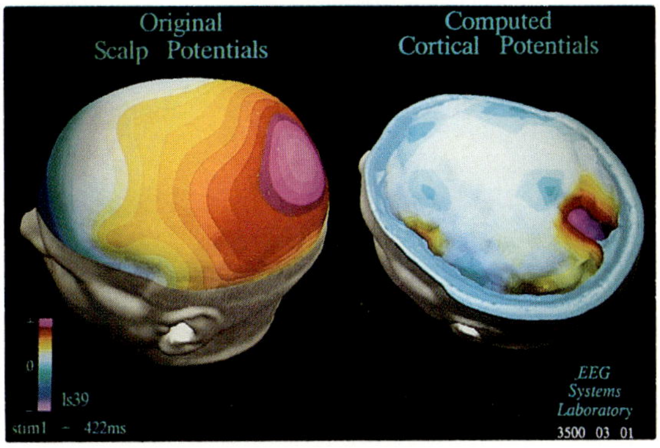


**7.5.** High-resolution EPs during linguistic encoding; data correspond to an EP peak in a subject 422 msec after the first visual stimulus in a grammatic condition of a linguistic judgment task. The stimulus was a closed-class word, known to involve processing in Broca's language area. The distinguishing feature of activation of the grammatic condition is evident as a positive potential over a circumscribed area of left prefrontal cortex (right), a feature that is not apparent in the conventional linked-mastoid EP map (left). The focus of activation over the posterior visual cortex in response to the visually presented stimuli occurred for both closed- and open-class words.

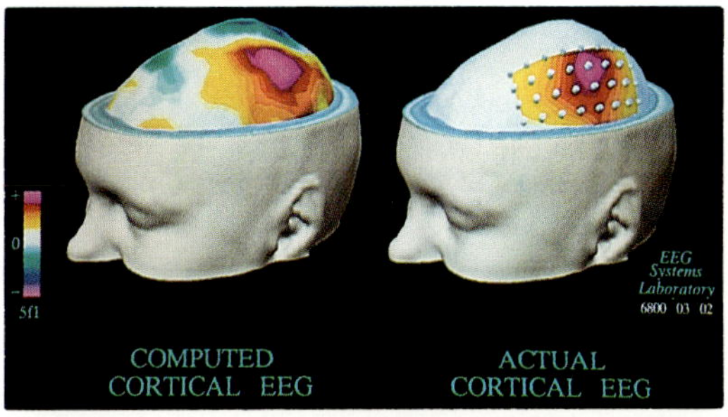


**7.6.** Validation of deblurring method by intracerebral recording, with response to steady-state electrical stimulation to the fingers of the right hand. Deblurred (left) and actual (right) cortical EPs from an epilepsy patient who had a 64-channel electrode grid implanted are shown. The left focus of cortical activation in the deblurred potentials is very similar in topography to the actual recorded cortical EP. (Scalp EP data were recorded prior to implantation of the grid.)

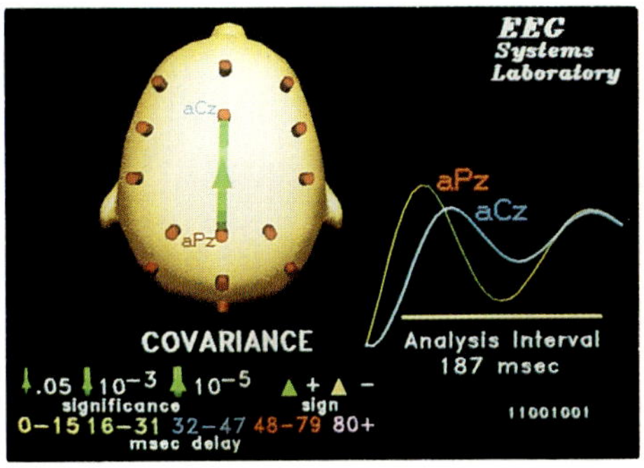


**7.7.** The EPC method. The EPC line of a model head is computed from EP segments recorded from positions aPz and aCz. EPs were spatially sharpened with a Laplacian derivation and filtered to extract activity in a 4–7 Hz band; the EPC analysis interval was 187 msec wide. EPC line widths indicate significance of the covariance between the two EP segments; scale is above the word "significance." Arrow color indicates the sign of the covariance (same color as line, positive; skin color, negative); the arrow points from the leading to the lagging electrode. Here, the EPC between aPz and aCz has a $p$ value < $10^{-5}$, and the aPz waveform leads the aCz waveform by 16–31 msec. The covariance is positive.

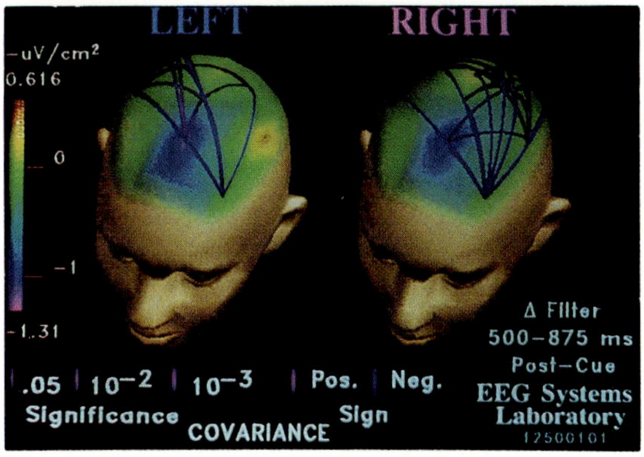


**7.8.** Preparatory EPC patterns preceding accurate responses. Patterns were computed in an interval from 500 to 875 msec after seven subjects were cued to prepare to make a response with the index finger of either the right or left hand when the next numeric stimulus was presented. Data are shown only for those trials for which the response was subsequently accurate. Preparatory patterns preceding accurate responses have covariances of the left prefrontal electrode with electrodes overlying the same motor, somatosensory, and parietal areas that were involved in actual response execution.

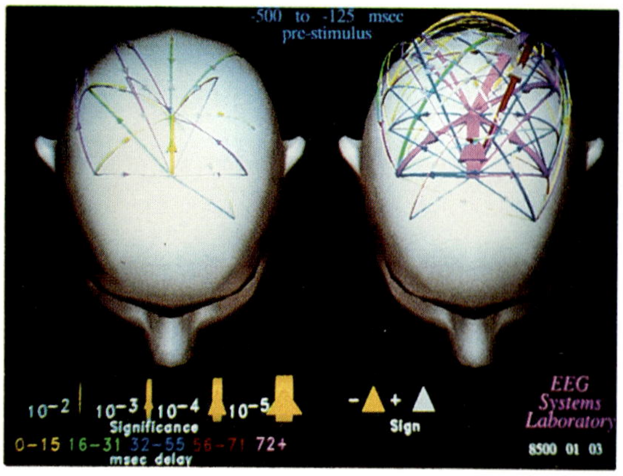


**7.9.** Preparatory EPC patterns during working memory and control tasks. EPCs are of a 375-msec interval when five subjects were staring at a blank screen awaiting the next stimulus number. In the working memory task (right), subjects had to remember the two stimulus numbers from the preceding two trials, while in the control task (left), they did not. The working memory task shows stronger EPCs overall and strong left-sided EPCs among anterior occipital, anterior parietal, and anterior central sites, which overlie cortical areas that may be involved in the maintenance of the numbers in working memory. (From Gevins & Cutillo, 1993.)

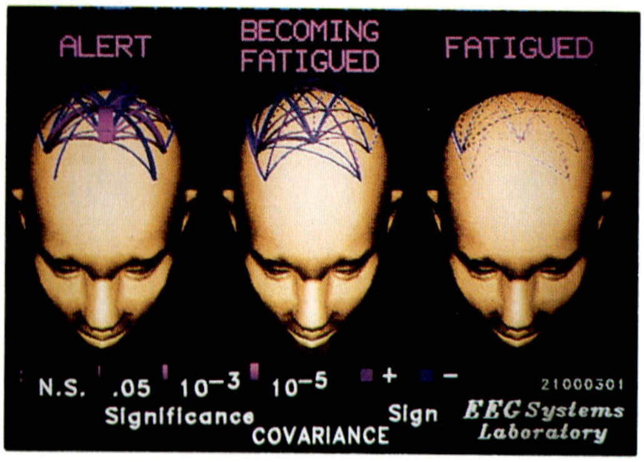


**7.10.** Preparatory EPC patterns during sustained performance of a working memory task. Shown are changes in EPC patterns in the prestimulus interval of the working memory task as five subjects performed working memory and control tasks for 10–14 hours [same subjects and task as Figure 7.9; left pattern (alert) is the same as that on the right of Figure 7.9]. The fatigued pattern (right) occurred later in the session after a significant decline in task performance. The middle pattern (becoming fatigued) occurred just prior to the appearance of significant signs of performance decrement after 7–8 hours of work. The EPC magnitude declined greatly from the alert to the becoming fatigued condition; the EPC pattern also changed, the emphasis shifting from midline central and precentral and left parietal sites to the right anterocentral and parietal and left central sites.

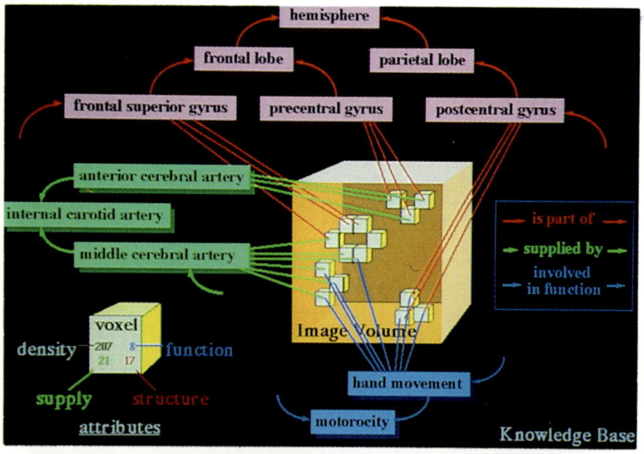


**9.1.** Structure of the three-dimensional anatomical model. Each voxel of the basic volume bears, besides a density value, an attribute describing its membership to a basic region (pictorial level). Relations between the basic regions are described in the knowledge base (descriptive level).

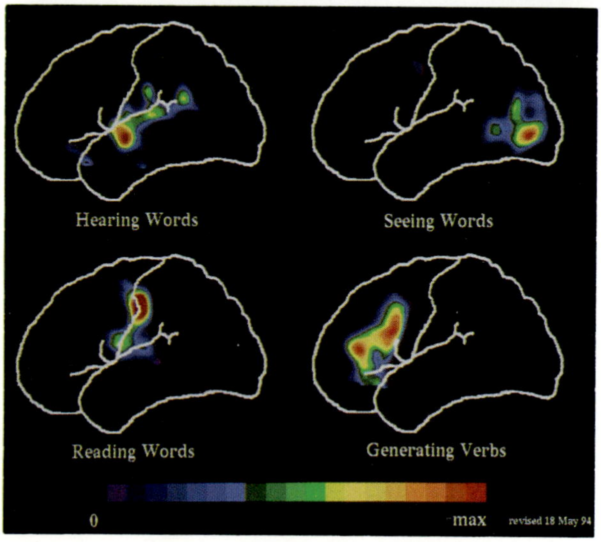


**10.5.** Subtraction of a control state from a task state. For "hearing words" and "seeing words," the control state was visual fixation. For "speaking words," the control state was "seeing words." For "generating verbs," the control state was "speaking words." Images represent averages of nine subjects.

pared using a series of dependent, two-tailed *t*-tests with the rejection level set at 0.05. No comparisons were made for the P7m source since only three subjects displayed that component in the frequent tone MEFs and only four in the rare tone MEFs. A significant difference in source location was obtained for the P2m component, with the source of the response to the rare 2 kHz tone being medial to that of the response to the frequent 1 kHz tone. This difference most likely reflects the tonotopic organization of the auditory cortex and replicates previous MEG studies of this effect. (e.g., Romani *et al.*, 1982; Pantev *et al.*, 1988). Finally, the responses to the two tones differed in the orientation of the equivalent dipoles of the P3m. In summary, aside from a substantial difference in the strength of the sources and a smaller one in the orientation of the P3m sources, the responses to both frequent and rare tones appear quite similar and contrast sharply with their EP counterparts.

From this study, we can make two conclusions: (1) The P3m occurs within the same latency range as the P3, and for most of its temporal course, it can be accounted for by a series of sources located in the floor of the Sylvian fissure and beyond the insula (in the thalamus and/or the hippocampus). (2) The P3m is not specific to rare and unpredictable target stimuli, as the EP waveforms suggest. Rather, P3m is a feature of the brain's response to both frequent and rare stimuli and the difference in the brain's response to the two stimuli is basically a difference in source strength and orientation. Taken together these results suggest that the differences in electrically recorded EPs to rare and frequent stimuli may not be due to additional cognitive processing of the former by a brain module separate from that which mediates sensory processing reflected in the earlier two EP components. Rather, the results suggest that stimulus events of increasingly greater "startle" value result in stronger responses of the same kind to which "cognitive" or "sensory" labels may be equally appropriately or inappropriately applied.

## Concluding Remarks

The overview of the MEG procedures attempted here and the few examples of their application to basic neuroscience research were meant to convey a general idea about the nature of this functional imaging modality and some of its most direct uses. Consequently, neither the overview nor the examples provide an exhaustive account of MEG and its applications. We expect, however, that the interested reader, having gained some insights about the fundamental principles of MEG, will be prepared to appreciate more detailed accounts of special issues in this field that continue to appear in the literature. These special issues fall into three main categories. The first includes technical issues of instrumentation. In the present overview we described the instruments consisting of either seven or 19 axial gradiometers operating at liquid helium temperatures. There are, however, systems available with more channels which can cover an entire hemisphere (e.g., the BTI 37-channel biomagnetometer). Using such systems, one avoids the problem of recording parts of the evoked, or spontaneous, magnetic field distribution at successive sessions and then treating the entire distribution as if it were simultaneously recorded. Moreover, there are prototypes being built using different types of gradiometers (i.e., planar). There are also prototypes under study with a sufficient number of channels to cover the entire head and still others that would be operating at higher temperatures, thus overcoming problems associated with the use of bulky hard dewars to contain liquid helium. The importance of improved instrumentation for MEG applications hardly needs further comment.

The second category of issues is the process of arriving at a valid description of the intracranial sources on the basis of magnetic field distributions recorded at the head surface. In the present overview we described one particular source model that is commonly used, namely the single equivalent current dipole, and one particular procedure of estimating the parameters of such model sources, namely one assuming that the head is a homogeneously conductive sphere with one spherical boundary (i.e., the skull). However, alternative source models have also been developed (e.g., Hämäläinen, 1992; Scherg, 1992; Mosher, Lewis, & Leahy, 1992; Baumgartner, Sutherling, Di, & Barth, 1991) as have procedures involving more realistic assumptions regarding the shape and relevant conductive properties of the head (e.g., Cuffin, 1993).

The third category of issues includes the range of MEG applications in research and the prospect of MEG as a clinically useful modality. In this overview, in the interest of expositional clarity, we limited the description of MEG applications to familiar (from the EEG/EP literature), well-defined evoked response components whose intracranial origin is already known or surmised. However, MEG has been creatively used to address issues in cognitive neuroscience (for a review, see Papanicolaou, Rogers, & Baumann, 1991; Alho *et al.*, 1993; Lü, Williamson, & Kaufman, 1991), explore frontal lobe function (e.g., Basile, Rogers, Bourbon, & Papanicolaou, 1993), address issues of brain plasticity (e.g., Mogilner *et al.*, 1993), and identify the sources of spontaneous epileptic activity (e.g., Barth *et al.*, 1984; Barth, Sutherling, Engle, & Beatty, 1982; Ricci *et al.*, 1987; Tiihonen *et al.*, 1990; Eisenberg, Papanicolaou, Baumann, Rogers, & Brown, 1991; Rose, Smith, & Soto, 1987; Stefan *et al.*, 1992; Sutherling, Crandall, Levesque, Darcey, & Barth, 1992).

It is often assumed that the outcome of the latter type of research will decide the issue of the clinical relevance of MEG. Whether or not this assumption is shown to be correct, we believe there is ample evidence now for the value of MEG as a means of advancing our understanding of the functional organization of the brain.

*Acknowledgments*

This work was supported in part by a grant # NS39540-02 to the first author from the National Institutes of Health. A version of this paper is also prepared for a special issue of *Brain and Cognition*. We wish to thank Dr. W. T. Bourbon for his helpful comments during the preparation of this manuscript.

## References

Alho, K., Huotilainen, M., Tiitinen, H., Ilmoniemi, R., Knuutila, J., & Näätänen, R. (1993). Memory-related processing of complex sound patterns in human auditory cortex: A MEG study. *Neuroreport, 7,* 391–394.

Allison, T., McCarthy, G., Wood, C. C., Darcey, T., Spencer, D. D., & Williamson, P. D. (1989). Human cortical potentials evoked by stimulation of the median nerve. I. Cytoarchitectonic areas generating short-latency activity. *Journal of Neurophysiology, 62,* 694–710.

Barth, D. S., Sutherling, W., Engle Jr., J., & Beatty, J. (1982). Neuromagnetic localization of epileptiform spike activity in the human brain. *Science, 281,* 891–894.
Barth, D. S., Sutherling, W., Engle Jr., J., & Beatty, J. (1984). Neuromagnetic evidence of spatially distributed sources underlying epileptiform spikes in the human brain. *Science, 223,* 293–306.
Basile, L. F. H., Rogers, R. L., Bourbon, W. T., & Papanicolaou, A. C. (1994). Slow magnetic flux from human frontal cortex. *Electroencephalography and Clinical Neurophysiology, 90,* 157–165.
Baule, G. M., & McFee, R. (1963). Detection of the magnetic field of the heart. *Journal of Applied Physics, 36,* 2066–2073.
Baumgartner, C., Sutherling, W., Di, S., & Barth, D. S. (1991). Spatiotemporal modeling of cerebral evoked magnetic fields to median nerve stimulation. *Electroencephalography and Clinical Neurophysiology, 79,* 27–35.
Bromm, B. (1989). Laboratory animal and human volunteer in the assessment of analgesic efficacy. In R. C. Chapman & H. Loeser (Eds.), *Issues in pain measurement.* New York: Raven Press.
Bromm, B., Laudahn, R., & Tarkka, I. M. (1992). Late magnetic field components evoked by auditory and pain inducing stimuli. In M. Hoke, S. N. Erne, Y. C. Okada, & G. L. Romani (Eds.), *Biomagnetism: Clinical aspects* (pp. 265–268). Amsterdam: Elsevier.
Brown, W. S., Marsh, J. R., & LaRue, A. (1983). Exponential electrophysiological aging: P3 latency. *Electroencephalography and Clinical Neurophysiology, 55,* 277–285.
Cohen, D. (1972). Magnetoencephalography: Detection of the brain's electrical activity with a superconducting magnetometer. *Science, 175,* 664–666.
Cohen, D., Edelsack, E., & Zimmerman, J. E. (1970). Magnetocardiograms taken inside a shielded room with a superconducting point-contact magnetometer. *Applied Physics Letter, 16,* 278–280.
Cuffin, B. N. (1993). Effects of local variations in skull and scalp thickness on EEG's and MEG's. *IEEE Transactions in Biomedical Engineering, 40*(1), 72–78.
Desmedt, J. E., & Debecker, J. (1979). Waveform and the neural mechanisms of the decision P350 elicited without prestimulus CNV or readiness potential in random sequences at near threshold auditory clicks and finger stimulation. *Electroencephalography and Clinical Neurophysiology, 47,* 648.
Dössel, O., David, B., Fuchs, M., Krüger, J., Kullmann, W. H., & Ludeke, K. M. (1991). A modular approach to multichannel magnetometry. *Clinical Physical and Physiological Measurement, 12* (Suppl. B), 75–79.
Eisenberg, H. M., Papanicolaou, A. C., Baumann, S. B., Rogers, R. L., & Brown, L. M. (1991). Reliability and validity of MEG localization of interictal spike sources: A case study. *Journal of Neurosurgery, 74,* 660–663.
Fuchs, M., & Dössel, O. (1992). On-line head position determination for MEG measurements. In M. Hoke, S. N. Erne, Y. C. Okada, & G. L. Romani (Eds.), *Biomagnetism: Clinical aspects* (pp. 869–874). Amsterdam: Elsevier.
Goodin, D. S., Squires, K. C., & Star, A. (1978). Long latency event-related components of the auditory evoked potentials in dementia. *Brain, 101,* 635.
Gordon, E., Sloggett, G., Harvey, I., Kraiuhin, C., Pennie, C., Yiannikas, C., & Meares, R., (1987). Locating the source of P300 via magnetic field recording. *Clinical Experimental Neurology, 23,* 101.
Halgren, E., Squires, N. K., Wilson, C. S., Rohrbaugh, J. W., Babb, T. L., & Crandall, P. H. (1980). Endogenous potentials generated in the human hippocampal formation and amygdala by infrequent events. *Science, 210,* 803.
Hämäläinen, M. S. (1992). Magnetoencephalography: A tool for functional brain imaging. *Brain Topography, 5*(2), 95–102.
Hämäläinen, M. S., Ilmonemi, R. J., Knuutila, J., & Reinikainen, K. (1987). Analysis of magnetoencephalographic data obtained with a four channel SQUID magnetometer. In K. Atsumi, M. Kotani, S. Llenos, T. Katila, & S. J. Williamson (Eds.), *Biomagnetism* (pp. 299–303). Tokyo: Tokyo Denki UP.
Hansch, E. C., Syndulko, K., Cohen, S. N., Goldberg, I. Z., Potvin, A. R., & Tourtellotte, W. W. (1982). Cognition in Parkinson's disease: An event-related potential perspective. *Annals of Neurology, 11,* 599.
Hari, R., Aittoniemi, K., Järvinen, M. D., Katila, T., & Varpula, T. (1980). Auditory evoked transient and sustained magnetic fields of the human brain. *Experimental Brain Research, 40,* 237–240.
Hari, R., Pelizzone, M., Mäkelä, J. P., Hällström, J., Leinonen, L., & Lounasmaa, O. V. (1987). Neuromagnetic responses of the human auditory cortex to on- and offsets of noise bursts. *Audiology, 25,* 31–43.
Huttunen, J., Hari, R., & Leinonen, L. (1987). Cerebral magnetic responses to stimulation of ulnar and median nerves. *Electroencephalography and Clinical Neurophysiology, 66,* 391–400.

Johnson Jr., R. (1989a). Development evidence for modality-dependent P300 generators: A normative study. *Psychophysiology, 26*(6), 651.

Johnson Jr., R. (1989b). Auditory and visual P300s in temporal lobectomy patients: Evidence for modality dependent generators. *Psychophysiology, 26*(6), 633.

Kimura, J. (1989). *Electrodiagnosis in diseases of nerve and muscle: Principles and practice* (pp. 375–408). Philadelphia: F. A. Davis Company.

Knight, R., Donatello, S., Woods, D., & Clayworth, C. (1989). Contributions of temporal parietal junction to the human auditory P3. *Brain Research, 502,* 109.

Kutas, M., McCarthy, G., & Donchin, E. (1977). Augmenting mental chronometry: The P300 as a measure of stimulus evaluation time. *Science, 197,* 792.

Lewine, J. S. (1990). Neuromagnetic techniques for the noninvasive analysis of brain function. In S. E. Freeman, E. Fukushima, & E. R. Greene (Eds.), *Noninvasive techniques in biology and medicine* (pp. 33–74). San Francisco: San Francisco Press.

Lewine, J. S., Roeder, S. B. W., Oakey, M. T., Arthur, D. L., Aine, C. J., George, J. S., & Flynn, E. R. (1990). Localization of the generators of the magnetic P3. In L. Kaufman, S. Williamson, & G. Stroink (Eds.), *Advances in biomagnetism* (p. 299). New York: Plenum Press.

Lü, Z.-L., Williamson, S. J., & Kaufman, L. (1992). Human auditory primary and association cortex have differing lifetimes for activation traces. *Brain Research, 572,* 236–241.

McCarthy, G., & Woods, D. (1987). Intracranial recordings and endogenous ERPs in humans. In R. J. Ellingson, N. M. F. Murray, & A. M. Halliday (Eds.), *The London symposia, Electroencephalography and Clinical Neurophysiology,* Suppl. 39, 331–337.

Mogilner, A., Grossman, J. A. I., Ribary, U., Joliot, M., Wolkmann, J., Rapaport, D., Beasley, R. W., & Llinas, R. R. (1993). Somatosensory cortical plasticity in adult humans revealed by magnetoencephalography. *Proceedings of the National Academy of Sciences, 90,* 3593–3597.

Mosher, J. C., Lewis, P. S., & Leahy, R. M. (1992). Multiple dipole modeling and localization from spatio-temporal MEG data. *IEEE Transactions in Biomedical Engineering, 39,* 541–557.

Okada, Y. C., Kaufman, L., & Williamson, S. J. (1983). The hippocampal formation as a source of the slow endogenous potentials. *Electroencephalography and Clinical Neurophysiology, 55,* 417.

Pantev, C., Hoke, M., Lehnertz, K., Lutkenhoner, B., Anogianakis, G., & Wittkowski, W. (1988). Tonotopic organization of the human auditory cortex revealed by transient auditory magnetic fields. *Electroencephalography and Clinical Neurophysiology, 69,* 160–170.

Papanicolaou, A. C., Baumann, S. B., Rogers, R. L., Saydjari, C., Amparo, E. G., & Eisenberg, H. M. (1990). Localization of auditory response sources using MEG and MRI. *Archives of Neurology, 47,* 33–37.

Papanicolaou, A. C., Levin, H. S., Eisenberg, H. M., Moore, B. D., Goethe, K. E., & High, W. M. (1984). Evoked potential correlates of posttraumatic amnesia after closed head injury. *Neurosurgery, 14,* 676–678.

Papanicolaou, A. C., Rogers, R. L., & Baumann, S. B. (1991). Applications of magnetoencephalography to the study of cognition. In R. A. Zappula (Ed.), *Windows on the brain: Neurophysiology's technological frontiers, Annals of the New York Academy of Sciences,* 118–127.

Papanicolaou, A. C., Rogers, R. L., Baumann, S. B., Saydjari, C., & Eisenberg, H. M. (1990). Source localization of two evoked magnetic field components using two alternative procedures. *Experimental Brain Research, 80,* 44–48.

Pritchard, W. S. (1981). Psychophysiology of P300. *Psychological Bulletin, 89,* 506.

Regan, D. (1989). *Human brain electrophysiology.* New York: Elsevier.

Ricci, G. B., Romani, G. L., Pizzella, V., Salustir, C., Torrioli, G., Buonomo, S., Peression, M., & Modena, I. (1987). Study of focal epilepsy by multichannel neuromagnetic measurements. *Electroencephalography and Clinical Neurophysiology, 66,* 358–368.

Rif, J., Hari, R., Hämäläinen, M. S., & Sams, M. (1991). Auditory attention affects two different areas in the human supratemporal cortex. *Electroencephalography and Clinical Neurophysiology, 79,* 464–472.

Rogers, R. L., Baumann, S. B., Papanicolaou, A. C., Bourbon, T. W., Alagarsamy, S., & Eisenberg, H. M. (1991). Localization of the P3 sources using magnetoencephalography and magnetic resonance imaging. *Electroencephalography and Clinical Neurophysiology, 79,* 308–321.

Rogers, R. L., Papanicolaou, A. C., Baumann, S. B., Eisenberg, H. M., & Saydjari, C. (1990a). Spatially distributed cortical excitation patterns of auditory processing during contralateral and ipsilateral stimulation. *Journal of Cognitive Neuroscience, 2,* 44–50.

Rogers, R. L., Papanicolaou, A. C., Baumann, S. B., Saydjari, C., & Eisenberg, H. M. (1990b). Neuromagnetic evidence of a dynamic excitation pattern generating the N100 auditory response. *Electroencephalography and Clinical Neurophysiology, 77,* 237–240.

Romani, G. L., Williamson, S. J., Kaufman, L., & Brenner, D. (1982). Characterization of the human auditory cortex by the neuromagnetic method. *Experimental Brain Research, 47*, 381.

Rose, D. F., Smith, P. D., & Soto, S. (1987). Magnetoencephalography and epilepsy research. *Science, 238*, 329–335.

Rösler, Fl. (1983). Endogenous ERPs and cognition. Probes, prospects, and pitfalls in matching pieces of the mind-body puzzle. In A. W. K. Gaillard & W. Ritter (Eds.), *Tutorials in ERP research: Endogenous components* (pp. 9–35). Amsterdam: North Holland.

Scherg, M. (1992). Functional imaging and localization of electromagnetic brain activity. *Brain Topography, 5*(2), 103–111.

Squires, K. C., Wickens, C., Squires, N. K., & Donchin, E. (1976). The effect of stimulus sequence on the waveform of the cortical event-related potential. *Science, 193*, 1142.

Stefan, H., Schneider, S. Feistel, H., Pawlic, G., Schuler, P., Abraham-Fuchs, K., Schlegel, T., Neubauer, U., & Huk, W. J. (1992). Ictal and interictal activity in partial epilepsy recorded with multichannel magneto-encephalography: Correlation of electroencephalography/electrocorticography, magnetic resonance imaging, single photon emission computed tomography and positron emission tomography findings. *Epilepsia, 33*, 874–887.

Sutherling, W., Crandall, P., Levesque, M., Darcey, T., & Barth, D. (1992). Physical interpretation of frontal lobe seizures. The dipole approximation and sensorimotor cortex. *Advances in Neurology, 57*, 339–347.

Talbot, J. D., Marrett, S., Evans, A. L., Meyer, E., Bushnell, M. D., & Duncan, G. H. (1991). Multiple representations of pain in human cerebral cortex. *Science, 251*, 1355–1358.

Tarkka, I. M., Laudahn, R., Braeutigam, S., & Bromm B. (1993). Comparison of source localization of short latency somatosensory evoked activity by electric and magnetic methods. *Neurology, 43*(4), 422.

Tiihonen, J., Hari, R., & Hämäläinen, M. S. (1989). Early deflections of cerebral magnetic responses to median nerve stimulation. *Electroencephalography and Clinical Neurophysiology, 74*, 290–296.

Tiihonen, J., Hari, R., Kajola, M., Nousiainen, U., & Vapaalahti, M. (1990). Localization of epileptic foci using a large area magnetometer and functional brain anatomy. *Annals of Neurology, 27*, 283–290.

Weinberg, H., Stroink, G., & Katila, T. (1987). *Biomagnetism*. New York: Wiley.

Williamson, S. J., & Kaufman, L. (1987). Analysis of neuromagnetic signals. In A. S. Gevins & A. Remond (Eds.), *Handbook of Electroencephalography and Clinical Neurophysiology:* Vol. 1 (pp. 405–448). New York: Elsevier.

Yingling, O. D., & Hosobuchi, Y. (1984). A subcortical correlate of P300 in man. *Electroencephalography and Clinical Neurophysiology, 59*, 2076.

# 9

# A Computerized Three-Dimensional Atlas of the Human Skull and Brain

U. TIEDE, M. BOMANS, K. H. HÖHNE, A. POMMERT, M. RIEMER, TH. SCHIEMANN, R. SCHUBERT, AND W. LIERSE†

Recently, it has become possible through the methods of digital image processing and computer graphics to reconstruct from magnetic resonance (MR) imaging or computer tomography (CT) the shape of anatomic objects in three dimensions (Herman & Liu, 1979; Höhne *et al.*, 1987, 1992; Höhne, Riemer, & Tiede, 1987; Vannier, Marsh, & Warren, 1983). Such reconstructions have proved to be useful in therapy (especially surgery) planning (Fishman, Ney, & Magid, 1990; Witte *et al.*, 1986; Zonneveld *et al.*, 1989). However, widespread application is limited by a problem peculiar to three-dimensional image processing: unlike two-dimensional cross-sectional images, objects within a three-dimensional scene may obscure each other. Therefore, any visualization must be preceded by a

This article is reprinted from the *American Journal of Neuroradiology* (1993), Volume 14, pages 551–559 by the American Society of Neuroradiology. Reprinted with permission of the authors and the American Society of Neuroradiology.

†*Deceased.*

U. TIEDE, M. BOMANS, K. H. HÖHNE, A. POMMERT, M. RIEMER, TH. SCHIEMANN, AND R. SCHUBERT Institute of Mathematics and Computer Science in Medicine (IMDM), Martinistrasse 52, 20246 Hamburg, Germany. W. LIERSE Department of Neuroanatomy, University of Hamburg, and University Hospital Eppendorf, Martinistrasse 52, 20246 Hamburg, Germany.

*Neuroimaging I: Basic Science*, edited by Erin D. Bigler. Plenum Press, New York, 1996.

segmentation step in which three-dimensional regions belonging to an organ must be identified. Only if this is done can they be visualized or removed. This segmentation step is the most difficult one in three-dimensional visualization. At present, only very few objects can be segmented automatically. The educational application of three-dimensional visualization techniques has been overlooked. In this case, only a limited number of specimens have to be segmented into their constituents; it is feasible for this work to be done by a person rather than by a computer. With this motivation, we have developed an interactive three-dimensional atlas of the human head. It is the purpose of this chapter to show the usefulness of this approach both for education and clinical work.

## Materials and Methods

### Basic Data

The basic idea is to describe the anatomy in two levels (Figure 1). The lower level is a volume data set from cross-sectional imaging such as MR imaging and/or CT (pictorial level). In addition to an intensity value, each volume element contains attributes describing its membership to basic *regions.* There are different kinds of basic regions depending on the feature considered, such as structural regions (e.g., the precentral gyrus of the brain); functional regions (such as the sensorimotor cortex); and regions supplied by a given blood vessel.

The upper level is a knowledge base that contains *relations* between the basic regions (descriptive level). These relations may concern structure (e.g., precentral gyrus is part of the frontal lobe); function (e.g., a sensory stimulus goes through the spinothalamic tract); blood supply (e.g., the internal carotid artery

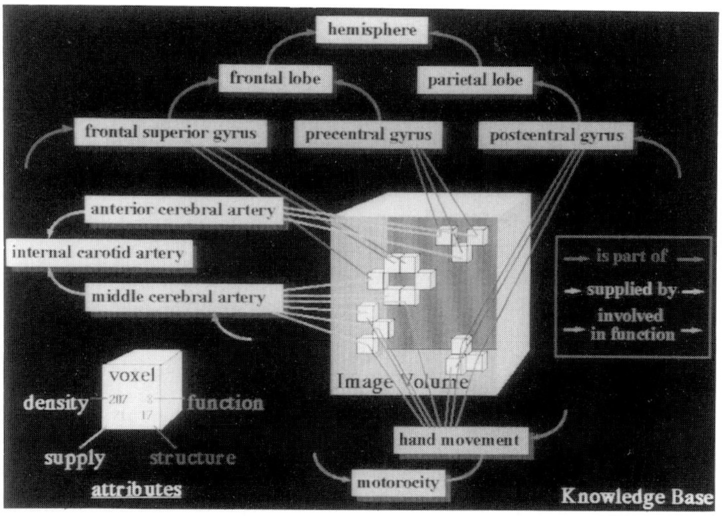


Figure 1. Structure of the three-dimensional anatomical model. Each voxel of the basic volume bears, besides a density value, an attribute describing its membership to a basic region (pictorial level). Relations between the basic regions are described in the knowledge base (descriptive level). For a color representation of this figure, see the color insert facing page 179.

supplies the anterior cerebral artery and the middle cerebral artery); and topology (e.g., the thalamus is next to the third ventricle).

### Construction of the Three-Dimensional Head Model

As a basis for the pictorial level, we use an MR image and a CT data set of a head as basic image volumes. The original cross-sectional images are interpolated to achieve an image volume with cubic voxels. This is performed by linear scaling in the direction perpendicular to the original slices. The resulting MR imaging volume contains 150 × 200 × 192 voxels with a width of 1.5 mm. The CT volume contains 240 × 222 × 246 voxels with a width of 1 mm.

The most critical step is the assignment of the anatomic knowledge to the image volume. This is done in several steps. Initially the gross constituents, such as skull, brain, and ventricular system, are segmented by an interactive segmentation program. An expert selects the appropriate regions using the intensity window control and special image processing tools (Bomans *et al.*, 1990; Höhne & Hanson, 1992; Schiermann, Bomans, Tiede, & Höhne, 1992). The subdivision of the coarse components into detailed structural regions (like the gyri) is done by the neuroanatomist using a "volume editor." For example, a gyrus would first be identified by painting its extent on the three-dimensional brain surface image. The gray matter (voxels below the painted surface) is included by specifying an intensity threshold. Refinements can now be made by repeating the painting procedure from different viewing angles and finally on the cross-sectional images. Although we have developed a set of special tools, this labeling step is tedious. The final step is naming the identified regions.

Once the image volume is labeled in the described way, the relations between the basic regions are entered into the descriptive level of the knowledge base. Two hundred basic structural regions have been labeled, and their structural relations are stored in the knowledge base. In addition, the database contains the paths of the major tracts and the sulci, which can be visualized as polygonal lines in three-dimensional space.

### Exploration Tools

The tools for the exploration of the acquired head model may be divided into four groups: visualization, composition, exploration, and simulation. For *visualization,* computer graphics provide all effects that a painter would use to paint a three-dimensional scene onto a planar screen. Depth impression is produced by obscuration and by perspective parallax. Proper impression of shape is achieved by simulating light reflection from the surfaces according to surface inclination and reflectivity. In this context, highlights play an important role. To achieve all this, we use tools that we have developed in previous projects (Tiede *et al.*, 1990). A viewer can generate arbitrary perspective views by choosing, for example, a certain viewpoint or light direction. Cutting tools are provided to reveal interior structures. The user may move a plane in any position and direction (Figure 2) in order to specify a cut. He may repeat this procedure in order to dissect the head in various sections. A particularly useful feature is selective cutting, which excludes objects from cutting. The colors are chosen arbitrarily.

In the *display composition* mode, components may be displayed via selection of objects from the list of names in the knowledge base, such as add/remove

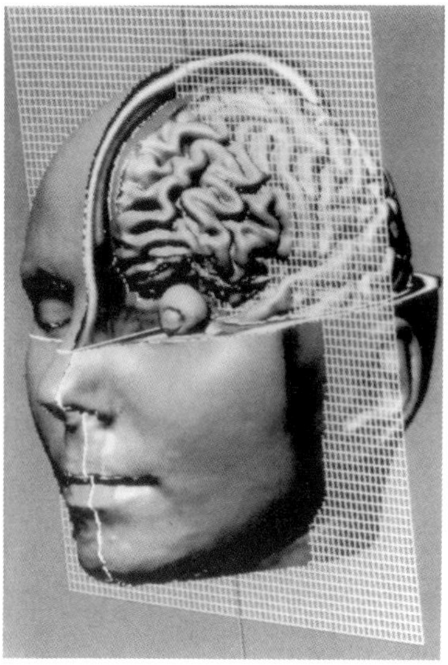

Figure 2. Dissection of the head model at the computer screen. A wire mesh is used to define cut planes. Within the sections generated this way, objects can be removed, providing the look and feel of an actual dissection.

objects (e.g., a gyrus); paint objects with a chosen color; and show objects involved in a certain function.

In the *exploration* mode, the visible image can be explored by pointing on a region of interest according to the location of the cursor on the visible three-dimensional surfaces or cut planes. New images can be created, for example, by coloring the chosen region, coloring parent regions (e.g., the lobe to which a gyrus belongs), removing a chosen region, or adding an object with a certain relation to the selected one (e.g., in front).

One can create output of text that can be associated with the displayed objects, including annotating the object, listing the parent names, and displaying the description attached to the object.

Finally, application-oriented tools for the *simulation* of medical procedures can be derived from the basic tools. Three of them are presently available: simulation of craniotomy, simulation of the path of a stereotactic device, and simulation of x-ray images from computer tomographic volumes.

The program is implemented in the C programming language under the UNIX (Bell Laboratories, Murray Hill, New Jersey) operating system using the OSF/Motif (Open Systems Foundation, Inc.) standard for the user interface. With these standard tools, it is run on standard UNIX workstations as they are used in connection with state-of-the-art MR and CT devices, such as DECstation 5000 (Digital Equipment Corporation, Maynard, Massachusetts), IBM RS/6000 (International Business Machines Corporation, Armonk, New York), and Sun SPARCstation (Sun Microsystems, Mountain View, California). The speed of the atlas is vital for practical application. For the most frequently used exploration actions such as coloring or annotation, the reaction time is less than 1 second. For exploration actions such as rotation and cutting, the reaction time depends on the quality chosen. The small images in Figure 3 (256 × 256 image matrix) appear in about 5 seconds on a DECstation 5000/200, a quality that is sufficient in many instances. Development of a high-quality image such as the large one in Figure 3 (1017 × 604 image matrix) may take 10 minutes.

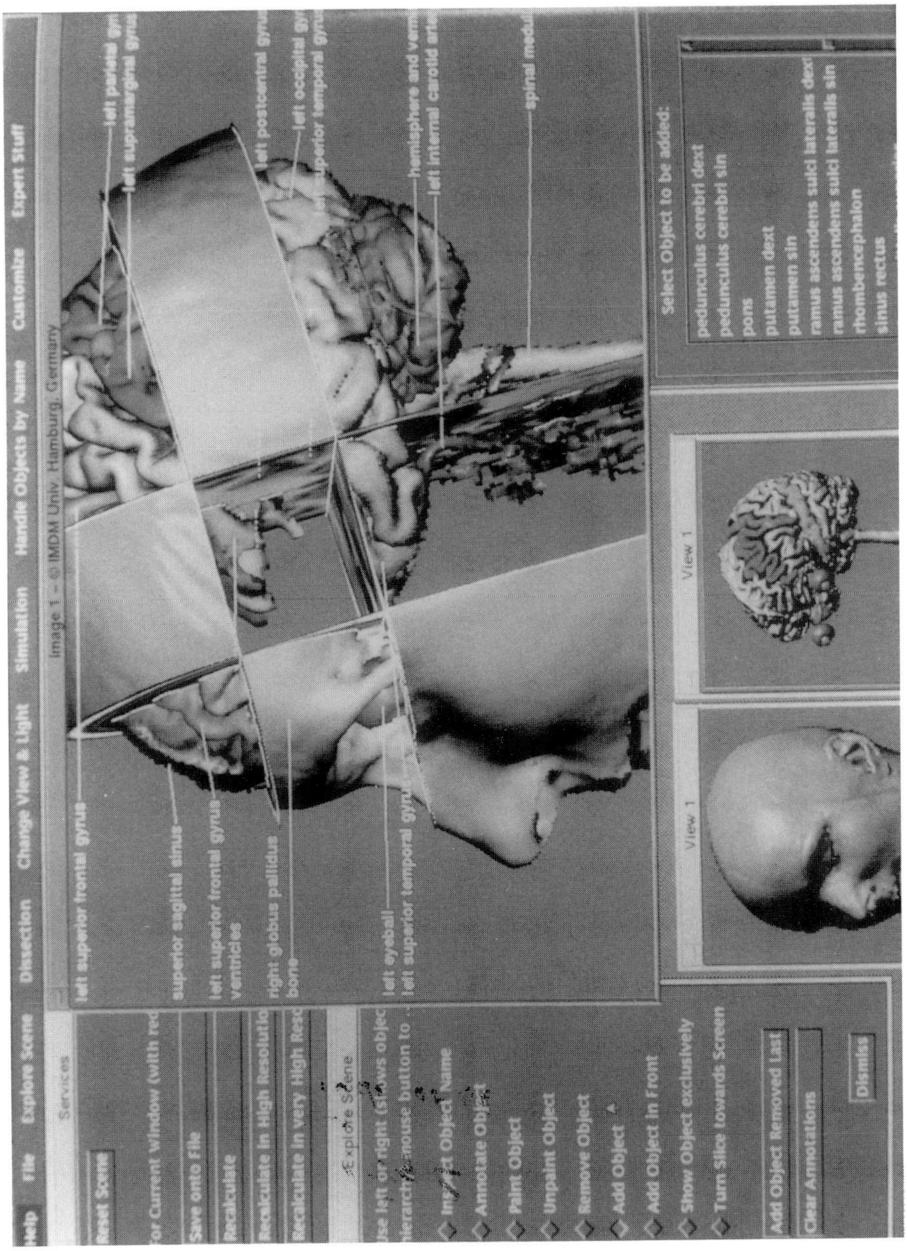


Figure 3. “User interface” of the atlas. A viewer can address the exploration tools (left) via a mouse and generate annotated views with the desired direction, size, and resolution. Objects to be displayed may be selected from the knowledge base (lower right corner).

# RESULTS

A student could begin to explore the brain model with a skin surface image of the head. As the first step, one could simply cut into the head and explore the appearing section. However, an anatomist would prefer to remove the tissue layer by layer. This can be simulated easily by "selective cutting" as shown in Figure 2. Here, cut planes have been specified and the objects "brain" and "eye" excluded from cutting. Since the cut surfaces exhibit the original intensity values, the scene has the appearance of an actual dissection.

The crucial advantage of the computer model is that, for any visible voxel, all related information can be queried. As shown in Figures 4, 5, and 6, the corresponding object names (such as the name of a gyrus) can be annotated at the cursor position, whether there are sections or multiple surfaces or both. The corresponding region can be colored as well. Thus, questions about the nature of the visible objects can be answered easily. The language for the annotations can be chosen (English, Latin, and German are currently available options).

Questions about the relationships between basic regions can be answered as far as they are described in the knowledge base. In the pilot project, for example, the structural parents of a basic region (e.g., the lobe to which a gyrus belongs) can be queried and colored, displayed, removed, etc. (Figure 4). A query addressing topologic knowledge might be "show all neighboring objects."

We can in the reverse case browse through a textual display of the knowledge base and select an object or group of objects that can be treated as shown

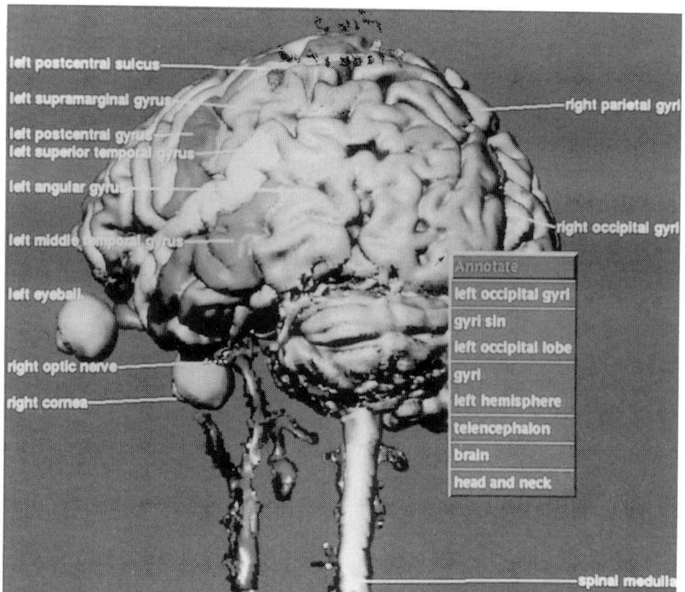


Figure 4. Views from any direction, such as this oblique dorsolateral one, can be produced. At any instant, for every visible voxel, information such as object names with their parents in the structural hierarchy can be queried (shown in the "pop-up" window). Regions may be annotated and/or colored by a "mouse" click.

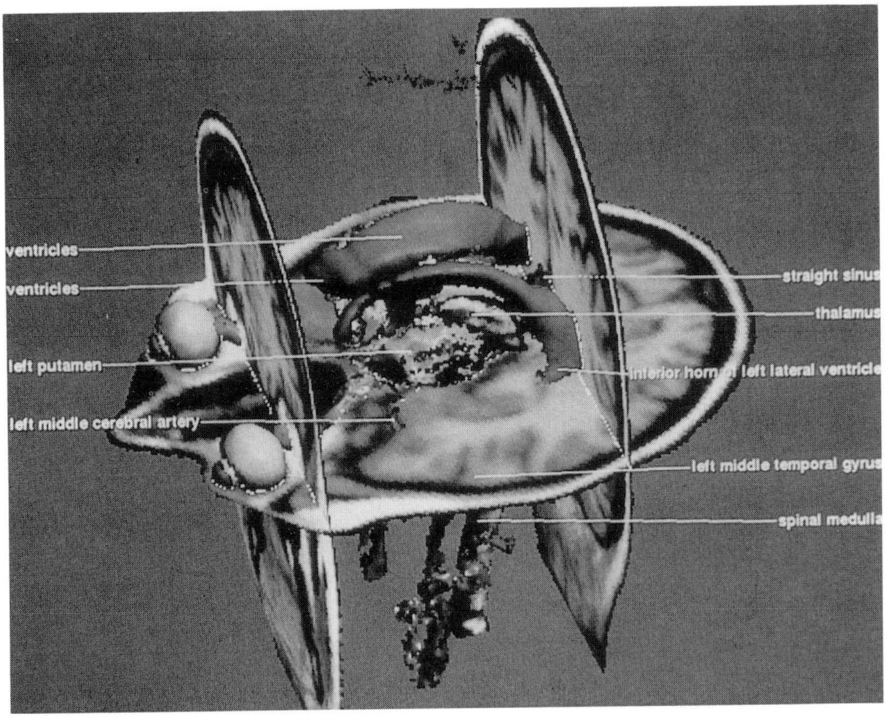


Figure 5. Integration of cross-sectional imaging with basic 3-D anatomy. Cross-sections and anatomic constituents may be chosen freely by the viewer.

above. For example, we could ask for the exclusive display of the thalamus and neighboring objects as shown in Figure 5. In this case, it is advisable to add the ventricular system for orientation. Unlike in an actual dissection, we can display it as a solid object. While in the previous examples radiologic imagery served to teach basic anatomy, the same data can be used for teaching and reference purposes in radiology itself. As shown in Figure 5, this can be done by presenting cross-sectional images in the context of basic anatomy. Whether we are on a section or an object surface, the exploration tools such as inquiry of names can be used.

A structure that is not well represented on MR imaging is bone. Therefore, we derive the anatomy of the skull from a CT data set. Unfortunately, we currently do not have the data sets form the same person. As an example of the use of the skull model, Figure 6 shows the transnasal access to the pituitary gland.

The previously described operations enable us to explore freely an anatomic object in a style comparable to dissection. Furthermore, the basic tools can be used for the simulation of certain clinical procedures. We demonstrate two of them in the following sections.

## Simulation of an X-Ray Exposure and Neurosurgery Planning

A CT data volume represents the absorption values. Thus it is easy to simulate an x-ray projection from a chosen direction with any beam geometry.

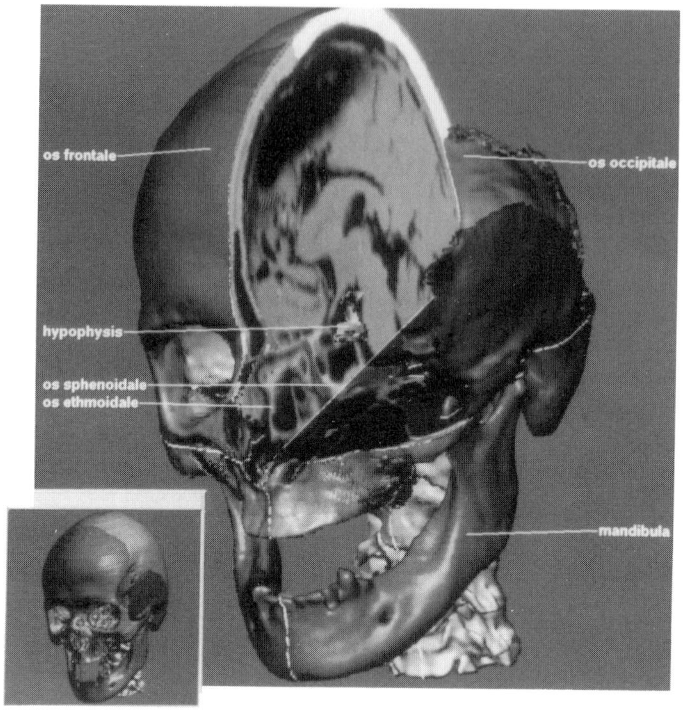


Figure 6. View showing transnasal access to the pituitary gland.

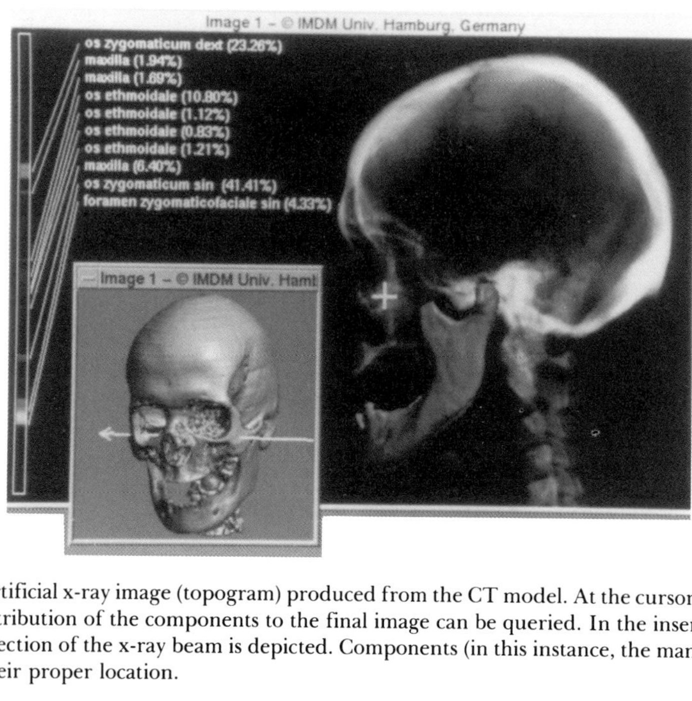


Figure 7. Artificial x-ray image (topogram) produced from the CT model. At the cursor position, the relative contribution of the components to the final image can be queried. In the inserted localized view the direction of the x-ray beam is depicted. Components (in this instance, the mandible) can be added to their proper location.

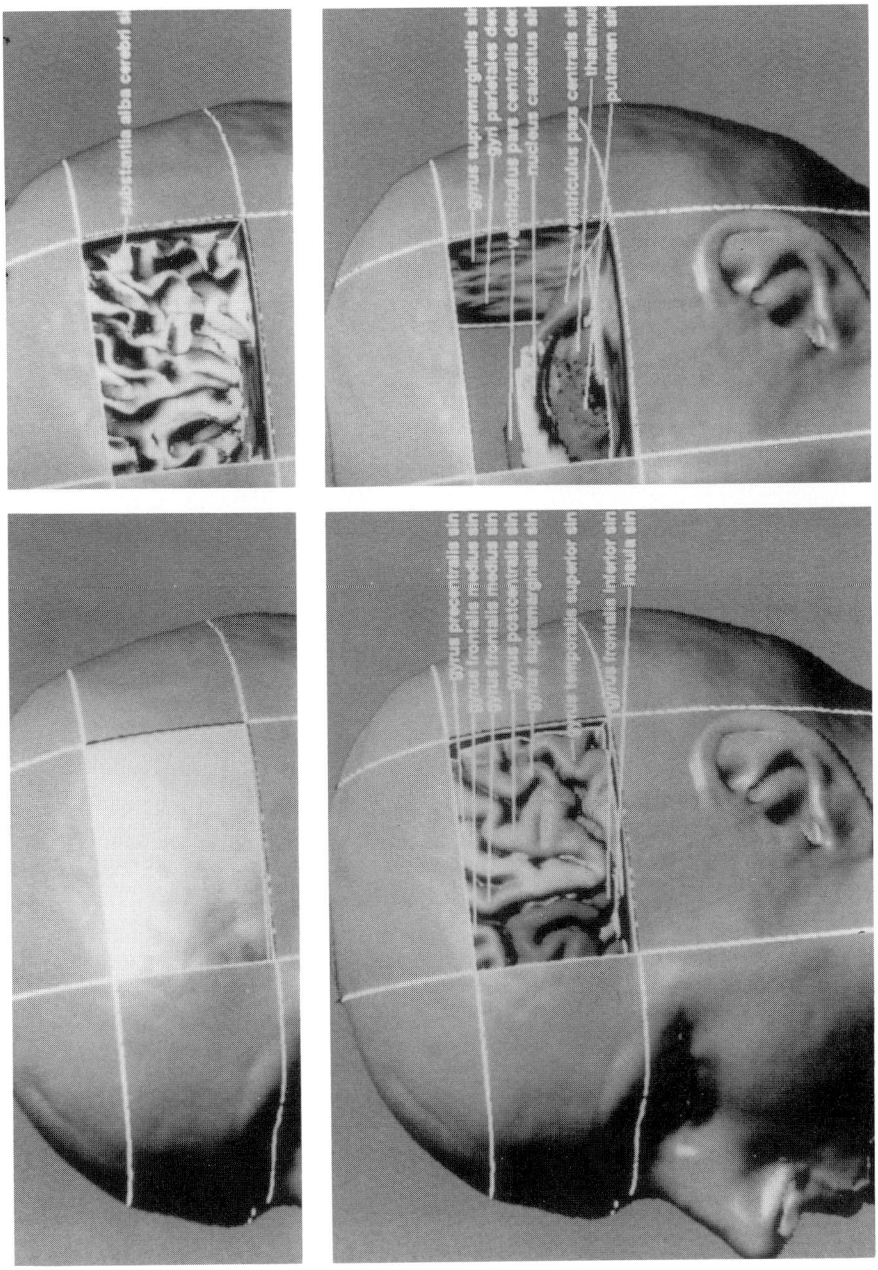


Figure 8. Simulation of craniotomy. Layers can be removed successively and information queried about the encountered objects.

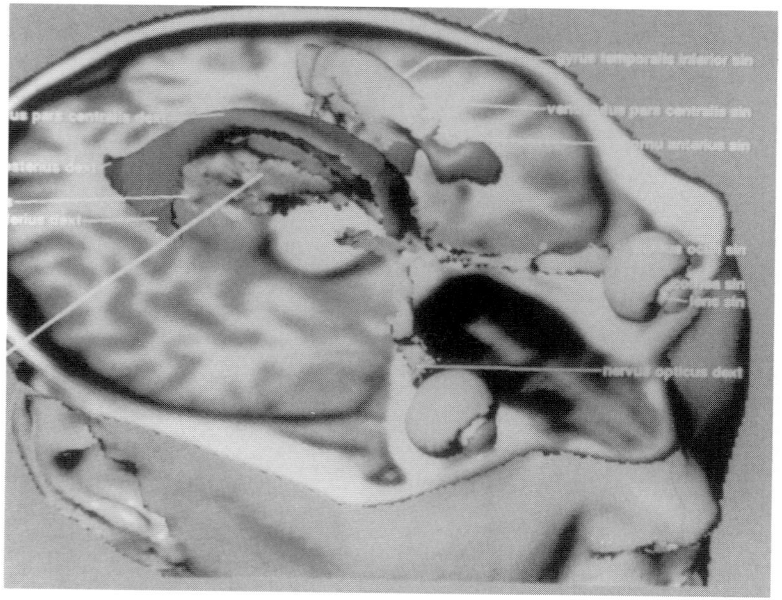


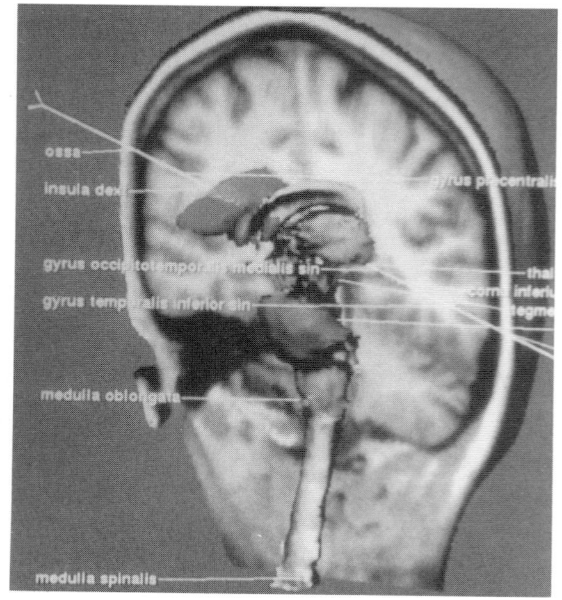


Figure 9. Stereotactic simulation. (A) Target (in this instance, the thalamus) and needle direction (arrow) can be defined. (B) A cut plane along the needle unveils the encountered objects and the puncture site at the skin surface. In this example, the precentral gyrus is involved.

Since we know the objects that cause the projection image, we can determine the contribution of the individual objects to the final intensity at any location (Figure 7). Thus, a student can conveniently check knowledge in radiologic image interpretation.

In neurosurgery, it is vital that the access to a lesion such as a tumor is optimal in that it avoids damage to vulnerable regions (like blood vessels or sensorimotor cortex). With the tools described, surgical operations and stereotactic interventions can be simulated in order to enhance the necessary surgical skills. Figure 8 shows the way a craniotomy can be simulated and the access path to a region of interest explored. As shown in Figure 9, stereotactic surgery can be simulated as well. Objects encountered using a certain needle position and direction can be identified easily.

## Discussion

Over centuries, textbooks and atlases were the only media for learning anatomy. With the development of so-called hypermedia techniques in the past decade, both pictorial and descriptive knowledge can be organized in such a way that a learner can browse through the knowledge on a computer screen in a sequence that depends on the teaching aim and individual needs (Jansen *et al.*, 1989). Still the underlying model is that of a book: hypermedia screens mostly exhibit a set of pages, in which the pictures are precomputed or digitized image matrices. Different aspects of the same object have to be available as different images. By having computerized anatomical models of the objects, different pictorial representations can be created at the user's will. Thus, it is also possible to integrate a dissection function into a learning program.

Computerized models of the human brain have been used for several purposes in the past. One purpose is the correlation of functional information (e.g., from positron emission tomography) with detailed structural information (e.g., from MR imaging) (Evans *et al.*, 1988; Greitz *et al.*, 1991; Lehmann *et al.*, 1991). A more academic application of a computer model is as a knowledge base for experiments in computer vision. Not much work has been done in the field of model-based atlases for medical education (Brinkley *et al.*, 1989; Mano *et al.*, 1990; Wahler-Lück, Schütz, & Kretschmann, 1991). The projects described until now have in common the representation of objects as surface layers. They do not allow the "look and feel" of a dissection, which is the main tool of exploring an anatomic object, because the objects are hollow. Cutting thus does not reveal realistic surfaces.

The approach presented here is based on a complete solid volume model. Through the linkage of descriptive information to sets of voxels, a variety of novel applications can be generated. Using state-of-the-art neuroradiologic imaging, it is now possible to generate computer models that allow dissections to explore human anatomy on a computer screen at any time and location. Thus, one can anticipate moving the dissection laboratory to places it is actually needed, such as the classroom, the examination room, the operating room, the x-ray reading room, or the library.

From the technical point of view, the volume model has proved to be a powerful data structure for constructing three-dimensional anatomic models. On one hand, this is because it is well suited to visualization of both surface and volume properties of the anatomy. It provides some of the look and feel which approach those of an actual dissection. On the other hand, it is flexible in that it easily allows the incorporation of extensions and more details on both the pictorial and descriptive levels. It must be admitted that three-dimensional editing of the spatial regions is a tedious task, but it need be done only once for a specimen. This can then be used in an unlimited number of copies. The required large computing power will be a less serious obstacle in the near future. A problem still remaining is the construction of a user interface that allows the easy and natural access to the exploration tools and the availability of computer power such that any reaction of the system could be in real time.

The spatial resolution achieved with MR imaging is not yet sufficient for all applications. Nonetheless, the pilot project has shown that an approach based on a solid volume model allows the user to choose, according to personal needs, a virtually infinite number of realistic scenes of the anatomy to be explored. Thus, the approach comprises the currently used atlas techniques in that it allows the generation of pictures in a form needed for both books and hypermedia documents. Beyond atlas generation, our approach even allows the generation of real solid models by using stereolithography. For general use, the database must still be refined. For completeness, image material, such as photographic images of cross-sections as representation for the actual anatomy of a specimen or positron emission tomography for functional aspects, must be included. The described techniques are being used for the construction of models of other parts of the body (abdomen) and other specimens (fetus). Their inclusion is not a problem in computer science, but a practical problem of data acquisition.

*Acknowledgments*

We are grateful to all members of our department who have supported this work. We thank Professor Zeumer and Dr. Siepmann, Department of Neuroradiology, for valuable discussions. The original MR image sequence from which the three-dimensional atlas has been generated has been kindly provided by Siemens Medical Systems (Erlangen). We are indebted to the Werner-Otto-Stiftung, Hamburg, Germany, which has supported this work in part.

## References

Bomans, M., Höhne, K. H., Tiede, U., & Riemer, M. (1990). 3D Segmentation of MR images of the head for 3D display. *IEEE Transactions on Medical Imaging, 9,* 177–183.

Brinkley, J. F., Prothero, J. S., Prothero, J. W., & Rosse, C. (1989). A framework for the design of knowledge-based systems in structural biology. In: *Proceedings of the Symposium on Computer Applications in Medical Care (SCAMC)* (pp. 61–65). Washington, DC: American Medical Informatics Association.

Evans, A. C., Beil, C., Marrett, S., *et al.* (1988). Anatomical-functional correlation using an adjustable MRI-based region of interest atlas with positron emission tomography. *Journal of Cerebral Blood Flow and Metabolism, 8,* 513–530.

Fishman, E. K., Ney, D. R., & Magid, O. (1990). Three-dimensional imaging: Clinical applications in orthopedics. In K. H. Höhne, H. Fuchs, & S. M. Pizer (Eds.), *3D-Imaging in medicine: Algorithms, systems, applications,* NATO ASI Series F (pp. 425–440). Berlin: Springer-Verlag.

Greitz, T., Bohm, C., Holte, S., & Eriksson, L. (1991). A computerized brain atlas: Construction, anatomical content, and some applications. *Journal of Computers and Tomography, 15,* 26–38.

Herman, G. T., & Liu, H. K. (1979). Three-dimensional display of human organs from computed tomograms. *Computer Graphics and Image Processing, 9,* 1–29.

Höhne, K. H., Bormans, M., Pflesser, B., Pommert, A., Riemer, M., Schiemann, T., & Tiede, U. (1992). Anatomic realism comes to diagnostic imaging. *Diagnostic Imaging, 1,* 115–121.

Höhne, K. H., DeLaPaz, R. L., Bernstein, R., & Taylor, R. C. (1987). Combined surface display and reformatting for the 3D-analysis of tomographic data. *Investigative Radiology, 22,* 658–664.

Höhne, K. H., & Hanson, W. A. (1992). Interactive 3D-segmentation of MRI and CT volumes using morphological operations. *Journal of Computer Assisted Tomography, 16,* 285–294.

Höhne, K. H., Riemer, M., & Tiede, U. (1987). Viewing operations for 3D-tomographic gray level data. In H. U. Lemke, M. L. Rhodes, C. C. Jaffe, & Felix, R. (Eds.), *Computer Assisted Radiology, (Proceedings in Computer Assisted Radiology '87).* Berlin: Springer-Verlag.

Jansen, W., Baak, J. P., Smeulder, A. W., & van Ginneken, A. M. (1989). A computer based handbook and atlas of pathology. *Pathology Research Practice, 185,* 652–656.

Lehmann, E. D., Hawkes, D. J., Hill, D. L., Bird, C. F., Robinson, G. P., Colchester, A. C., & Maisey, M. N. (1991). Computer-aided interpretation of SPECT images of the brain using an MRI-derived 3D neuro-anatomical atlas. *Medical Informatics, 16,* 151–166.

Mano, I., Suto, Y., Suzuki, M., *et al.* (1990). Computerized three-dimensional normal atlas. *Radiation Medicine, 8,* 50–54.

Schiemann, T., Bomans, M., Tiede, U., & Höhne, K. H. (1992). Interactive 3D-segmentation. In *Proceedings SPIE 1808: Visualization in biomedical computing II.* Chapel Hill, NC: The International Society for Optical Engineering.

Tiede, U., Höhne, K. H., Bomans, M., Pommert, A., Riemer, M., & Wiebecke, G. (1990). Investigation of medical 3D-rendering algorithms. *Computer Graphics and Applications, 10,* 41–53.

Vannier, M. W., Marsh, J. L., & Warren, J. O. (1983). Three-dimensional computer graphics for craniofacial surgical planning and evaluation. *Computer Graphics, 17,* 263–273.

Wahler-Lück, M., Schütz, T., & Kretschmann, H.-J. (1991). A new anatomical representation of the human visual pathways. *Graefe's Archive of Clinical and Experimental Ophthalmology, 229,* 201–205.

Witte, G., Höltje, W.-J., Tiede, U., & Riemer, M. (1986). Die dreidimensionale Darstellung computertomographischer Untersuchungen craniofacialer Miffbildungen. *Fortschritte auf dem Gebiete der Roentgenstrahlen, 144,* 400–405.

Zonneveld, F. W., Lobregt, S., van der Meulen, J. C. H., & Vaandrager, J. M. (1989). Three-dimensional imaging in craniofacial surgery. *World Journal of Surgery, 13,* 328–342.

10

# Neurobehavioral Probes as Applied in Physiological Neuroimaging Studies

## Methodological Considerations

RUBEN C. GUR, JUAN MANUEL GUTIERREZ, JAMES A. HOLDNACK, AND RICHARD N. MAHR

The emergence of neuroimaging technology has opened a window to researchers in neuropsychology into the *in vivo* functioning of the human brain. These techniques provide information regarding the neuroanatomical and physiological functioning of the brain and, coupled with what has been termed neurobehavioral probes (Gur, Erwin, & Gur, 1992) can yield insight into the neural networks that regulate human behavior. However, it is also important to recognize the limitations of this methodology, both current and inherent. The goal of this chapter is to outline the potential and the limitations of neuroimaging techniques and behavioral probes as they are applied to the study of brain–behavior relationships.

To accomplish this goal, we will first review the main techniques in use today and the principles on which they are based. We will then outline their evolving application to neuropsychological research. Differences in neurobehavioral research paradigms will be presented and their relative merits described. While

RUBEN C. GUR, JUAN MANUEL GUTIERREZ, JAMES A. HOLDNACK, AND RICHARD N. MAHR Brain Behavior Laboratory, Neuropsychiatry Section, Department of Psychiatry, University of Pennsylvania, Philadelphia, Pennsylvania 19104.

*Neuroimaging I: Basic Science*, edited by Erin D. Bigler. Plenum Press, New York, 1996.

the main avenue for linking neurobehavioral processes with regional brain function is likely to be in measures of physiologic activity during performance of tasks, there is also potential in studies correlating behavioral measure with resting neural activity. This issue will be reviewed. We will conclude by suggesting directions for future research.

## Neuroimaging Methods

Since the use of functional imaging techniques is not independent of neuroanatomical substrate, it is important to consider anatomy in the analysis and interpretation of physiologic data. We will therefore briefly present the main methods for anatomic imaging before we describe the physiologic techniques.

### Neuroanatomical Methods

The two major techniques currently available for imaging brain anatomy are x-ray computed tomography (CT) and magnetic resonance imaging (MRI). The latter technique can clearly display with exquisite detail brain structures and many lesions (Figure 1). If a lesion is detectable with CT or MRI, then we can state with considerable confidence that at least some of the behavioral deficits have been caused by atrophy or dystrophy of brain tissue.

Magnetic resonance imaging is superior to CT in many respects, particularly in brain areas near the skull or other bony tissue, where beam hardening effects of bone distort the CT scans. Some lesions can be missed by CT and detected by MRI. On the other hand, abnormal signal intensity may occur in MRI scans of normal individuals, and the field still lacks standard scanning procedures and normative data. Very small lesions, or diffuse neuronal loss such as that which occurs in Alzheimer's disease, can have deleterious effects on behavior and yet be undetectable by routine clinical reading of either CT or MRI. Computerized techniques for quantitating brain atrophy have been developed (Kohn *et al.*, 1991) and should prove more sensitive for detection of such tissue loss.

Aside from the obvious uses of MR scanning such as the examination of structural integrity in cerebral insult, various other applications are possible. Because MRI is not susceptible to bone artifacts, it can be readily used to image the temporal lobe regions of the brain as well as identify sulcal cerebrospinal fluid (CSF). The imaging of the temporal lobes by MR is considered critical for the investigation of diseases such as schizophrenia. Other advantages of MRI over CT are superior resolution and lack of ionizing radiation. Thus, there is minimal risk and therefore no ethical reason to limit its use in healthy individuals (Cleghorn, Zipursky, & List, 1991; Pfefferbaum, Lim, Rosenblum, & Zipursky, 1990; Gur & Pearlson, 1993).

### Methods for Physiologic Neuroimaging

The initial methods for physiologic neuroimaging had in common their use of radioisotopes. These techniques take advantage of the fact that active neurons have metabolic needs for oxygen and glucose and that cerebral blood flow rates change in response to these needs. Such measures can help identify regions of

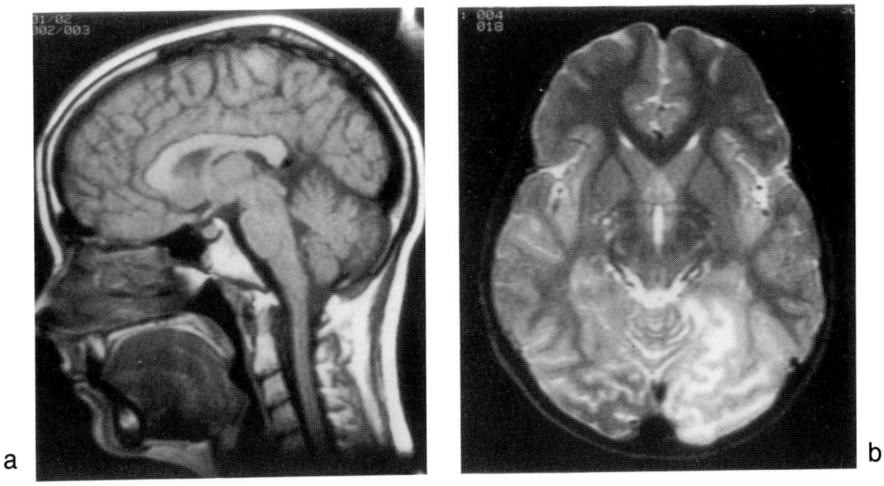


Figure 1. MRI. (a) Sagittal view of a normal brain. (b) Axial view of a cerebral infarct. Individuals with lesions in this area usually have difficulty understanding language.

abnormal physiologic activity associated with behavioral deficits. Furthermore, such measures obtained during the performance of cognitive tasks could help delineate brain regions necessary for regulating cognitive processes.

These isotopic techniques for measuring cerebral metabolism and blood flow can be traced to the pioneering method of Kety and Schmidt (1948) for measuring whole-brain metabolism and blood flow. The technique used intracarotid injection of nitrous oxide, and measurement of arterial-venous differences in concentration yielded accurate and reproducible data on brain metabolism and blood flow. However, this technique is limited by its invasiveness and provides only whole-brain values.

133 Xenon Clearance Methods. Safe regional measurements were first made possible by the introduction of the $^{133}$Xe clearance techniques for measuring regional cerebral blood flow (rCBF). The highly diffusible $^{133}$Xe can be administered as a gas mixed in air or in saline. Its clearance from the brain is measurable by stationary scintillation detectors. The rate of clearance enables considerably accurate quantitation of rCBF in the fast-clearing gray matter compartment, as well as calculation of mean flow of gray and white matter. Initial applications used carotid injections (Olesen, Paulson, & Lassen, 1971), which were invasive and only enabled measurements in one hemisphere at a time. In 1975, Obrist, Thompson, Wang, and Wilkinson (1975) reported the $^{133}$Xe inhalation technique and presented models for quantifying rCBF with this noninvasive procedure. The technique permits simultaneous measurements from both hemispheres, and the number of brain regions that can be measured depends on the number of detectors. Initial studies were performed with up to 16 detectors, 8 over each hemisphere, but commercially available systems now enable the placement of up to 254 detectors. The quantitative data can be displayed topographically as shown in Figure 2.

Note that rCBF is typically higher in the front of the brain. This "hyperfrontal" pattern has been observed routinely in normal subjects (Ingvar, 1979). The main limitation of the technique is that it is optimal for measuring rCBF only on

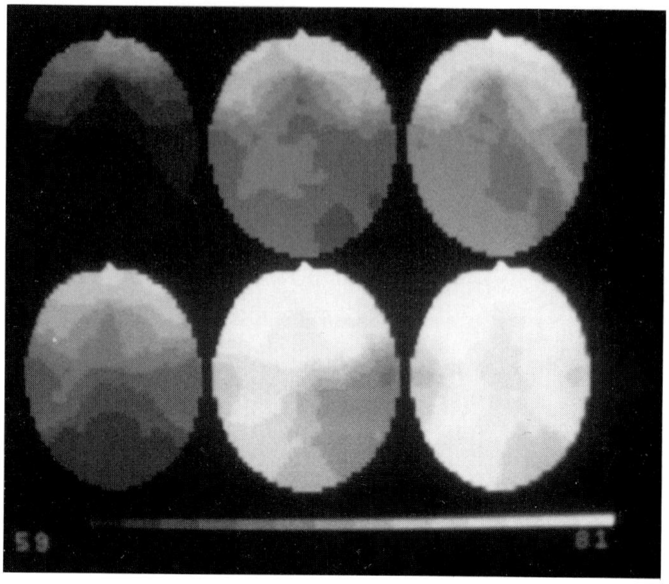


Figure 2. A topographic display of rCBF in a group of normal males (upper row) and females (lower row) during rest (first column), a verbal analogies task (middle column), and a spatial line orientation task (last column). Note the hyperfrontal pattern at rest, the higher CBF in females, and the increased CBF in the left for the verbal and the right for the spatial task.

the brain surface near the skull, and hence it is restricted to the study of cortical brain regions.

POSITRON EMISSION TOMOGRAPHY (PET). PET has made it possible to measure *in vivo* biochemical and physiological processes in the human brain with three-dimensional resolution. Initial work with animals used selectively labeled chemical compounds (radioisotopes) to measure the rate of the biochemical process. This has been extended to humans by principles of computed tomography (Reivich *et al.*, 1979), and now the technique has been adapted to use several radionuclides that decay through the emission of positrons. Subjects are administered radionuclides that are unstable because their nuclei have an excess positive charge. The radionuclides are usually given intravenously and are taken up by tissue. Through the emission of a positron they get rid of their energy and undergo the process of annihilation where the positrons interact with an electron (negatively charged). The two photons emitted from each annihilation travel in opposite directions and the energy generated is measured by detector arrays. By computed tomography principles, the coincidental counts are used to generate images reflecting the regional rate of radionuclide uptake. This information enables the calculation, depending on the specific radionuclide, of such varied physiologic parameters as oxygen and glucose metabolism, blood flow, drug pharmacokinetics, or receptor density of neurotransmitters. In order to relate this physiologic information to anatomic regions of interest (ROI), an atlas of brain anatomy is required. These can be based on computerized images of sliced brains or on x-ray CT or MRI scans. Multiple brain "slices" can be obtained with PET (Figure 3).

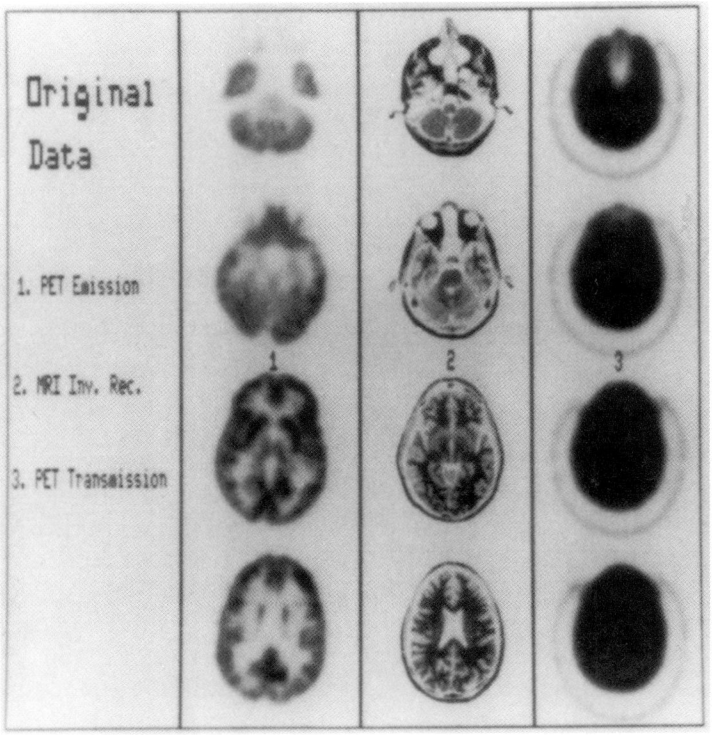


Figure 3. Raw data from MRI and PET used as input for merging. MRI data were acquired at 4.2-mm intervals with a slice thickness of 4.0 mm using an inversion recovery pulse sequence (TE = 30 msec, TI = 300 msec, RT = 1276 msec). Transmission and emission PET data were also acquired, although only the emission data are shown in this image set. Note the alignment of sections in terms of scaling, angle, and level. From Mazziotta J.C., *et al.* Region of interest insures: The relationship between structure and function in the brain. *J. Cereb Blood Flow Metab* 1991;11:A51–A56.

Single Photon Emission Computed Tomography (SPECT). SPECT is another technique for three-dimensional imaging of rCBF. The technique uses radionuclides, which unlike positron emitters, do not require the availability of a dedicated cyclotron for their production. There are advantages and disadvantages in applying SPECT compared to PET technology. The advantages of SPECT include its relatively lower startup cost and cost person and its greater availability (Innis, 1992). However, PET maintains some decided advantages over SPECT technology in image resolution (5 mm compared to 7–8 mm for SPECT; Innis, 1992) and established quantification methods. Although SPECT technology has been available for approximately ten years, FDA approval of ligands for use in neuroimaging occurred only more recently in 1988 (Devous, 1992).

A family of amines including $^{123}$IMP and $^{99}$Te hexamethyl oxime (HMPAO), which accumulate in the brain in proportion to cerebral blood flow, were developed to overcome the limitations of $^{133}$Xe. However, no methods for computing absolute blood flows have been established for these ligands (Beats, Burns, & Levy, 1991; Miller *et al.,* 1991; Van Heertum, 1992).

While perfusion agents have received the most attention, ligands that have an affinity for specific neuroreceptors such as benzodiazepine and D2 receptors are also available. However, their application in humans has yet to be fully explored (Hall *et al.*, 1992; Innis, 1992).

Improvements in SPECT scanners have reduced scan time, increased resolution, and allowed for coregistration with MRI (Van Heertum, 1992). Advances in this imaging technology and in quantification methods continue such that one day it may rival the image quality of PET.

FUNCTIONAL MRI. Recent studies have led to much interest in functional imaging with MR scanners. Early uses of MR presupposed that the tissue nuclei were static and could be easily located through the procedure of excitation and re-emission of energy. However, functional imaging with MR has been hampered by its sensitivity to movement by the patient being examined. These movements include heartbeat, bowel peristalsis, breathing, voluntary movements by uncooperative patients, and, of most interest to the current discussion, blood flow. These difficulties were generally considered as a group and referred to as the motion artifact (Wood & Henkelman, 1985; Stehling, Turner, & Mansfield, 1991). Cerebral blood flow (CBF) is extremely fast (in major arteries of the brain, blood moves at the speed of 10 cm/second; Olendorf & Olendorf, 1989). Thus, despite even brief intervals between pulses of a conventional scan, the blood moves a substantial distance within the time of a scan. Therefore, these factors can easily be understood to mean that CBF will not respond to conventional MR scanning in the same manner as more static tissue. Consequently, there are various problems in MR imaging of CBF and hence the use of MRI in functional imaging. For example, spin echo is difficult to achieve because blood movement may not allow for the absorption of all radio wave pulses or because the blood will enter a region other than that being scanned by the time the scan is complete. Further complicating functional imaging is the fact that within blood vessels, blood travels at different rates (blood in the center of the vessel moves faster), and this phenomenon contributes to dephasing (Olendorf & Olendorf, 1989). Lastly, because the human arterial system is complex, CBF will travel in various directions, further contributing to dephasing. Nevertheless, subsequent study of various types of blood flow measurement procedures continues (Gore *et al.*, 1992) and it is hoped that functional MRI can find great success and continue to expand the depth of investigation in the assessment of the neurologically impaired patient.

### OVERALL ISSUES RELATED TO THE TECHNOLOGY

Much of the research utilizing neuroimaging has overlooked the validity of various assumptions made related to the technology. Such oversight may have contributed to inconsistent or often contradictory findings. The validity of the tracer model, the precision and accuracy of the data acquisition, the propagation of errors, and the assumptions used for image reconstruction and quantitation are probably as important as appropriate experimental design and statistical analysis (Volkow, Brodie, & Bendriem, 1991). The precision of isotopic measurements can be affected by several variables, including background noise radiation ("random coincidences"), scatter, accuracy of anatomical correlation of metabolic function (coregistration), definition of the region of interest (ROI) under study, accuracy of reconstruction, and analysis of the images generated, as well as head movement.

It is important to keep in mind these sources of possible confound. Moreover, as Gur and Gur (1992) note, some of this research has been further hindered by methodological problems such as small samples, insufficient attention to normal human variability of brain function, and ignoring the high intercorrelations of data across brain regions (Warach *et al.*, 1982).

## Application of Neuroimaging in Neuropsychology

Physiologic neuroimaging methods provide information on the rate of activity of neural systems, and this raises the question of measurement conditions. Measures of brain anatomy would not be influenced by the subject's current state of mind, whether he is frightened by the procedure, or whether she is disturbed by conversation among staff members. By contrast, should physiologic measures be insensitive to such effects we would probably lose interest in them as indicators of brain function. Hence, when contemplating physiologic neuroimaging studies the experimental conditions require careful consideration.

### Resting Baseline Measures

An ongoing debate in the area of physiologic neuroimaging is whether resting baseline conditions are feasible or desirable. Several investigators have maintained that physiologic neuroimaging studies would be altogether uninformative if subjects were not provided with a task or a uniform stimulation condition. Thus, Buchsbaum *et al.*, (1982) have used electric shocks and the continuous performance task during measurement of cerebral glucose metabolism with PET. A resting baseline condition, it is argued, is too unstructured, hence unreliable, and would unduly increase intersubject variability.

Contrary to this we have argued that a standard resting baseline condition would be invaluable for interpreting activation data and for comparison across studies and research centers. It is possible to structure a resting baseline condition with standard ambient stimulation and instructions (e.g., to rest, not to make any serious mental efforts, and refrain from falling asleep). We have described such a condition (Gur *et al.*, 1982) that yields reproducible data (Gur, *et al.*, 1983, 1987; Warach *et al.*, 1992). Furthermore, the variability in the resting values and the regional landscape did not seem random, since it correlated with age (Gur, Gur, Obrist, Skolnick, & Reivich, 1987) and gender (Gur *et al.*, 1983). Finally, several studies have reported systematic correlations between resting baseline measures of cerebral metabolism or blood flow and neuropsychological performance (Gur *et al.*, 1988, 1994).

### Activated Measures

Regardless of one's position in the debate over the value of obtaining resting baseline measures, it became apparent rather early that measures of CBF and metabolism during the performance of cognitive tasks can accentuate differences between patients and controls. For example, in the first study in which we compared medicated patients with schizophrenia to sociodemographically balanced healthy controls, we found no differences in overall or hemispheric CBF using the $^{133}$Xe clearance technique, but distinct abnormalities were seen in the

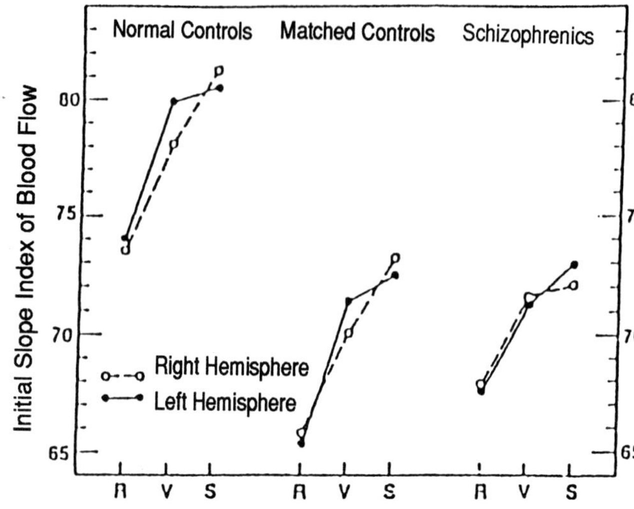


Figure 4. Regional cerebral blood flow in two hemispheres of schizophrenics and matched controls when resting (R), solving verbal analogies (V), and performing spatial task (S). Normal control values from an earlier study (AU-PROVIDE THIS) for comparison with latest matched sample.

pattern of hemispheric changes induced by verbal (analogies) and spatial (line orientation) tasks. In contrast to controls, who showed greater left hemispheric increase for the verbal and greater right hemispheric increase for the spatial task, patients with schizophrenia had a bilaterally symmetric activation for the verbal task, thus failing to show left hemispheric dominance for this task, and instead showed greater left than right hemispheric activation for the spatial task (Figure 4). Similarly, Weinberger, Berman, and Zec (1986) found no regional abnormalities in resting CBF of patients with schizophrenia. However, distinct abnormalities were demonstrated in frontal regions during activation with the Wisconsin Card Sorting Test of abstraction and mental flexibility, which is putatively sensitive to frontal lobe damage.

Neurobehavioral Probes. Arguably the most crucial aspect in designing activated studies is the selection of stimuli and procedures to be used as neurobehavioral probes. The selection of particular tasks to be used for activation is ultimately in the hands of the investigator and tailored to test specific hypotheses. However, several considerations could be useful in trying to maximize the yield of such research (Gur, Erwin, & Gur, 1992). First and foremost, it is important to begin with well-defined behavioral dimensions, where there is psychometric evidence for convergent and divergent "construct validity" and some neurobehavioral evidence for regional specificity. It is then necessary to find a task which measures this construct and can be adapted for use with neuroimaging. This can be rather trivial for some tasks, but a challenge for others. When changes are extensive it would be prudent to find out whether this produces changes in the performance and its construct validity.

The ability to obtain performance measures concomitant with the physiologic parameters can be helpful in the interpretation of regional effects. For example, in our first experiment with verbal and spatial tasks we used a Gestalt completion task for the latter. While the verbal task reliably increased left hemispheric CBF relative to right, the spatial task produced, on the average, bilaterally symmetric changes. Indeed, of the 36 subjects, all right-handed males, 17 showed greater right and 17 greater left hemispheric increase. The performance data, however, were different between these two groups, with right hemispheric increase associated with better performance on the spatial task. In general, we found performance data quite essential for evaluating relationships between mental effort and neural activity. For example, in a study of normal young volunteers who were given an easy and a difficult version of a verbal analogies task (within-subject design), we found effects of task difficulty on CBF measured with the $^{133}$Xe clearance method. We also found a relationship between the activated CBF and performance, which was mediated by the subject's anxiety (Gur *et al.*, 1988).

In addition to selecting a unitary behavioral dimension and tasks that yield performance measures with established psychometric properties of reliability and validity, task selection can be guided by reference to its relevance to the disorder. In trying to understand neural substrates of a psychiatric disorder using the physiologic neuroimaging approach, it would make sense to apply tasks for which there is evidence of impairment associated with the disorder. Our initial choice of verbal and spatial tasks was guided by evidence from behavioral measures suggesting dysfunction in cognitive domains associated with the left hemisphere. We have since learned to appreciate Chapman and Chapman's (1973) arguments for establishing differential deficit, and this can be incorporated into the neuroimaging paradigm by selecting control tasks with comparable psychometric properties. The past decade has also seen the development of efficient and well-standardized neuropsychological testing "batteries". Such batteries provide "neuropsychological profiles" of populations and can be used for establishing areas of differential deficit. These domains can then be selected for further scrutiny using activation studies.

In this respect, we have learned that computerized testing confers substantial advantages. It assures more uniform administration, can be conveniently "locked in" with the physiologic measures, and provides more accurate data as well as data that would be almost impossible to obtain manually (such as item-by-item reaction times). This also facilitates integration of behavioral data with neuroimaging since the data are available on media conveniently accessed by computers.

Cognition versus Emotion. So far we have restricted the discussion of behavior to cognitive variables, partly because this is the area where most progress has been made. There is reason, however, to believe that the realm of emotion is perhaps just as pertinent to the study of psychiatric disorders such as schizophrenia and schizotypy. This work is more difficult because there are less established paradigms, and data are meager. Yet there is evidence that application of activation tasks that probe emotional processing may yield unique information on the pathophysiology of schizophrenia.

A simple emotion discrimination task was developed which requires the subject to indicate whether a face is happy, sad, or neutral and rate the emotional intensity. The faces were the same as in the memory task except that they expressed happy and sad emotions. The posers were actors and actresses, and expressions were used which could be rated reliably by a normative sample. The task yields reliable measures of sensitivity and specificity of discriminating happy from neutral and sad from neutral and was found to be sensitive to gender differences (Erwin *et al.*, 1992). The task also discriminated patients with depression from comparable controls (Heimberg, Gur, Erwin, Shtasel, & Gur, 1992). When applied to patients with schizophrenia, this task revealed differential impairment relative to an age discrimination task, using the same stimuli (Gur *et al.*, 1992). This suggests that patients with schizophrenia are impaired in emotional discrimination. We have administered the task to a sample of healthy young adults and found right hemispheric activation for both happy and sad discrimination. However, subtracting happy from sad showed lateralized differences in the frontal region, with happy discrimination associated with relatively greater left hemispheric activation (Gur, Skolnick, & Gur, 1994). The same stimuli were used for standardized mood induction. The effects on mood were robust and reproducible (Schneider, Gur, Gur, & Muenz, 1994), and differential effects were observed on CBF measured with the $^{133}$Xe clearance method (Schneider, Gur, Jaggi, & Gur, 1994) and with PET (Schneider, Gur, Harper-Mozley, *et al.*, 1995). Patients with schizophrenia showed a blunted response to the induction. Thus, it seems that applying emotional probes in addition to the cognitive probes that are being currently applied would be of value in evaluating patients with schizophrenia and schizotypy.

ANALYSIS OF ACTIVATION DATA: QUANTITATIVE VERSUS SUBTRACTION METHODS. In recognizing the value of activated measurements, an issue that comes immediately to the fore is whether it is important to obtain quantitative parameters of activity. By this we mean the ability to measure with physiological units the rate of blood flow or metabolism. This is an important decision, since quantitation requires the availability of reliable estimates of the isotope arterial concentrations for the input function, which in the case of PET involves arterial catheterization (no reliable methods of CBF quantitation are currently available for SPECT).

A strong argument can be made that the additional effort and risk of quantitative measurements are unnecessary for examining task activation effects since all we are interested in are the task-induced changes in regional landscapes of activity. For this purpose it is sufficient to measure the activity during different task conditions which control for elements of the task and subtract the topography of control conditions from that of the condition of interest. The feasibility of this approach was demonstrated elegantly in a series of experiments (Posner, Petersen, Fox, & Raichle, 1988; Fox, Mintun, Reiman, & Raichle, 1988). They used rapid bolus injections of $^{15}$O-labeled water with PET, where voxel concentration levels were transformed to proportions of whole slice and regions were then identified across subjects using a computerized stereotactic atlas. They showed in healthy subjects how attentional and linguistic subsystems could be isolated by the "subtraction" method (Figure 5). Judging from these results it is fair to conclude that the nonquantitative subtraction method is likely to yield many important insights on brain behavior relations.

However, the subtraction method may have several limitations and perhaps may involve some risks. Most importantly, without quantitation it is impossible to tell whether activity in a region of interest has increased, decreased, or stayed the same across conditions. A "hot spot" on the subtraction image only refers to relative amount of change in the number of count compared to the other regions. This may address some hypotheses which do not include statements about the relationship between brain activity and mental effort, but will have limited

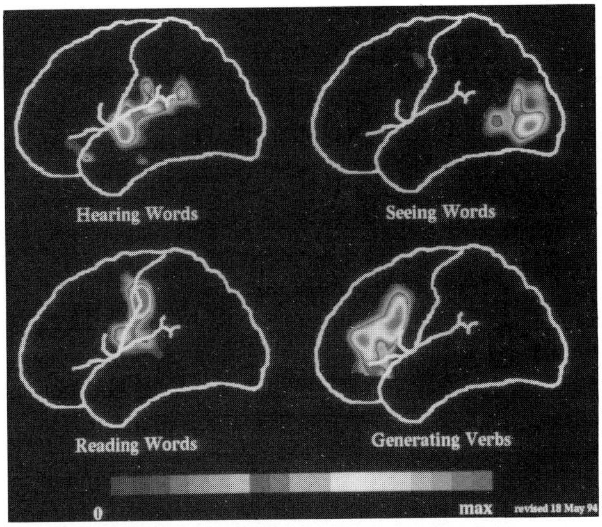


Figure 5. Subtraction of a control state from a task state. For "hearing words" and "seeing words," the control state was visual fixation. For "speaking words," the control state was "seeing words." For "generating verbs," the control state was "speaking words." Images represent averages of 9 subjects. Courtesy of Dr. Karl Friston, from Friston, Frith, Liddle, & Frackowiak (1993). For a color representation of this figure, see the color insert facing page 179.

value for linking these constructs. It should be pointed out further that even for hypotheses which are satisfied with statements about relative activation, the subtraction method makes a further questionable assumption of linear relationship between count density and CBF or metabolism.

Lack of quantitation could also lead to error when interpreting effects. Recently, Drevets *et al.* (1992) from the St. Louis group retracted an earlier report of increased left temporal CBF in anxiety patients during lactate-induced panic attacks. When PET data were cross-registered with MRI-based neuroanatomy, it became clear that the "hot spot" was on facial muscles, not brain. Drevets *et al.* (1992) point out that the mistake could have been avoided by anatomic cross-registration, but even without cross-registration a quantitative approach would have revealed that the values are uncharacteristic of brain parenchyma. Cross-registration in its present state, when applied to "aggregate brains" as is done in the subtraction method, would not safeguard against other artifacts that could be detected by quantitation (e.g., large vessel effects). Thus, while accepting the heuristic value of the subtraction method for identifying neural networks related to behavioral dimensions, we believe it is worthwhile to proceed with caution and use methods which permit quantification in physiologic units.

On the positive side, proponents of the subtraction method at the Hammersmith Institute in London (Friston, Frith, Liddle, & Frackowiak, 1991) have developed and refined a set of algorithms for standardized analysis of subtraction data. The method transforms pixel counts into probability maps based on both the magnitude and variability of changes induced by task conditions. These statistical probability maps (SPM) are then placed in a standard stereotactic coordinate system and overlayed on an anatomic image, which helps illustrate the topographic location and distribution of the changes in activity (Figure 6). A genre of studies has been using SPM to demonstrate task-related changes in

First PC

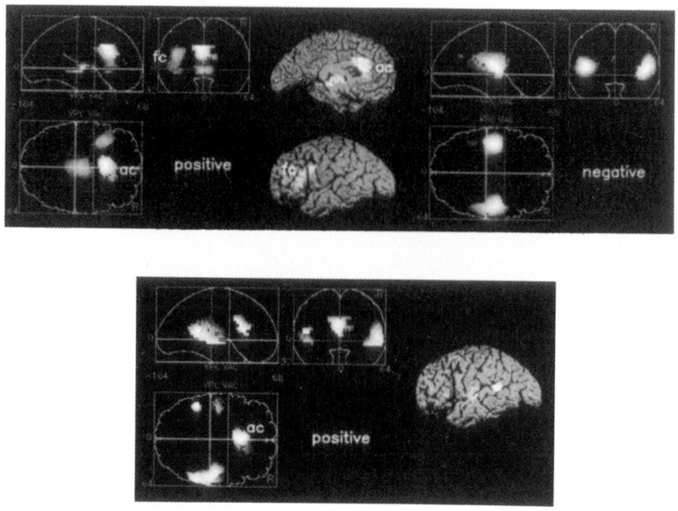


Second PC

Figure 6. Results of a word generation PET activation study of 6 normal subjects, each studied 12 times, during either a word generation task or a baseline word shadowing task. The essential difference between these tasks is the intrinsic generation of word representations and the "working memory" for words already produced. These cognitive components are seen to activate the anterior cingulate, the left dorsolateral prefrontal cortex, and the thalamus. Profound deactivations are seen bilaterally in the periauditory regions and related superior temporal cortex (top). Other important time-dependent changes in these data involved a montonic decrease in regional cerebral blood flow in the anterior cingulate and left temporal cortices [including Wernicke's area (bottom)] that were not related to the task being performed. The results are displayed in the form of statistical parametric maps (SPMs). In this instance the statistical parameter is derived from a principal component analysis of the PET time-series (averaged over all subjects). The display formats are standard and represent maximum intensity projections on viewing the brain from the side, from the top, and from the back. The same data are rendered on drawings of the cortical surface to aid interpretation. See Friston, Frith, Liddle, & Frackowiak (1992) for a fuller description of this study. Courtesy of Karl Friston.

brain activity, and this field is likely to expand in the near future. The great advantage of SPM is that it fosters comparability among studies and corrects to some extent for some obvious problems such as Type I (experimenter-wise) error associated with multiple statistical comparisons. However, the sample sizes used in many of these studies are small and we believe the results should be considered with extreme caution.

Cross-Modality Integration. Throughout this chapter we have attempted to show how hypotheses generated in this area of inquiry can be tested using integration of measures across modalities. This is probably the key for making inroads into our subject matter, yet it is not easy to do.

Cross-modality integration requires the joint examination of very different types of data, each with its own unreliability and uncertainty. There is need to reduce the data to some parameters which can be interpreted across modalities. Thus, comparing the magnitude of regional task activation with performance may require definition of physiologic activation parameters (such as task-rest per region divided by global changes) that could be theoretically linked to performance parameters (such as memory relative to general intellectual abilities).

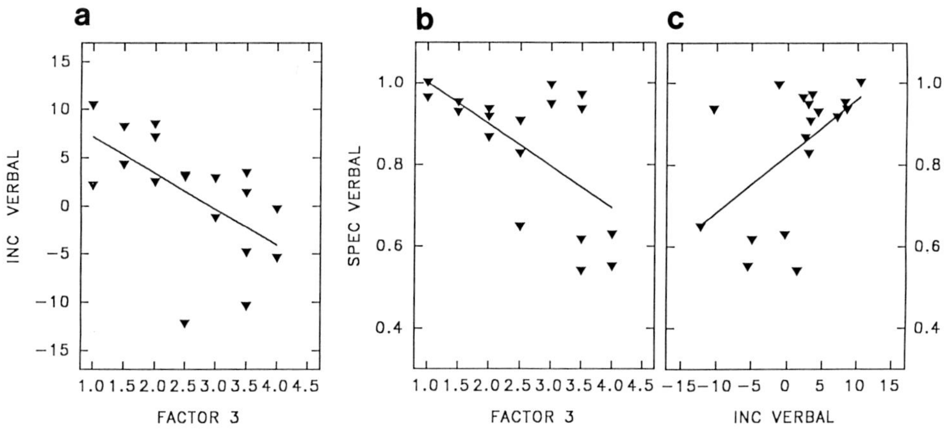


Figure 7. Scatterplots showing the relationships among (a) rCBF increase and severity of hallucinations and delusions (Factor 3), (b) verbal specificity scores and the hallucinations-delusions factor, and (c) verbal specificity scores and CBF increase for the verbal memory task.

The multitude of data necessarily involved in such integration is also staggering, and utmost care is required to safeguard against excessive Type I error. Sufficiently large samples for exploratory analyses are prohibited by the high costs of the methods, and hence it is particularly important to be guided and driven by hypotheses. On the other hand, the temptation to explore with a new technology is natural, and given the novelty of the method current hypotheses would not lead us very far.

We can divide the data into clinical (self and clinical rating scales, outcome measures, quality of life), neurobehavioral (neuropsychological testing, performance during activation studies), neuroanatomic (volumetric MRI), and neurophysiologic. An abnormality at the neuronal level becomes more credible as a potential substrate for the psychiatric disorder if it is correlated with clinical severity. This has motivated our search for neurophysiologic correlates of symptom severity (e.g., Gur *et al.,* 1987; Gur, Resnick, & Gur, 1989). More recently we found that, using the memory activation paradigm, clinical performance and neurophysiologic data can be integrated to link a specific abnormality in the regional neurophysiologic response to severity of a cluster of symptoms (Figure 7). This is obviously a complicated and laborious process, but it is possible to conceive of its success in continuously and systematically refining our hypotheses on brain-behavior relationships.

## Limitations and Future Directions

While neuropsychologists can be legitimately excited by the potential of physiologic neuroimaging to shed light on hitherto inaccessible issues concerning brain-behavior relationships, it is equally necessary to keep in mind the current as well as inherent limitations of these methods. At present they provide limited spatial resolution and even more limited time resolution. Thus, the most advanced PET scanners capable of imaging the entire brain have spatial resolution greater than 5 mm. There is a single-slice system which approaches 2 mm resolution (Budinger, Derenzo, Huesman, Jagust, & Valk, 1991), but this is probably the theoretical limit because of the distance traveled by positrons prior to annihilation. This will limit the range of anatomic detail we can study. Functional MRI has yet to be standardized and shown reliable and valid for tapping significant physiologic processes. The time resolution of the currently fastest CBF procedure is 40 seconds, which while impressive, is still far from enabling us to observe important computations believed to require fractions of this epoch. We are still far from imaging methods that can break the time barrier, and almost certainly functional neuroimaging will not resolve molecular changes or events taking place in milliseconds.

These limitations notwithstanding, there is now available an impressive array of methods for measuring regional brain physiology, and the application of these methods in neuropsychology and neuropsychiatric disorders is becoming routine. Whether the results to date can be considered of lasting importance is still debatable. However, with the experience we have gained in "taming" this technology we should be on our way to making more robust and theoretically heuristic discoveries. This can be buttressed by the application of well-defined and structured "neurobehavioral probes" that provide psychometrically reliable performance data. Such performance data are particularly helpful in examining correlations with quantitative physiologic parameters, and we have tried to argue in favor of such methods as superior to the nonquantitative "subtraction" procedures.

The most challenging aspect of this work is the integration of neuroimaging data with clinical and neurobehavioral data. This process is still in its infancy and is quite difficult to implement, but successful integration will inevitably lead to better understanding of the multifaceted nature of brain-behavior regulation and the dysfunction characteristic of neuropsychiatric disorders.

In applying physiologic neuroimaging procedures to the study of neuropsychology and neuropsychiatric disorders we have emphasized the need for careful and deliberate choice of activation tasks with respect to the behavioral dimension relevant to the disorder being investigated. Such tasks would preferably have established psychometric properties with acceptable levels of reliability and construct validity. The role of neuropsychologists in this research is, therefore, not only to supply knowledge on regional brain regulation of behavior but also to enlist their training in psychometrics and test methodology to safeguard these principles of sound behavioral research. This is probably an area where we need to work hardest to convince our colleagues from the many other disciplines involved in this research who are not sensitive to such issues.

*Acknowledgments*

This work was supported by grants MH43880 and MH19112. We thank Helen Mitchell-Sears for her help in preparing this manuscript.

## References

Beats, B., Burns, A., & Levy, R. (1991). Single photon emission tomography in dementia. *International Journal of Geriatric Psychiatry, 6,* 57–62.

Buchsbaum, M. S., Ingvar, D. H., Kessler, R., Waters, R. N., Cappelletti, J., van Kammen, D. P., King, C., Johnson, J. L., Manning, R. G., Flynn, R. W., Mann, L. S., Bunney, W. E., Sokoloff, L. (1982). Cerebral glucography with positron tomography. *Archives of General Psychiatry, 39,* 251–259.

Budinger, T. F., Derenzo, S. E., Huesman, R. H., Jagust, W. J., & Valk, P. (1991). High resolution positron emission tomography for medical science studies. *Acta Radiologia 376* (Suppl.), 15–23.

Chapman, L. J., & Chapman, J. P. (1973). Problems in the measurement of cognitive deficit. *Psychological Bulletin, 79,* 380–385.

Cleghorn, J. M., Zipursky, R. B., & List, S. J. (1991). Structural and functional brain imaging in schizophrenia. *Journal of Psychiatry Neuroscience, 16* (2), 54–74.

Devous, M. D. (1992). Comparison of SPECT applications in neurology and psychiatry. *Journal of Clinical Psychiatry, 53* (Suppl. 11), 29–34.

Drevets, W. C. Videen, T. O., MacLeod, A. K., Haller, J. W., & Raichle, M. E. (1992). PET images of blood flow changed during anxiety: Correction. *Science, 256,* 1696.

Erwin, R. J., Gur, R. C., Gur, R. E., Skolnick, B. E., Mawhinney-Hee, M., & Smailis, J. (1992). Facial emotion discrimination: I. Task construction and behavioral findings in normals. *Psychiatry Research, 42,* 231–240.

Fox, P. T., Mintun, M. A., Reiman, E. M., & Raichle, M. E. (1988). Enhanced detection of focal brain response using intersubject averaging and change-distribution analysis of subtracted PET images. *Journal of Cerebral Blood Flow and Metabolism, 8,* 642–653.

Friston, K. J., Frith, C. D., Liddle, P. F., & Frackowiak, R. S. (1991). Comparing functional (PET) images: The assessment of significant change. *Journal of Cerebral Blood Flow and Metabolism, 11,* 690–699.

Friston, K. J., Frith, C. D., Liddle, P. F., & Frackowiak, R. S. J. (1993). Functional connectivity: The principal component analysis of large (PET) data sets. *Journal of Cerebral Blood Flow and Metabolism, 13,* 5–14.

Gore, J. C., McCarthy, G., Constable, R. T., Anderson, A. W., Kennan, R. P., Rattner, Z., Zhong, J. (1992). Echo-planar imaging of the left inferior frontal lobe during word generation [Abstract], In Book of Abstracts, Eleventh annual meeting of the Society of Magnetic Resonance in Medicine, Berlin (p. 1834).

Gur, R. C., Erwin, R. J., & Gur, R. E. (1992). Neurobehavioral probes for physiologic neuroimaging studies. *Archives of General Psychiatry, 49,* 409–414.

Gur, R. C., Erwin, R. J., Gur, R. E., Zwil, A. S., Heimberg, C., & Kraemer, H. C. (1992). Facial emotion discrimination: II. Behavioral findings in depression. *Psychiatry Research, 42,* 241–251.

Gur, R. C., Gur, R. E., Obrist, W. D., Hungerbuhler, J. P., Younkin, D., Rosen, A. D., Skolnick, B. E., & Reivich, M. (1982). Sex and handedness differences in cerebral blood flow during rest and cognitive activity. *Science, 217,* 659–661.

Gur, R. C., Gur, R. E., Obrist, W. D., Skolnick, B. E., & Reivich, M. (1987). Age and regional cerebral blood flow at rest and during cognitive activity. *Archives of General Psychiatry, 44,* 617–621.

Gur, R. C., Gur, R. E., Resnick, S. M., Skolnick, B. E., Alavi, A., & Reivich, M. (1987). The effect of anxiety on cortical cerebral blood flow and metabolism. *Journal of Cerebral Blood Flow and Metabolism, 7,* 173–177.

Gur, R. C., Gur, R. E., Skolnick, B. E., Resnick, S. M., Silver, F. L., Chawluk, J. B., Muenz, L., Obrist, W. D., & Reivich, M. (1988). Effects of task difficulty on regional cerebral blood flow: Relationships with anxiety and performance. *Psychophysiology, 25,* 392–399.

Gur, R. C., Ragland, J. D., Resnick, S. M., Skolnick, B. E., Jaggi, J., Muenz, L., & Gur, R. E. (1994). Lateralized increases in cerebral blood flow during performance of verbal and spatial tasks: Relationship with performance level. *Brain and Cognition, 24,* 244–258.

Gur, R. C., Skolnick, B. E., & Gur, R. E. (1994). Effects of emotional discrimination tasks on cerebral blood flow: Regional activation and its relation to performance. *Brain and Cognition, 25,* 271–286.

Gur, R. E. (1983). Brain function in psychiatric disorders I. Regional cerebral blood flow in medicated schizophrenics. *Archives of General Psychiatry, 40,* 1253.

Gur, R. E., & Gur, R. C. (1992). Neurotransmitters are important, but so is metabolism. *Neuropsychopharmacology, 7,* 63–65.

Gur, R. E., & Pearlson, G. D. (1993). Neuroimaging in schizophrenia research. *Schizophrenia Bulletin, 19,* 337–353.

Gur, R. E., Resnick, S. M., & Gur, R. C. (1989). Laterality and frontality of cerebral blood flow and metabolism in schizophrenia: Relationship to symptom specificity. *Psychiatry Research, 27,* 325–334.

Gur, R. E., Skolnick, B. E., Gur, R. C., Caroff, S., Rieger, W., Obrist, W. D., Younkin, D., & Reivich, M. (1983). Brain function in psychiatric disorders: I. Regional cerebral blood flow in medicated schizophrenics. *Archives of General Psychiatry, 40,* 1250–1254.

Hall, H., Farde, L., Halldin, C., Hogberg, T., Larsson, S., & Sedvall, G. (1992). Imaging of dopamine receptors using PET and SPECT. *Neurochemistry International, 20* (Suppl.), 329s–333s.

Heimberg, C., Gur, R. E., Erwin, R. J., Shtasel, D. L., & Gur, R. C. (1992). Facial emotion discrimination: III. Behavioral findings in schizophrenia. *Psychiatry Research, 42,* 253–265.

Ingvar, D. H. (1979). "Hyperfrontal" distribution of the cerebral grey matter flow in resting wakefulness: On the functional anatomy of the conscious state. *Acta Neurologica Scandinavica, 60,* 12–25.

Innis, R. B. (1992). Neuroreceptor imaging with SPECT. *Journal of Clinical Psychiatry, 53* (Suppl. 11), 29–34.

Kety, S. S., & Schmidt, C. F. (1948). The nitrous oxide method for the quantitative determination of cerebral blood flow in man: Theory, procedure and normal values. *Journal of Clinical Investigation, 27,* 476–483.

Kohn, M. I., Tanna, N. K., Herman, G. T., Resnick, S. M., Mozley, P. D., Gur, R. E., Alavi, A., Zimmerman, R. A., & Gur, R. C. (1991). Analysis of brain and CSF volumes from magnetic resonance imaging: Methods, reliability, and validation. *Radiology, 178,* 115–122.

Mazziotta, J. C., Pelizzari, C. C., Chen, A. T., Bookstein, F. L., & Valentino, D. (1991). Region of interest issues: The relationship between structure and function in the brain. *Journal of Cerebral Blood Flow and Metabolism, 11,* 451–456.

Miller, B. L., Cummings, J. L., Villanueva-Meyer, J., Boone, K., Mehringer, C. M., Lesser, I. M., & Mena, I. (1991). *Neurology, 41,* 1374–1382.

Obrist, W. D., Thompson, H. K., Wang, H. S., & Wilkinson, W. E. (1975). Regional cerebral blood flow estimated by 133-Xenon inhalation. *Stroke, 6,* 245–256.

Olendorf, W., & Olendorf, W. (1989). *Basics of magnetic resonance imaging.* Boston: Martinus Nijhoff Publishing.

Olesen, J., Paulson, O. B., & Lassen, N. A. (1971). Regional cerebral blood flow in man determined by the initial slope of the clearance of intra-arterially injected 133 Xenon. *Stroke, 2,* 519–540.

Pfefferbaum, A., Lim, K. O., Rosenbloom, M., & Zipursky, R. B. (1990). Brain magnetic resonance imaging: Approaches for investigation of schizophrenia. *Schizophrenia Bulletin, 16* (3).

Posner, M. I., Petersen, S. E., Fox, P. T., & Raichle, M. E. (1988). Localization of cognitive operations in the human brain. *Science, 240,* 1627–1631.

Reivich, M., Kuhl, D., Wolf, A. P., Greenberg, J., Phelps, M., Ido, T., Casella, V., Fowler, J., Hoffman, E., Alavi, A., Som, P., & Sokoloff, L. (1979). The 18-F fluorodeoxyglucose method for the measurement of local cerebral glucose utilization in man. *Circulation Research, 44,* 127–137.

Schneider, F., Gur, R. E., Mozley, L. H., Smith, R. J., Mozley, P. D., Censits, D. M., Alavi, A., & Gur, R. C. (1995). Mood effects on cerebral blood flow correlate with emotional self-rating. A PET $H_2{}^{15}O$ study. *Psychiatry Research, 61,* 265–283.

Schneider, F., Gur, R. C., Gur, R. E., & Muenz, L. (1994). Standardized mood induction with happy and sad facial expressions. *Psychiatry Research, 51,* 19–31.

Schneider, F., Gur, R. C., Jaggi, J. L., & Gur, R. E. (1994). Differential effects of mood on cortical cerebral blood flow: A $^{133}$Xenon clearance study. *Psychiatry Research, 52,* 215–236.

Stehling, M. K., Turner, R., & Mansfield, P. (1991). Echo-planar imaging: Magnetic resonance in a fraction of second. *Science, 254* (5028), 43–50.

Van Heertum, R. L. (1992). Brain SPECT imaging and psychiatry. *Journal of Clinical Psychiatry, 53* (Suppl. 11), 7–12.

Volkow, N. D., Brodie, J., & Bendriem, B. (1991). Positron emission tomography: Basic principles and applications in psychiatric research. *Annals of the New York Academy of Science, 620,* 128–144.

Warach, S., Gur, R. C., Gur, R. E., Skolnick, B. E., Obrist, W. D., Hurtig, H. I., Goldberg, H. I., Younkin, D., & Reivich, M. (1982). Sex and handedness differences in interhemispheric correlations of human regional cerebral blood flow. *Society for Neuroscience Abstracts, 8,* 917.

Warach, S., Gur, R. C., Gur, R. E., Skolnick, B. E., Obrist, W. D., & Reivich, M. (1992). Decreases in frontal and parietal lobe regional cerebral blood flow related to habituation. *Journal of Cerebral Blood Flow and Metabolism, 12,* 546–553.

Weinberger, D. R., Berman, K. F., & Zec, R. F. (1986). Physiologic dysfunction of dorsolateral prefrontal cortex in schizophrenia: I. Regional cerebral blood flow evidence. *Archives of General Psychiatry, 43,* 114–125.

Wood, M. L., & Henkelman, R. M. (1985). Motion artifacts in MRI. *Medical Physics, 12,* 143.

III

# Appendix
## MRI Brain Atlas

Marc Steed and Erin D. Bigler

---

Marc Steed and Erin D. Bigler Department of Psychology, Brigham Young University, Provo, Utah 84602-5543.

# Axial Images

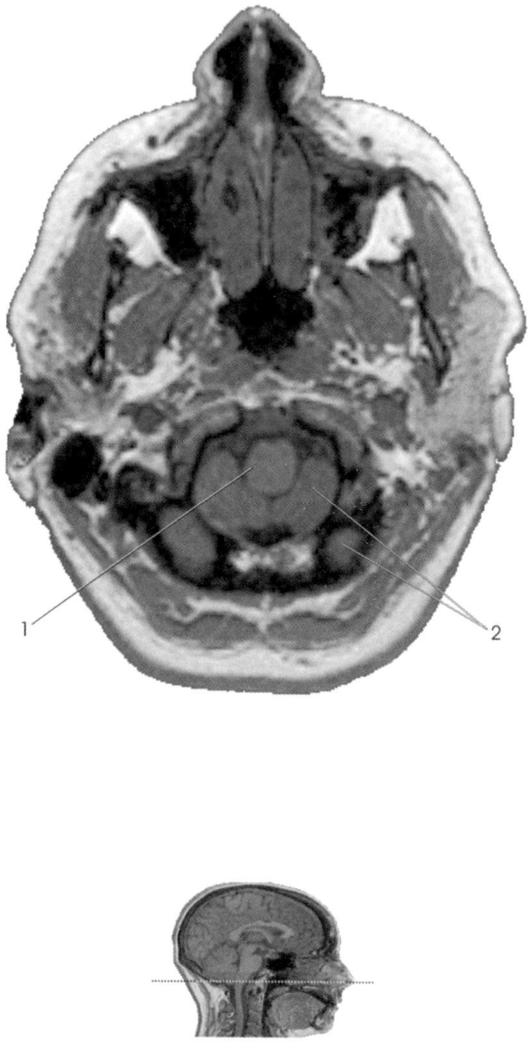


Image 1. (1) Medulla; (2) cerebellum.

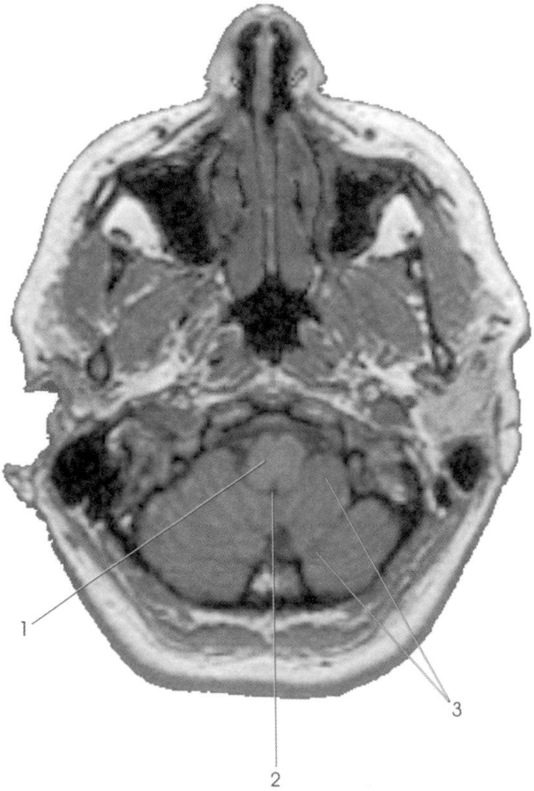


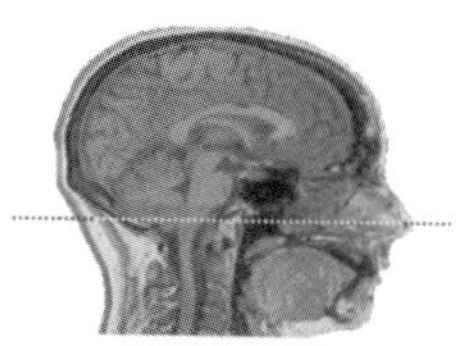

Image 2. (1) Medulla; (2) fourth ventricle; (3) cerebellum.

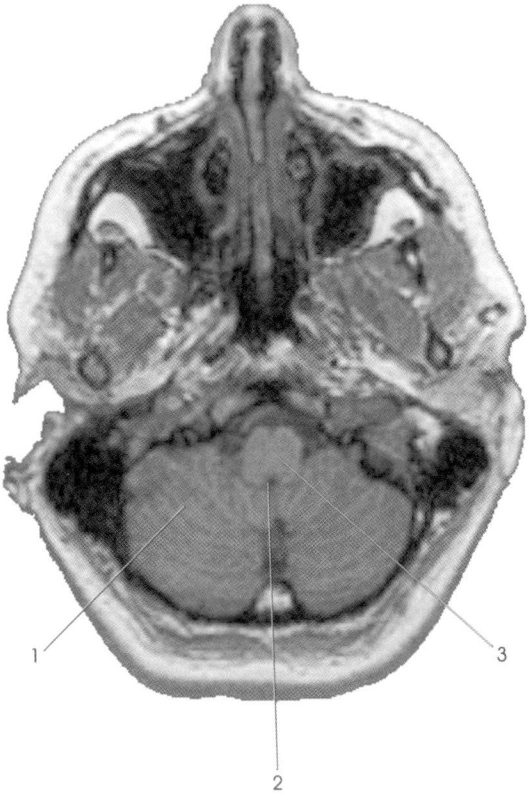


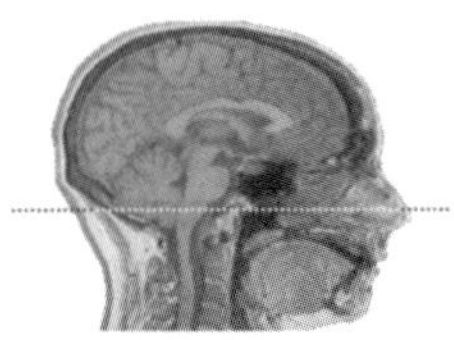

Image 3. (1) Cerebellum; (2) fourth ventricle; (3) medulla.

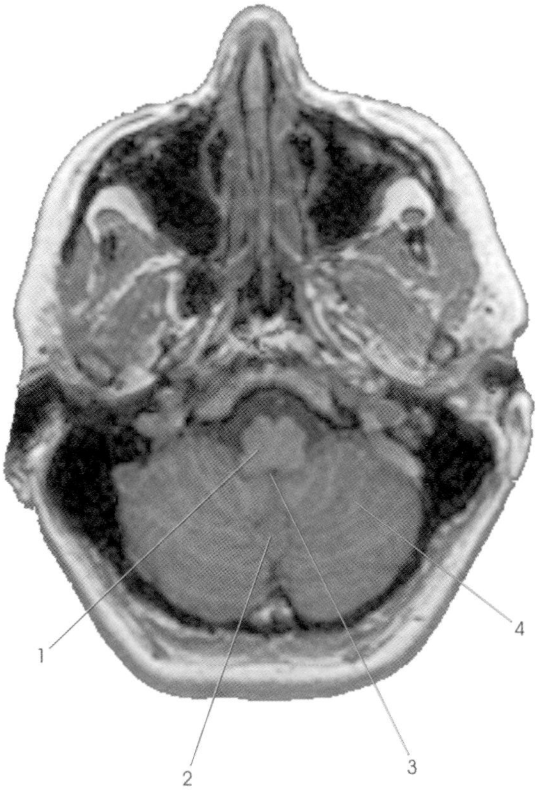


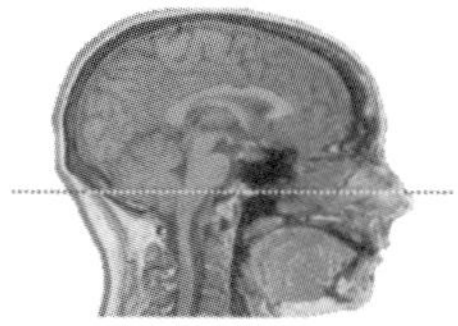

Image 4. (1) Medulla; (2) pyramis; (3) fourth ventricle; (4) cerebellum.

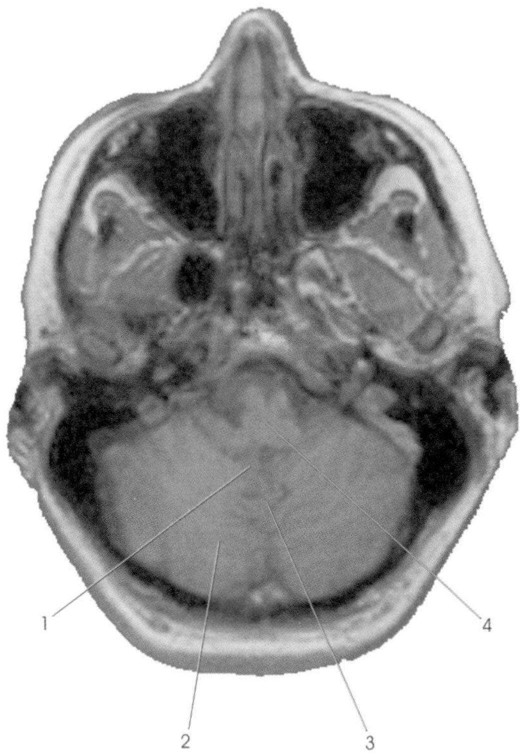


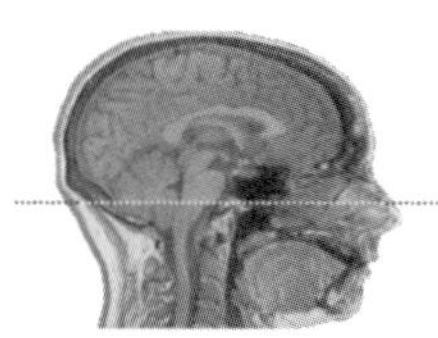

Image 5. (1) Nodulus; (2) cerebellum; (3) uvula; (4) pons.

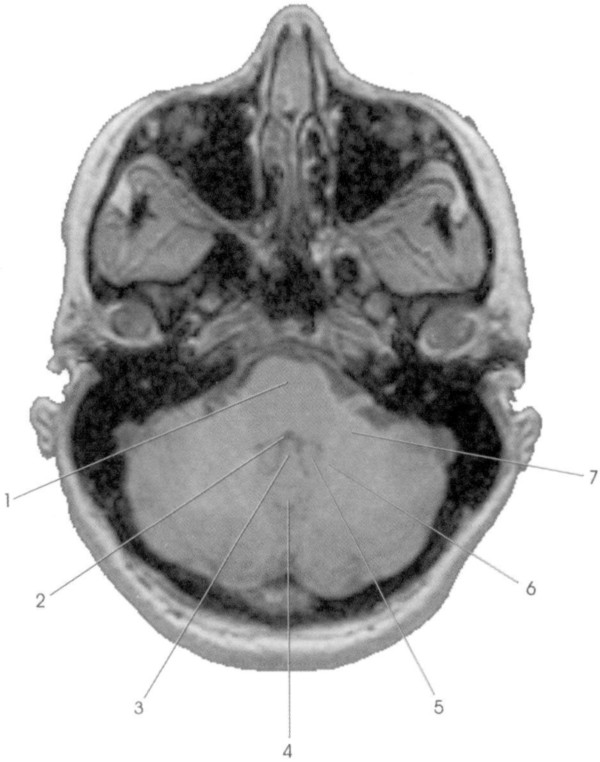


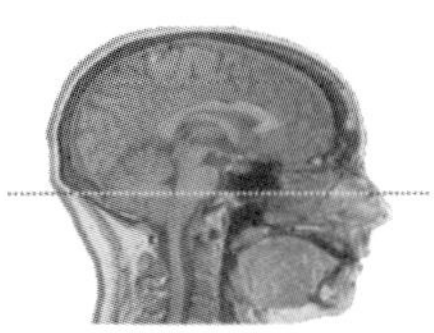

Image 6. (1) Pons; (2) fourth ventricle; (3) nodulus; (4) uvula; (5) cerebellar tonsil; (6) dentate nucleus; (7) middle cerebellar peduncle.

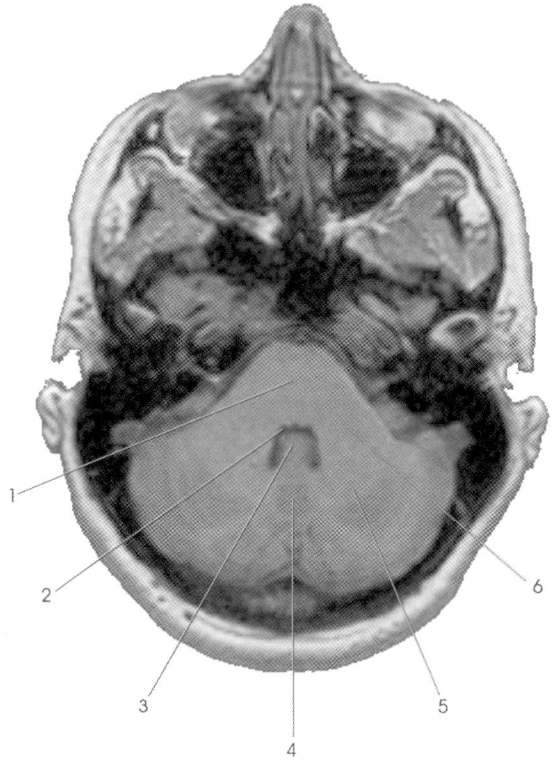


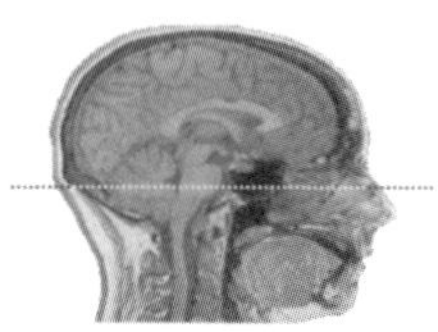

Image 7. (1) Pons; (2) fourth ventricle; (3) nodulus; (4) uvula; (5) cerebellum; (6) middle cerebellar peduncle.

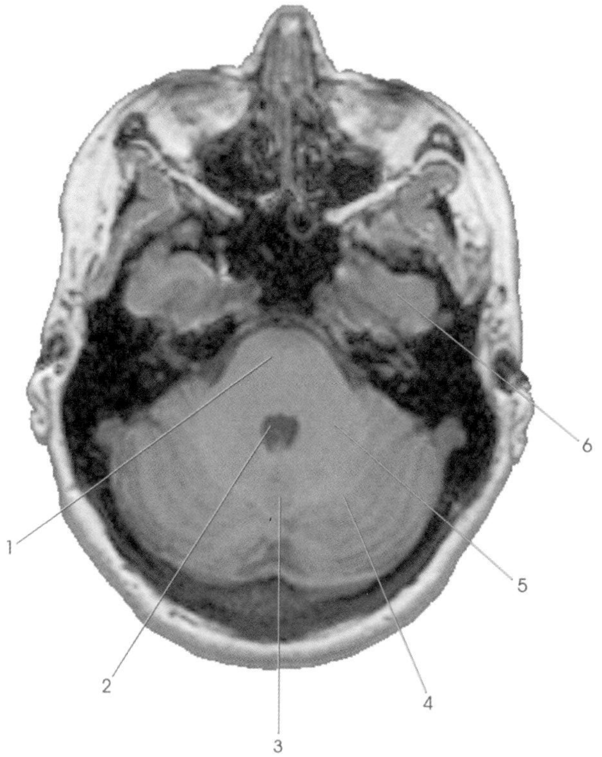


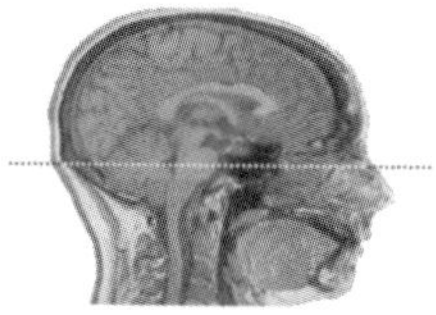

Image 8. (1) Pons; (2) fourth ventricle; (3) culmen; (4) cerebellum; (5) middle cerebellar peduncle; (6) temporal lobe.

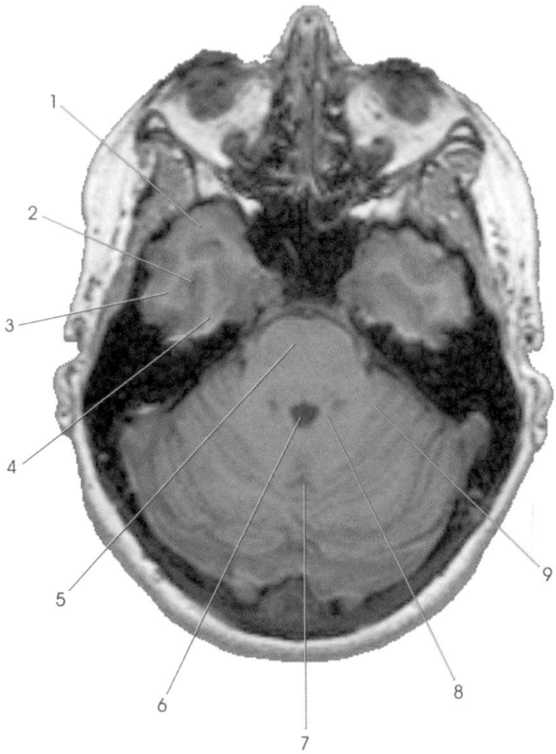


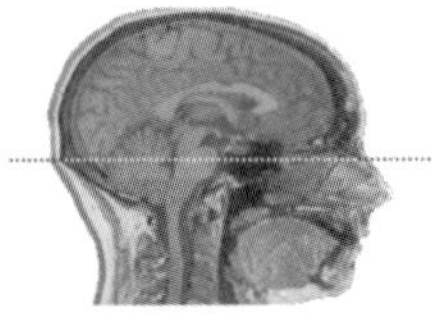

Image 9. (1) Middle temporal gyrus (T2); (2) occipitotemporal sulcus; (3) inferior temporal gyrus (T3); (4) temporal pole; (5) pons; (6) fourth ventricle; (7) culmen; (8) superior cerebellar peduncle; (9) middle cerebellar peduncle.

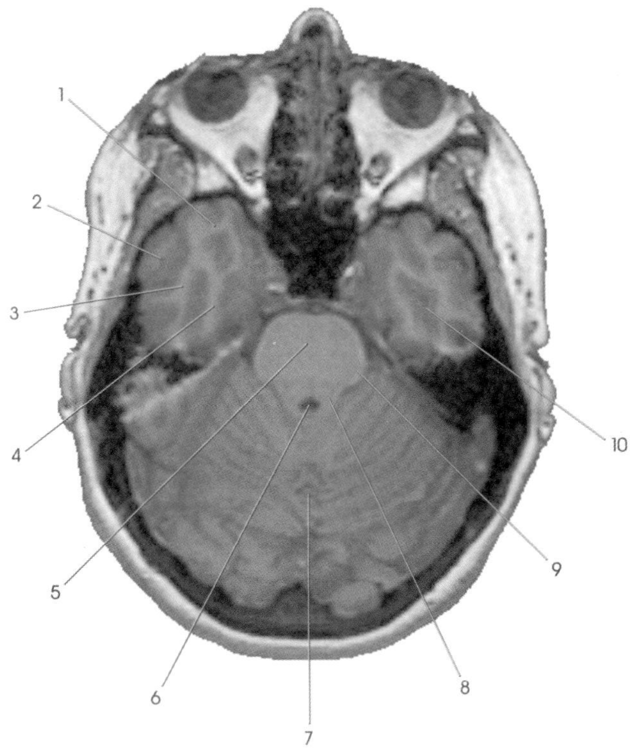


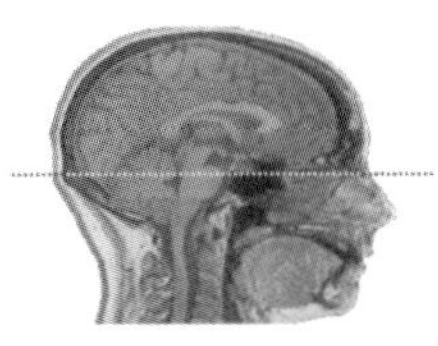

Image 10. (1) Middle temporal gyrus (T2); (2) occipitotemporal sulcus; (3) inferior temporal gyrus (T3); (4) temporal pole; (5) pons; (6) fourth ventricle; (7) culmen; (8) superior cerebellar peduncle; (9) brachium pontis; (10) occipitotemporal sulcus.

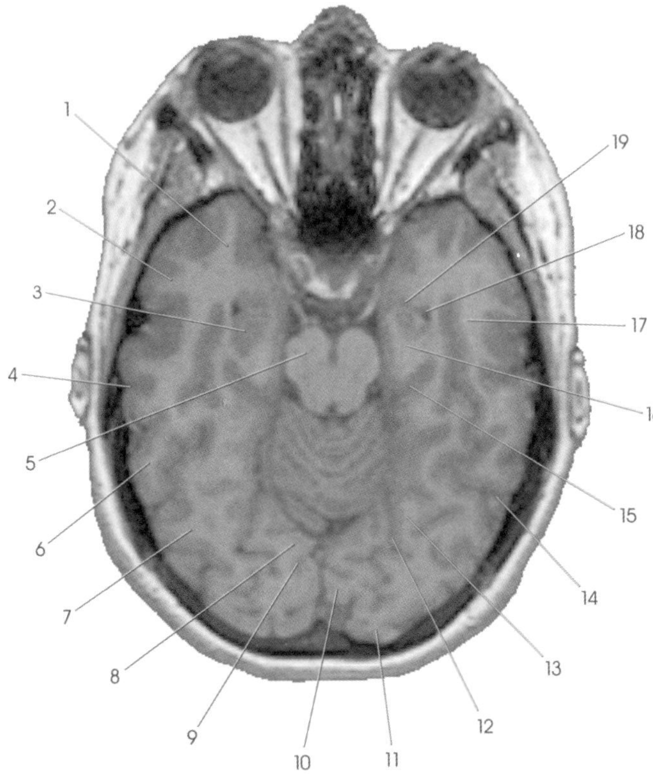


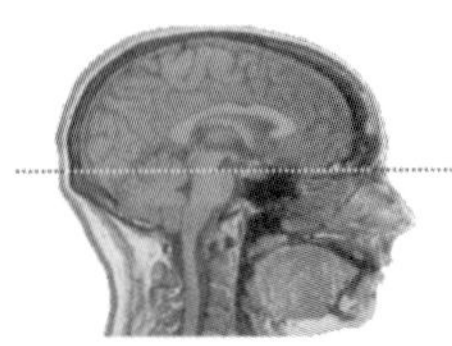

Image 11. (1) Superior temporal gyrus (T1); (2) middle temporal gyrus (T2); (3) head of hippocampus; (4) middle temporal gyrus (T2); (5) cerebral peduncle; (6) inferior temporal gyrus (T3); (7) inferior temporal gyrus (T3); (8) inferior lingual gyrus; (9) lingual sulcus; (10) superior lingual gyrus; (11) occipital pole; (12) collateral sulcus; (13) fusiform gyrus; (14) occipitotemporal sulcus; (15) collateral sulcus; (16) parahippocampal gyrus; (17) middle temporal gyrus (T2); (18) temporal horn of lateral ventricle; (19) amygdala.

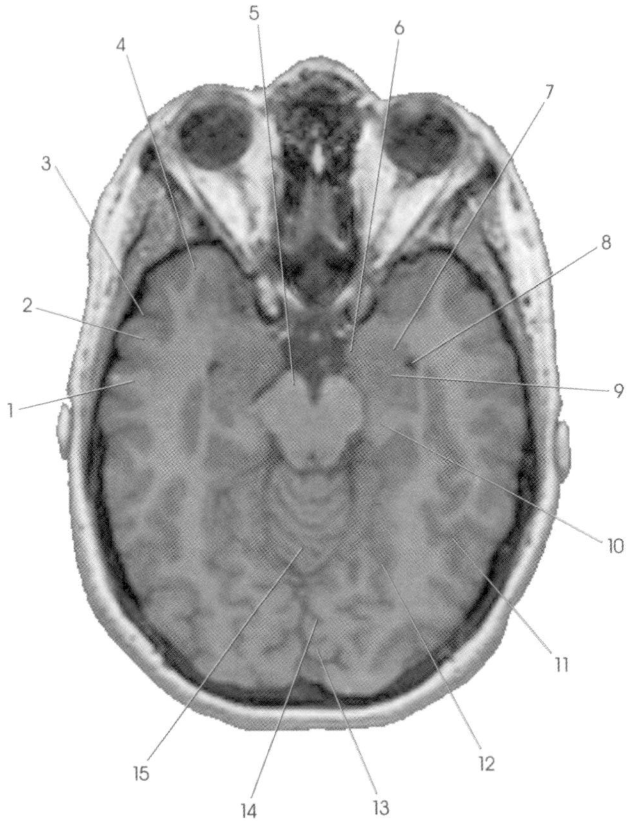


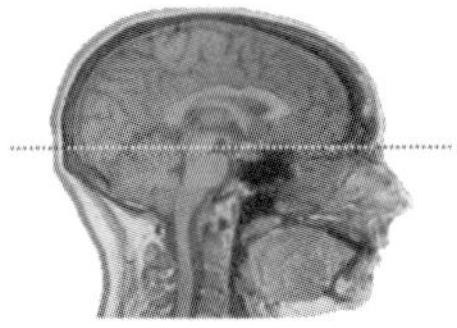

Image 12. (1) Inferior temporal gyrus (T3); (2) middle temporal gyrus (T2); (3) parallel sulcus; (4) superior temporal gyrus (T1); (5) cerebral peduncle; (6) uncus; (7) amygdala; (8) temporal horn of lateral ventricle; (9) hippocampus; (10) parahippocampal gyrus; (11) occipitotemporal sulcus; (12) collateral sulcus; (13) lingual gyrus; (14) lingual sulcus; (15) vermis of cerebellum.

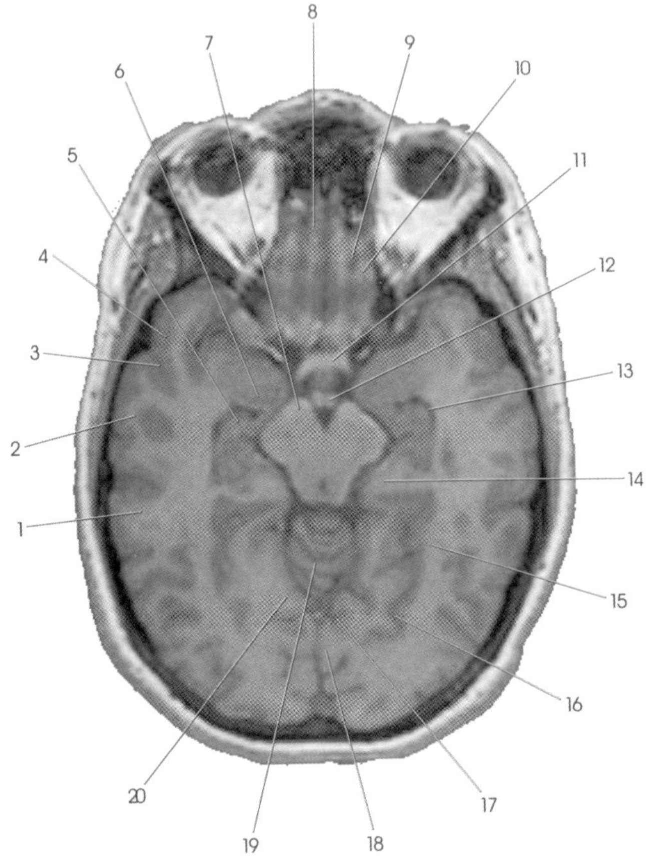


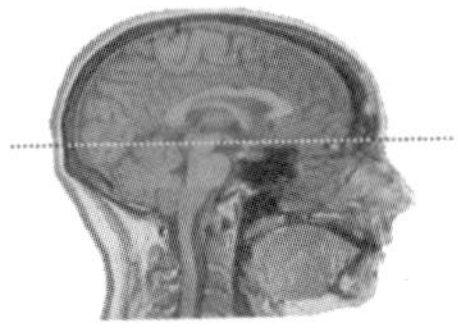

Image 13. (1) Inferior temporal gyrus (T3); (2) middle temporal gyrus (T2); (3) parallel sulcus; (4) superior temporal gyrus (T1); (5) hippocampus; (6) amygdala; (7) cerebral peduncle; (8) gyrus rectus; (9) olfactory sulcus; (10) medial orbital gyrus; (11) optic chiasm; (12) mamillary body; (13) temporal horn of lateral ventricle; (14) parahippocampal gyrus; (15) fusiform gyrus; (16) collateral sulcus; (17) ambient cistern; (18) lingual gyrus; (19) vermis of cerebellum; (20) inferior colliculus.

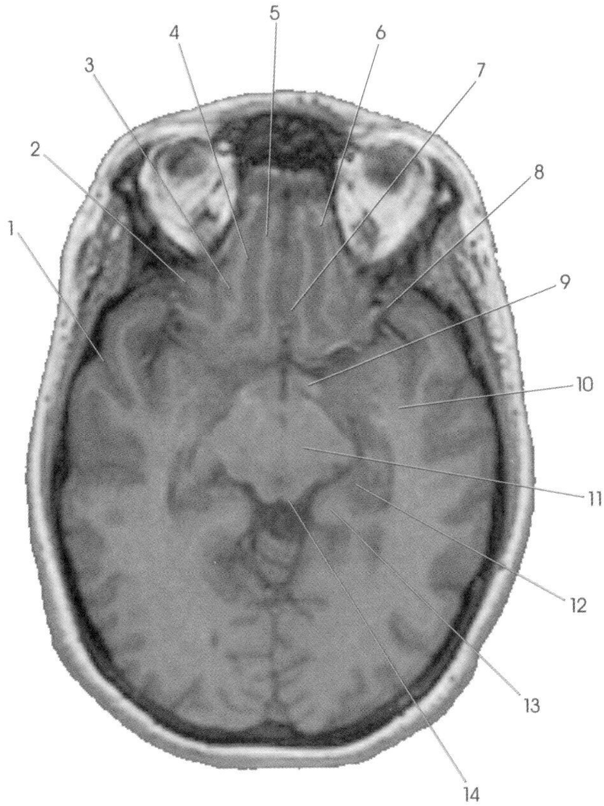


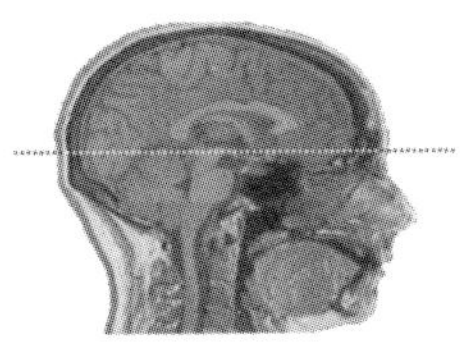

Image 14. (1) Parallel sulcus; (2) posterior orbital gyrus; (3) posterior orbital gyrus; (4) olfactory sulcus; (5) frontomarginal gyrus; (6) olfactory sulcus; (7) gyrus rectus; (8) basal part of lateral fissure; (9) optic tract; (10) temporal stem; (11) red nucleus; (12) hippocampus; (13) parahippocampal gyrus; (14) superior colliculus.

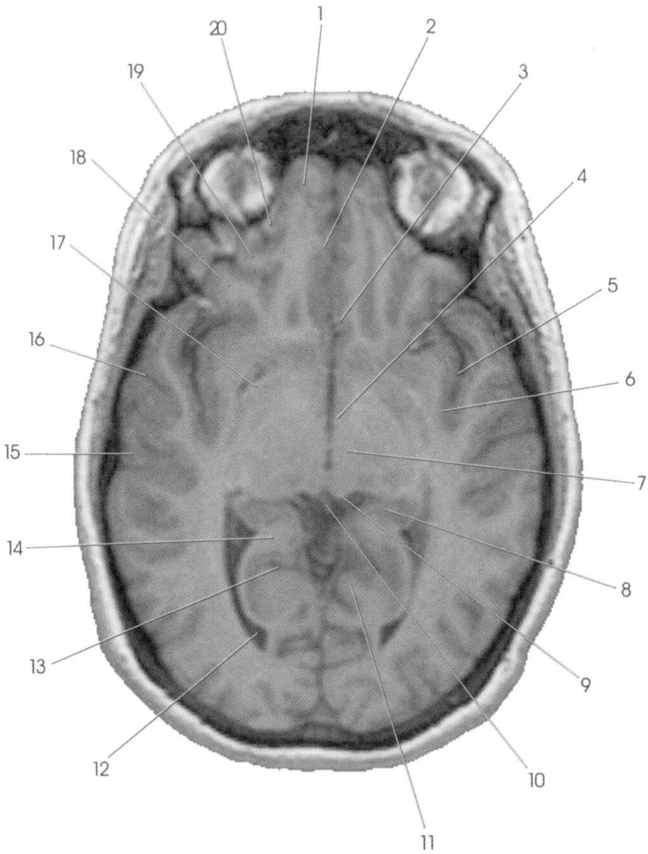


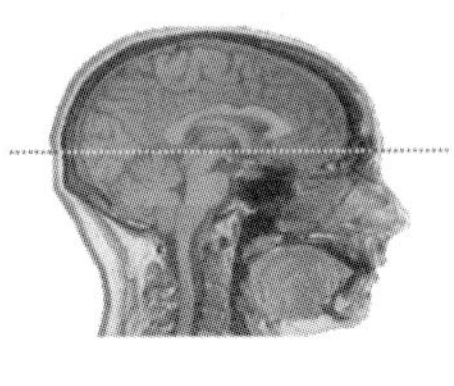

Image 15. (1) Frontomarginal gyrus; (2) gyrus rectus; (3) subcallosal gyrus; (4) third ventricle; (5) lateral fissure; (6) insular cortex; (7) red nucleus; (8) hippocampus; (9) superior colliculus; (10) pineal body; (11) lingual gyrus; (12) occipital horn of lateral ventricle; (13) calcarine sulcus; (14) isthmus of cingulum; (15) middle temporal gyrus (T2); (16) parallel sulcus; (17) anterior commisure; (18) lateral orbital gyrus; (19) middle orbital gyrus; (20) medial orbital gyrus.

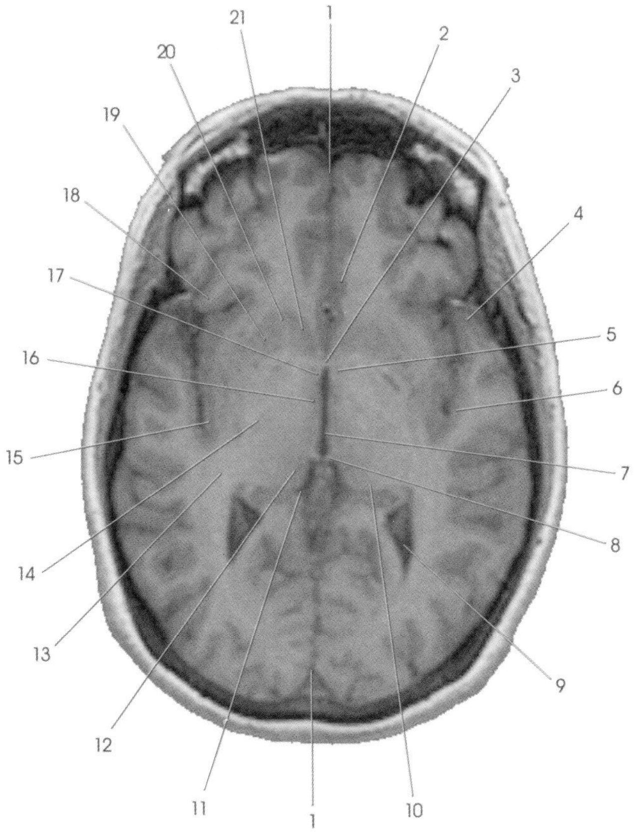


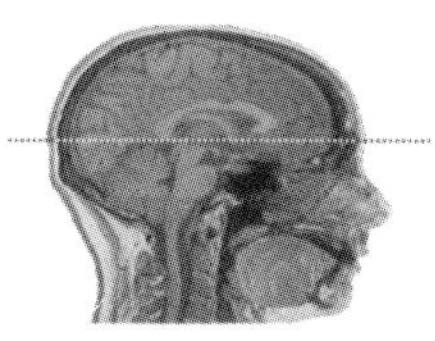

Image 16. (1) Interhemispheric fissure; (2) cingulate cortex; (3) anterior commisure; (4) temporal pole; (5) lateral preoptic nucleus; (6) circular sulcus of insular cortex; (7) third ventricle; (8) stria medullaris of thalamus; (9) atria of lateral ventricle; (10) lateral pulvinar nucleus of thalamus; (11) cistern of velum interpositum; (12) medial pulvinar nucleus of thalamus; (13) temporal stem; (14) ventrolateral nucleus of thalamus; (15) insular cortex; (16) dorsal medial nucleus of thalamus; (17) column of fornix; (18) inferior frontal gyrus (F3), pars orbitalis; (19) putamen; (20) anterior limb of internal capsule; (21) head of caudate nucleus.

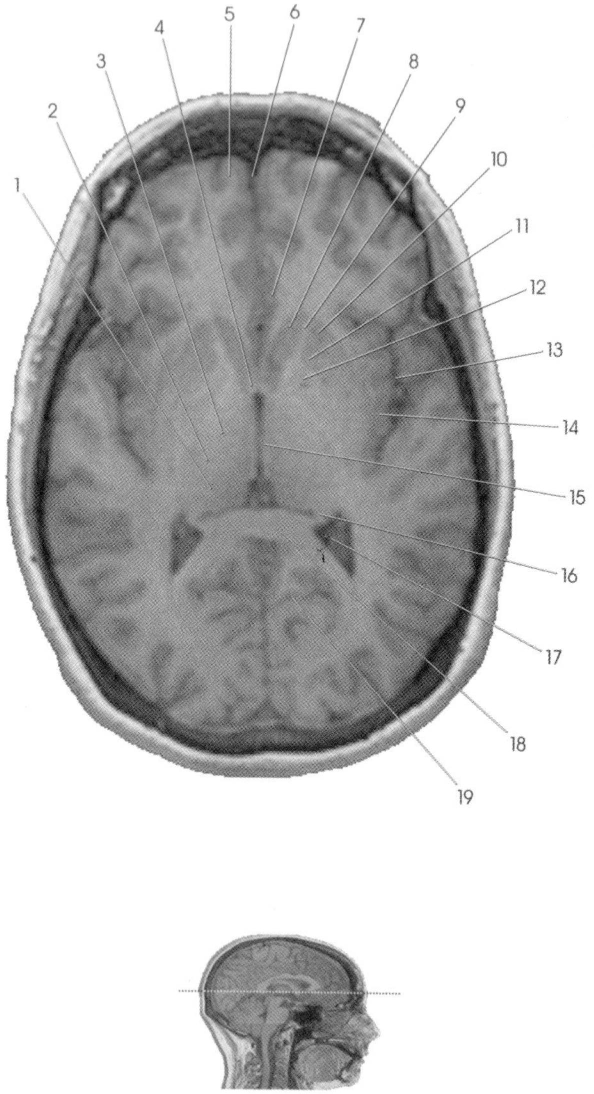


Image 17. (1) Thalamus; (2) lateral posterior thalamic nucleus; (3) ventral lateral thalamic nucleus; (4) fornix; (5) frontopolar gyrus; (6) interhemispheric fissure; (7) cingulate gyrus; (8) head of caudate nucleus; (9) anterior limb of internal capsule; (10) putamen; (11) pars lateralis of globus pallidus; (12) pars medialis of globus pallidus; (13) lateral fissure; (14) insular cortex; (15) third ventricle; (16) tail of hippocampus; (17) choroid plexus of lateral ventricle; (18) splenium of corpus callosum; (19) parieto-occipital sulcus.

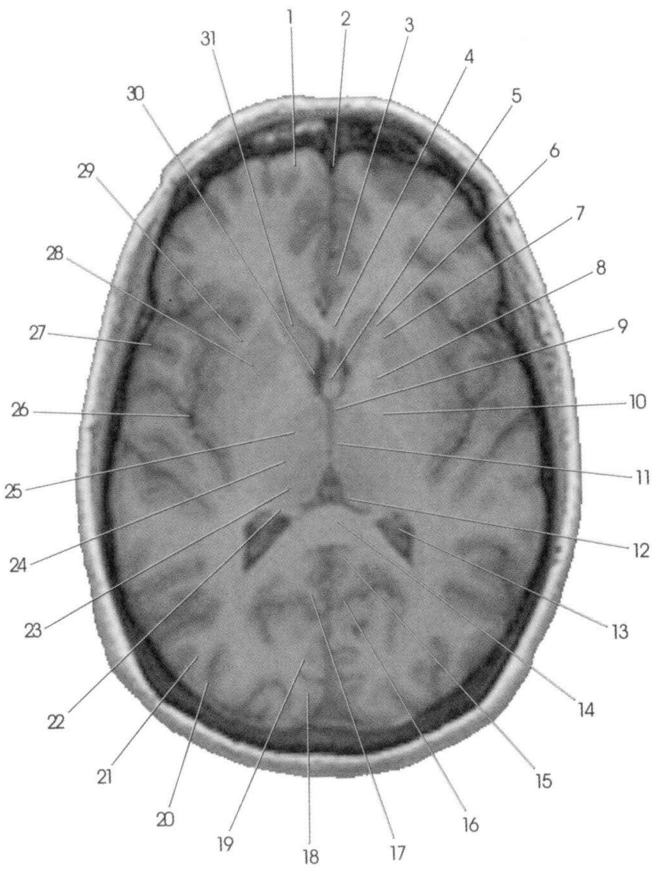


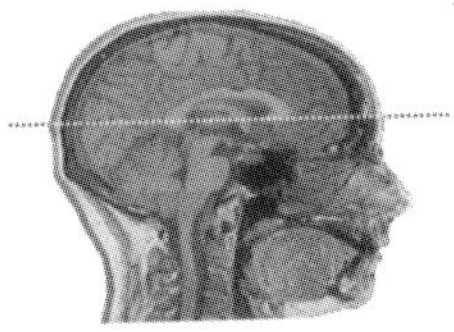

Image 18. (1) Superior frontal gyrus (F1); (2) interhemispheric fissure; (3) cingulate gyrus; (4) genu and rostrum of corpus callosum; (5) fornix; (6) anterior limb of internal capsule; (7) putamen; (8) genu of internal capsule; (9) anterior thalamic nucleus; (10) posterior limb of internal capsule; (11) dorsomedial thalamic nucleus; (12) stria medullaris; (13) occipital horn of lateral ventricle; (14) splenium of corpus callosum; (15) cingulate gyrus; (16) parieto-occipital sulcus; (17) cingulate gyrus; (18) superior occipital gyrus (O1); (19) cuneus; (20) transverse occipital sulcus; (21) middle occipital gyrus (O2); (22) crus of fornix; (23) thalamus; (24) lateral posterior thalamic nucleus; (25) ventral lateral thalamic nucleus; (26) lateral fissure; (27) inferior precentral sulcus; (28) putamen; (29) external capsule; (30) frontal horn of lateral ventricle; (31) head of caudate nucleus.

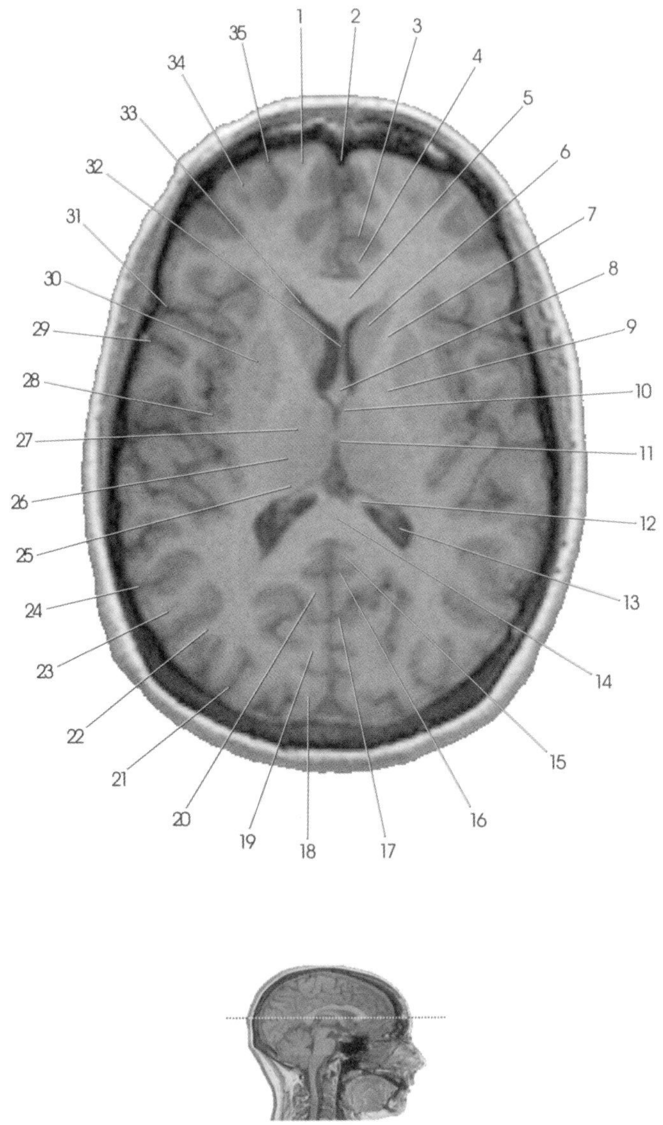


Image 19. (1) Superior frontal gyrus (F1); (2) interhemispheric fissure; (3) cingulate sulcus; (4) cingulate gyrus; (5) genu of corpus callosum; (6) head of caudate nucleus; (7) anterior limb of internal capsule; (8) fornix; (9) posterior limb of internal capsule; (10) foramen of munro; (11) dorsomedial thalamic nucleus; (12) fornix; (13) atria of lateral ventricle; (14) splenium of corpus callosum; (15) cingulate gyrus; (16) cingulate sulcus; (17) parieto-occipital sulcus; (18) superior occipital gyrus (01); (19) cuneus; (20) cingulate gyrus; (21) transverse occipital sulcus; (22) middle occipital gyrus (O2); (23) angular gyrus; (24) parallel sulcus; (25) pulvinar; (26) lateral posterior thalamic nucleus; (27) ventral lateral thalamic nucleus; (28) insular cortex; (29) inferior precentral sulcus; (30) putamen; (31) horizontal ramus of lateral fissure; (32) septum pellucidum; (33) frontal horn of lateral ventricle; (34) middle frontal gyrus (F2); (35) superior frontal sulcus.

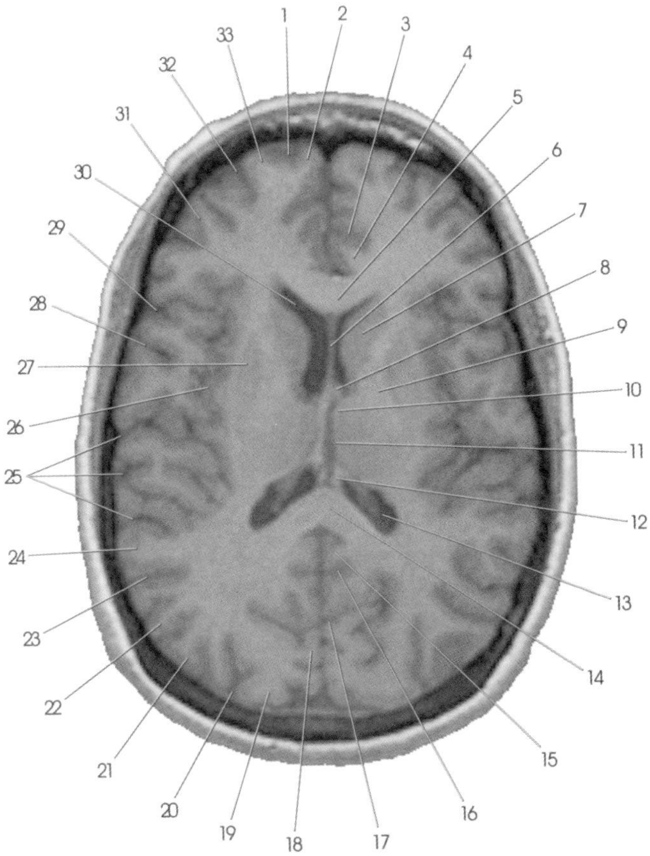


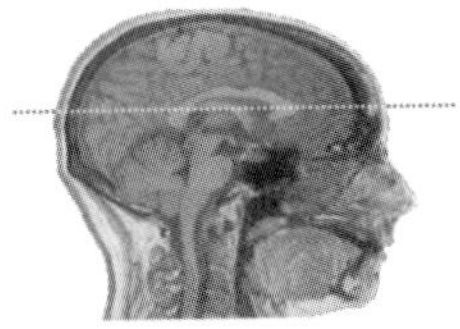

Image 20. (1) Superior frontal sulcus; (2) superior frontal gyrus (F1); (3) cingulate sulcus; (4) cingulate gyrus; (5) corpus callosum; (6) septum pellucidum; (7) head of caudate nucleus; (8) fornix; (9) genu of internal capsule; (10) anterior thalamic nucleus; (11) dorsomedial thalamic nucleus; (12) fornix; (13) atria of lateral ventricle; (14) corpus callosum; (15) cingulate gyrus; (16) cingulate sulcus; (17) parieto-occipital sulcus; (18) cuneus; (19) superior occipital gyrus (O1); (20) intra-occipital sulcus; (21) middle occipital gyrus (O2); (22) angular gyrus; (23) parallel sulcus; (24) supramarginal gyrus; (25) discontinuous lateral fissure; (26) insular cortex; (27) putamen; (28) inferior precentral sulcus; (29) horizontal ramus of lateral fissure; (30) frontal horn of lateral ventricle; (31) middle frontal gyrus (F2); (32) superior frontal sulcus; (33) superior frontal gyrus (F1).

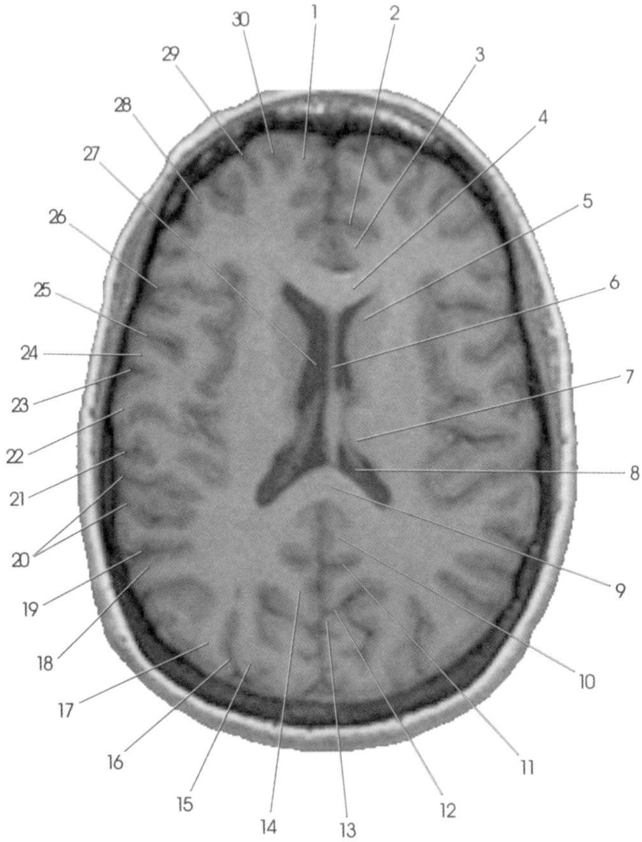


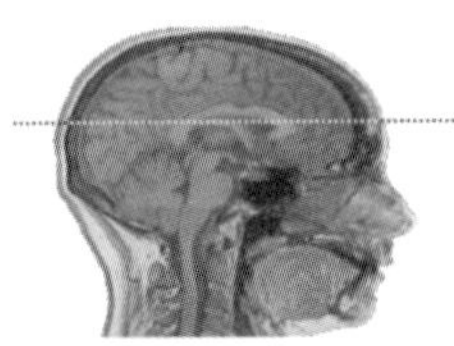

Image 21. (1) Superior frontal gyrus (F1); (2) cingulate sulcus; (3) cingulate gyrus; (4) genu of corpus callosum; (5) head of caudate nucleus; (6) septum pellucidum; (7) lateral dorsal thalamic nucleus; (8) atria of lateral ventricle; (9) splenium of corpus callosum; (10) cingulate gyrus; (11) cingulate sulcus; (12) parieto-occipital sulcus; (13) cuneus; (14) precuneus; (15) middle occipital gyrus (O2); (16) intra-occipital sulcus; (17) angular gyrus; (18) supramarginal gyrus; (19) parallel sulcus; (20) discontinuous lateral fissure; (21) postcentral sulcus; (22) postcentral gyrus; (23) central sulcus; (24) precentral gyrus; (25) inferior precentral sulcus; (26) horizontal ramus of lateral fissure; (27) body of lateral ventricle; (28) middle frontal gyrus (F2); (29) superior frontal sulcus; (30) superior frontal gyrus (F1).

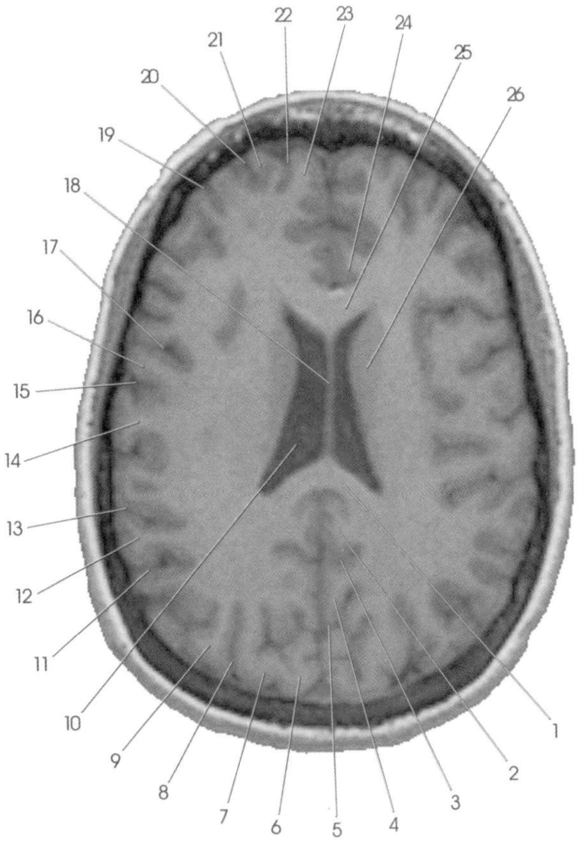


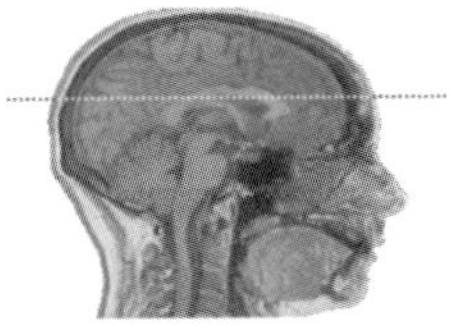

Image 22. (1) Corpus callosum; (2) cingulate gyrus; (3) cingulate sulcus; (4) precuneus; (5) parieto-occipital sulcus; (6) cuneus; (7) middle occipital gyrus (O2); (8) intra-occipital sulcus; (9) middle occipital gyrus (O2); (10) body of lateral ventricle; (11) parallel sulcus; (12) supramarginal gyrus; (13) lateral fissure; (14) postcentral gyrus; (15) central sulcus; (16) precentral gyrus; (17) inferior precentral sulcus; (18) septum pellucidum; (19) inferior frontal sulcus; (20) inferior frontal sulcus; (21) middle frontal gyrus (F2); (22) superior frontal sulcus; (23) superior frontal gyrus (F1); (24) cingulate gyrus; (25) corpus callosum; (26) caudate nucleus.

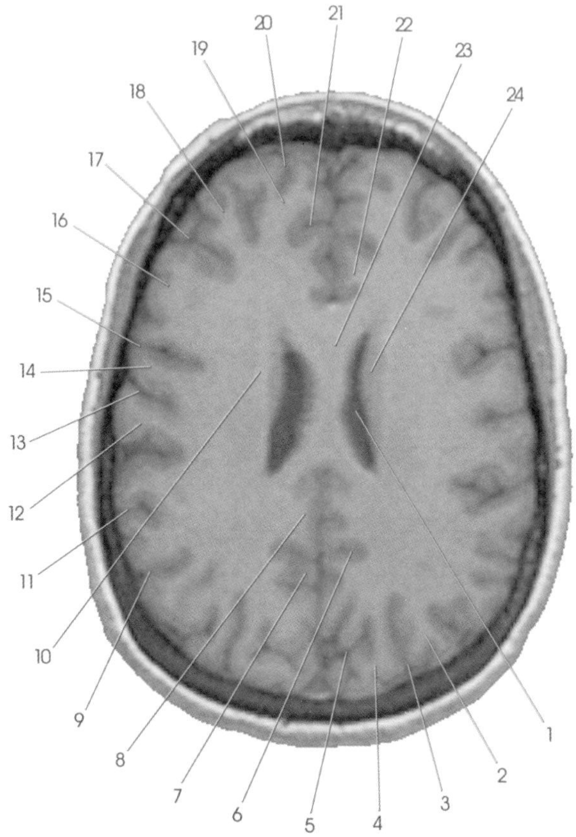


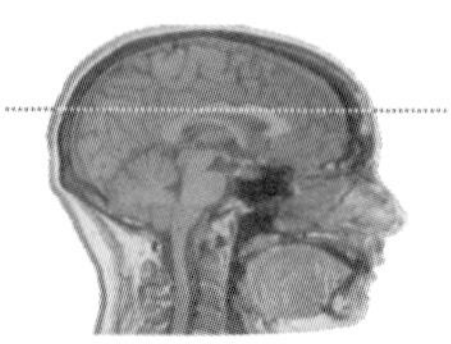

Image 23. (1) Body of lateral ventricle; (2) second parieto-occipital "pli de plassage"; (3) intra-occipital sulcus; (4) first parieto-occipital "pli de plassage"; (5) parieto-occipital sulcus; (6) cingulate sulcus; (7) precuneus; (8) cingulate gyrus; (9) parallel sulcus; (10) corona radiata; (11) lateral fissure; (12) postcentral gyrus; (13) central sulcus; (14) precentral gyrus; (15) inferior precentral sulcus; (16) inferior frontal gyrus (F3); (17) inferior frontal sulcus; (18) middle frontal gyrus (F2); (19) superior frontal gyrus (F1); (20) superior frontal sulcus; (21) cingulate sulcus; (22) cingulate gyrus; (23) corpus callosum; (24) caudate nucleus.

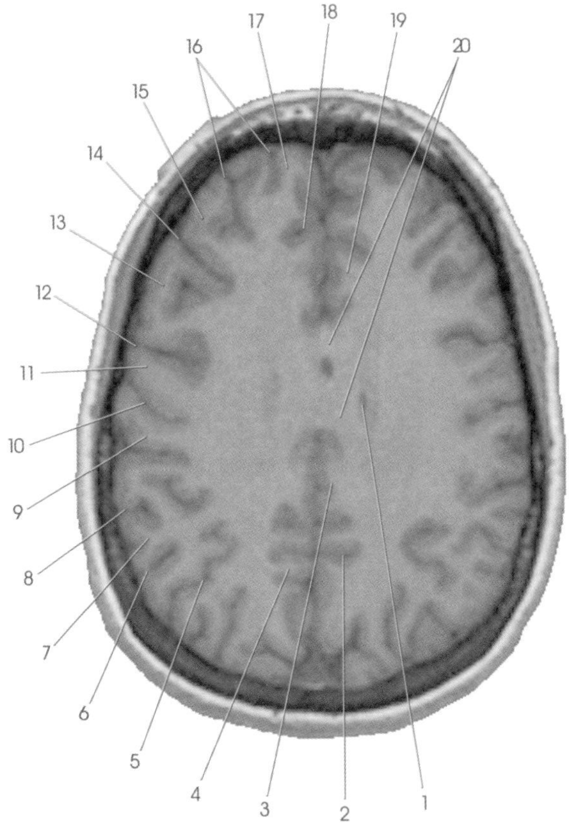


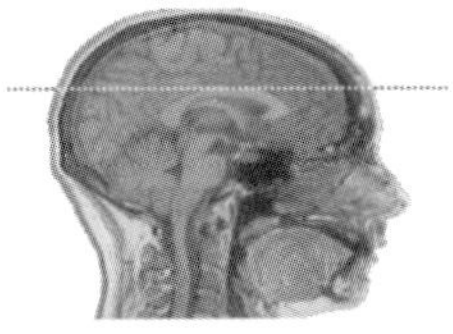

Image 24. (1) Body of lateral ventricle; (2) cingulate sulcus; (3) cingulate gyrus; (4) parietal lobule; (5) intraparietal sulcus; (6) parallel sulcus; (7) supramarginal gyrus; (8) lateral fissure; (9) postcentral gyrus; (10) central sulcus; (11) precentral gyrus; (12) inferior precentral sulcus; (13) inferior frontal gyrus (F3); (14) inferior frontal sulcus; (15) middle frontal gyrus (F2); (16) superior frontal sulcus; (17) superior frontal gyrus (F1); (18) cingulate sulcus; (19) cingulate gyrus; (20) corpus callosum.

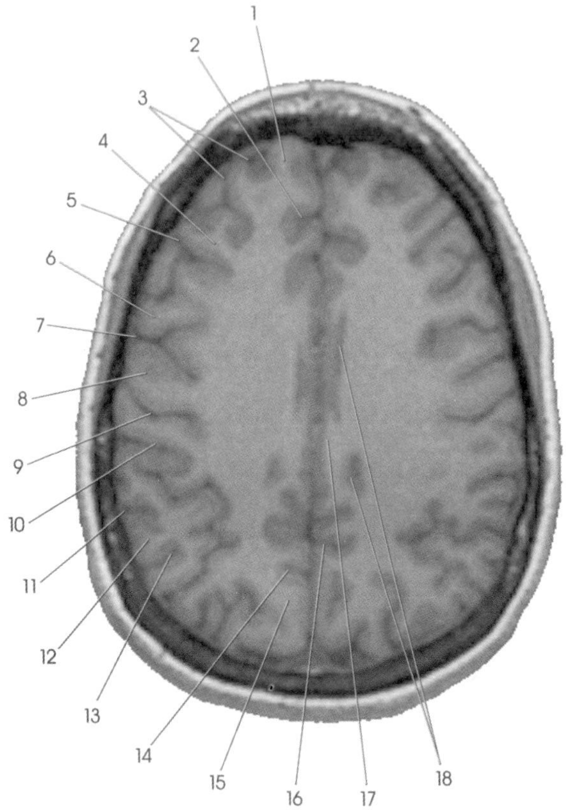


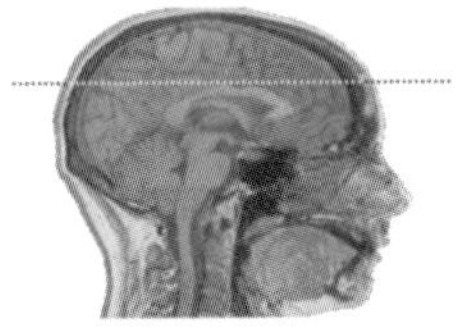

Image 25. (1) Superior frontal gyrus (F1); (2) cingulate sulcus; (3) superior frontal sulcus; (4) middle frontal gyrus (F2); (5) inferior frontal sulcus; (6) inferior frontal gyrus (F3); (7) inferior precentral sulcus; (8) precentral gyrus; (9) central sulcus; (10) postcentral gyrus; (11) lateral fissure; (12) supramarginal gyrus; (13) parallel sulcus; (14) parietal lobule; (15) cuneus; (16) cingulate sulcus; (17) cingulate gyrus; (18) cingulate sulcus.

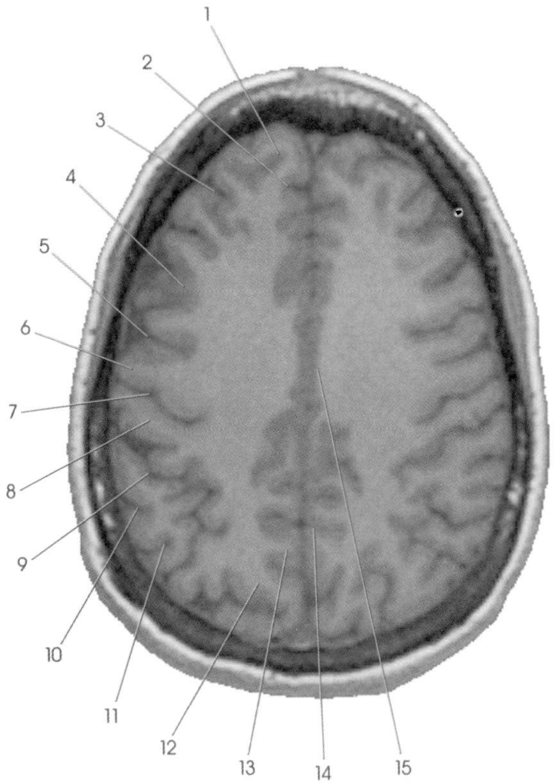


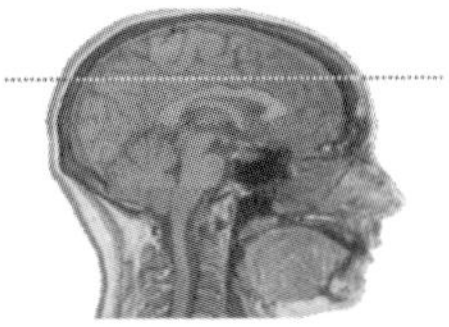

Image 26. (1) Superior frontal gyrus (F1); (2) superior frontal gyrus (F1); (3) superior frontal sulcus; (4) middle frontal gyrus (F2); (5) inferior precentral sulcus; (6) precentral gyrus; (7) central sulcus; (8) postcentral gyrus; (9) intraparietal sulcus; (10) lateral fissure; (11) parallel sulcus; (12) superior parietal gyrus (P1); (13) parietal lobule; (14) subparietal sulcus; (15) cingulate gyrus.

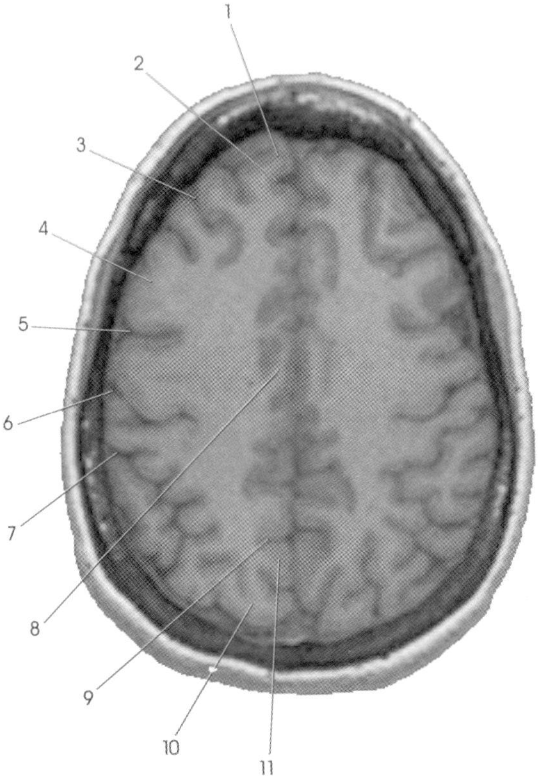


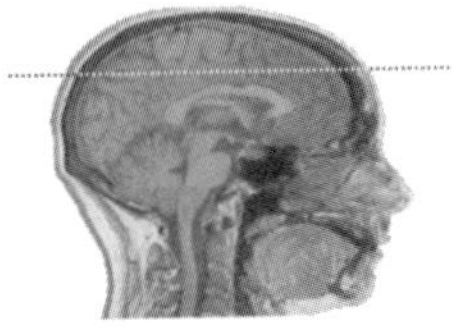

Image 27. (1) Superior frontal gyrus (F1); (2) interhemispheric fissure; (3) superior frontal sulcus; (4) middle frontal gyrus (F2); (5) inferior precentral sulcus; (6) central sulcus; (7) intraparietal sulcus; (8) cingulate gyrus; (9) transverse parietal sulcus; (10) precuneus; (11) precuneus.

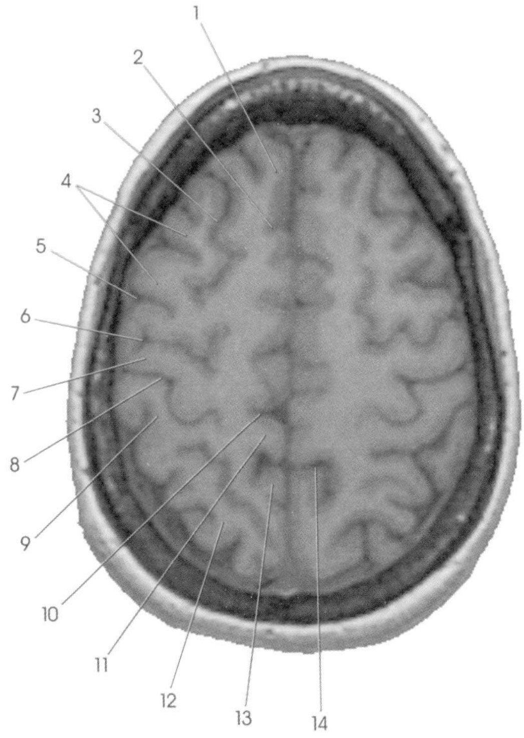


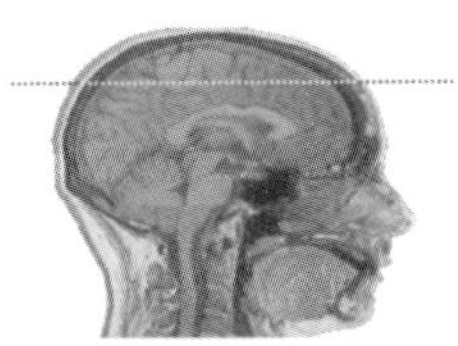

Image 28. (1) Superior frontal gyrus (F1); (2) superior frontal gyrus (F1); (3) superior frontal sulcus; (4) middle frontal gyrus (F2); (5) inferior precentral sulcus; (6) superior precentral sulcus; (7) precentral gyrus; (8) central sulcus; (9) postcentral gyrus; (10) paracentral sulcus; (11) paracentral lobule; (12) superior parietal gyrus (P1); (13) parietal lobule; (14) marginal ramus of cingulate sulcus.

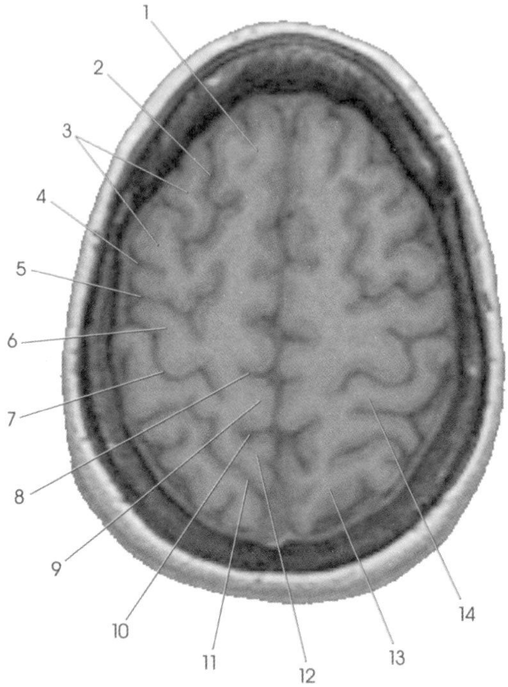


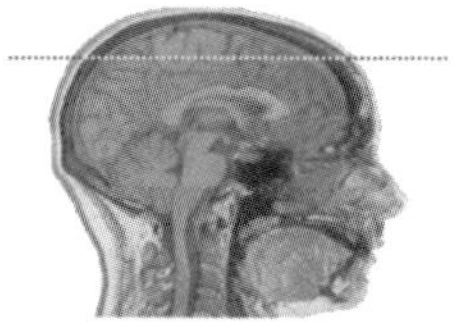

Image 29. (1) Superior frontal gyrus (F1); (2) superior frontal sulcus; (3) middle frontal gyrus (F2); (4) inferior precentral sulcus; (5) superior precentral sulcus; (6) precentral gyrus; (7) central sulcus; (8) paracentral sulcus; (9) paracentral lobule; (10) cingulate sulcus; (11) transverse parietal sulcus; (12) parietal lobule; (13) superior parietal gyrus (P1); (14) postcentral gyrus.

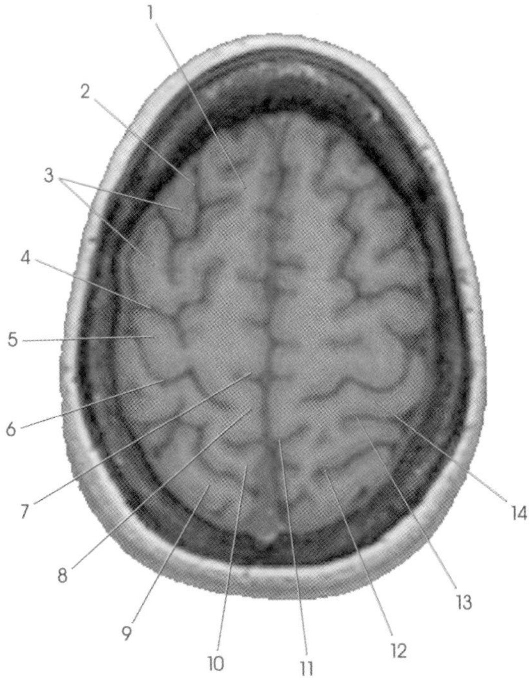


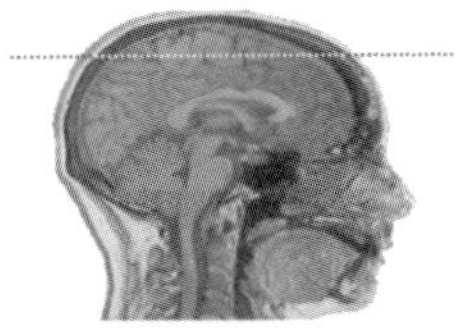

Image 30. (1) Superior frontal gyrus (F1); (2) superior frontal sulcus; (3) middle frontal gyrus (F2); (4) superior precentral sulcus; (5) precentral gyrus; (6) central sulcus; (7) paracentral sulcus; (8) paracentral lobule; (9) superior parietal gyrus (P1); (10) parietal lobule; (11) marginal ramus of cingulate sulcus; (12) transverse parietal sulcus; (13) postcentral sulcus; (14) postcentral gyrus.

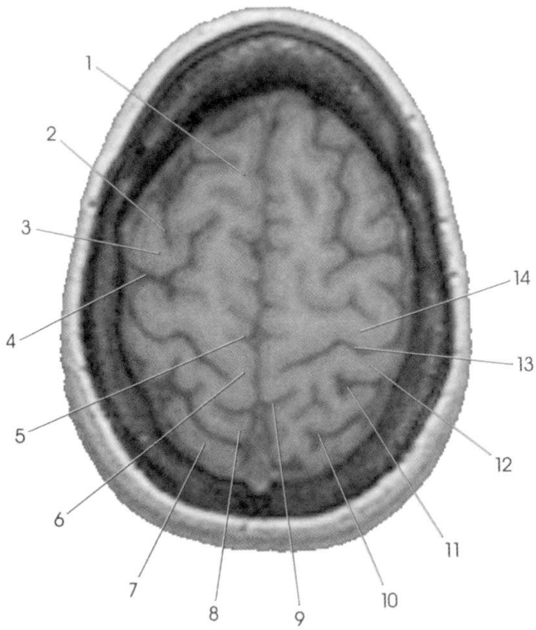


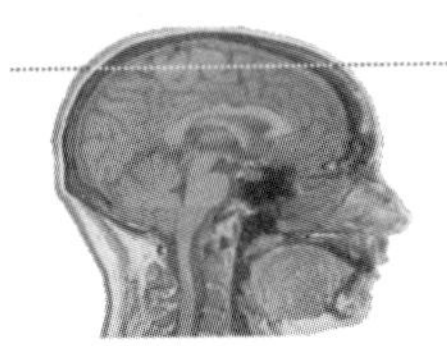

Image 31. (1) Superior frontal gyrus (F1); (2) superior frontal sulcus; (3) middle frontal gyrus (F2); (4) superior precentral sulcus; (5) interhemispheric fissure; (6) paracentral lobule; (7) superior parietal gyrus (P1); (8) parietal lobule; (9) marginal ramus of cingulate sulcus; (10) transverse parietal sulcus; (11) postcentral sulcus; (12) postcentral gyrus; (13) central sulcus; (14) precentral gyrus.

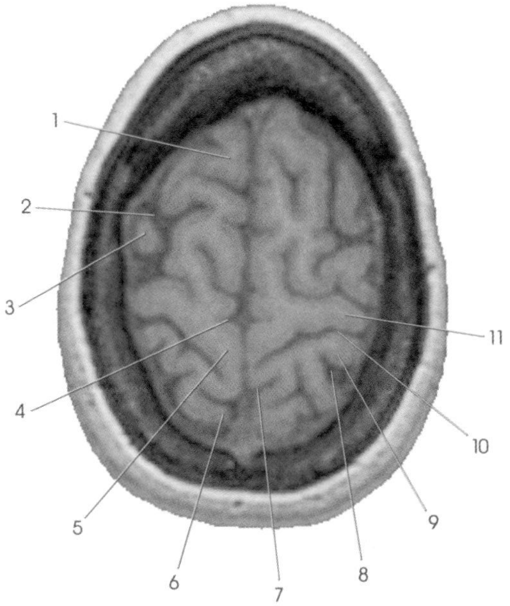


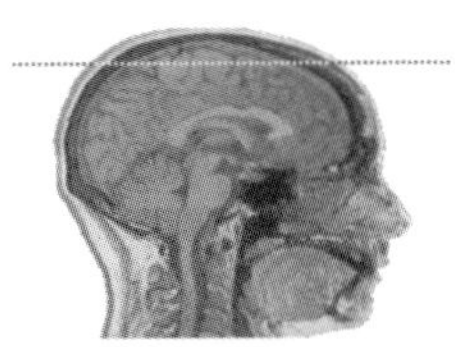

Image 32. (1) Superior frontal gyrus (F1); (2) superior frontal sulcus; (3) middle frontal gyrus (F2); (4) paracentral sulcus; (5) paracentral lobule; (6) superior parietal gyrus (P1); (7) marginal ramus of cingulate sulcus; (8) postcentral sulcus; (9) postcentral gyrus; (10) central sulcus; (11) precentral gyrus.

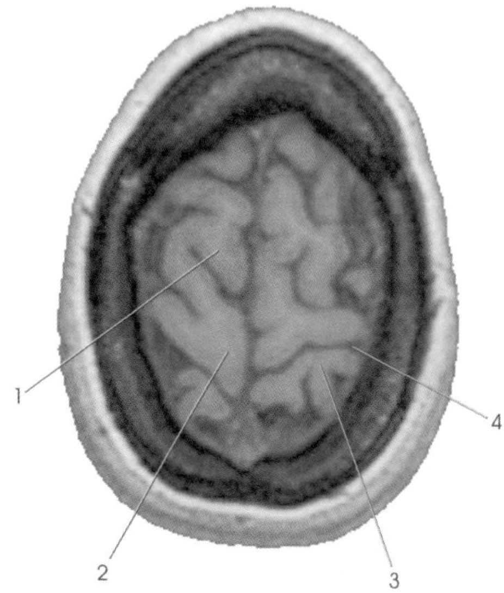


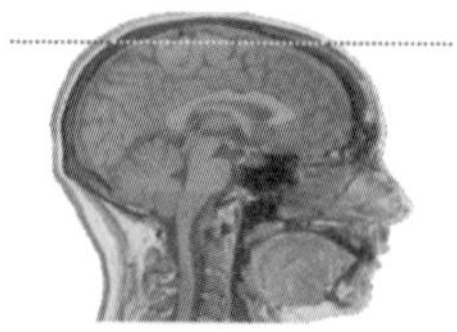

Image 33. (1) Superior frontal gyrus (F1); (2) paracentral lobule; (3) postcentral gyrus; (4) central sulcus.

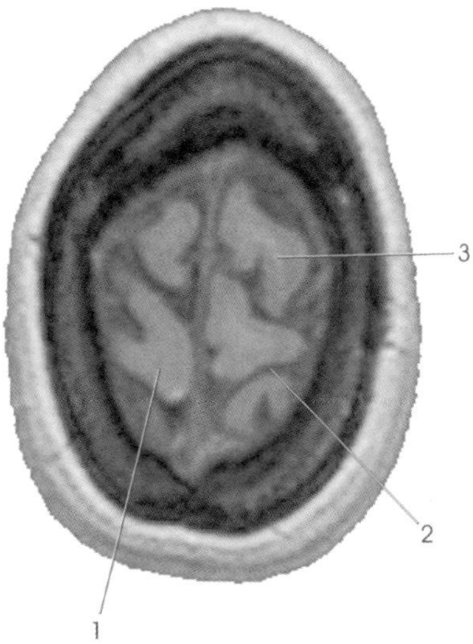


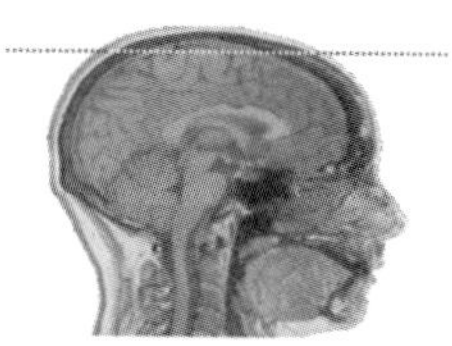

Image 34. (1) Paracentral lobule; (2) central sulcus; (3) superior frontal gyrus (F1).

# Sagittal Images

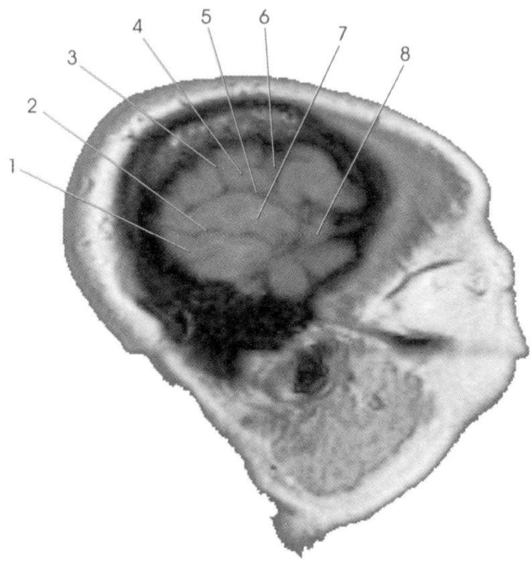


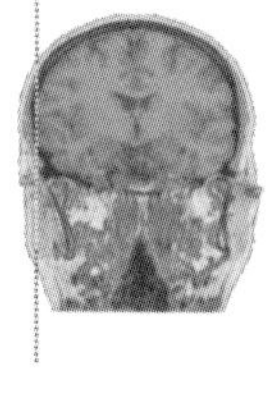

Image 1. (1) Middle temporal gyrus (T2); (2) parallel sulcus; (3) posterior segment of lateral fissure; (4) postcentral gyrus; (5) middle segment of lateral fissure; (6) intraparietal sulcus; (7) superior temporal gyrus (T1); (8) superior temporal gyrus (T1).

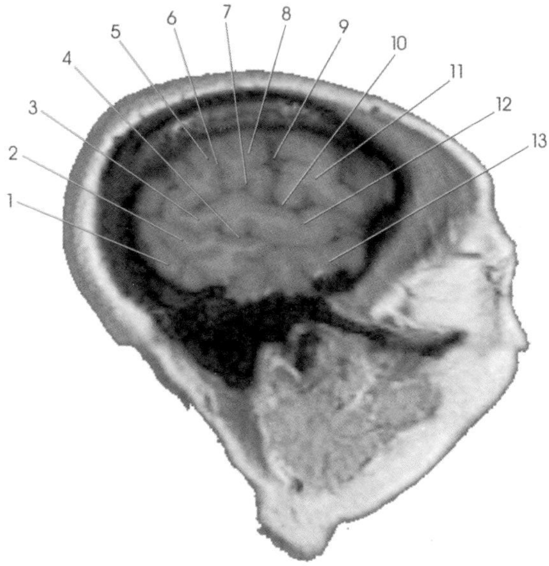


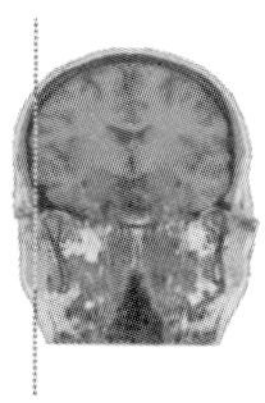

Image 2. (1) Inferior temporal gyrus (T3); (2) middle temporal gyrus (T2); (3) superior temporal gyrus (T1); (4) parallel sulcus; (5) supramarginal gyrus; (6) posterior segment of lateral fissure; (7) lateral fissure; (8) supramarginal gyrus; (9) intraparietal sulcus; (10) middle segment of lateral fissure; (11) subcentral gyrus; (12) superior temporal gyrus (T1); (13) middle temporal gyrus (T2).

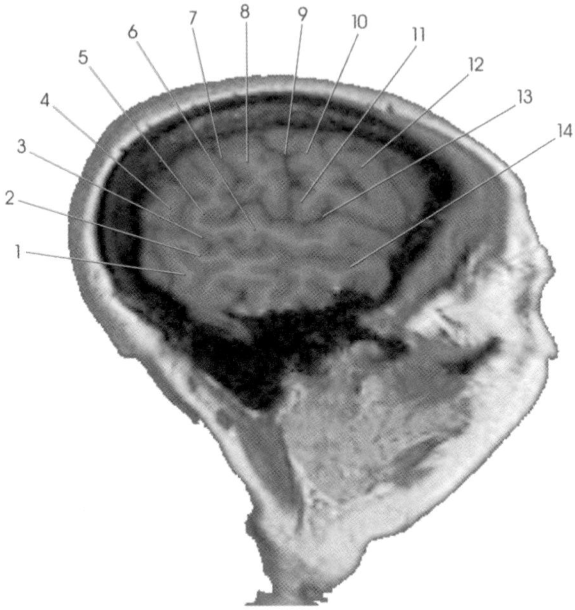


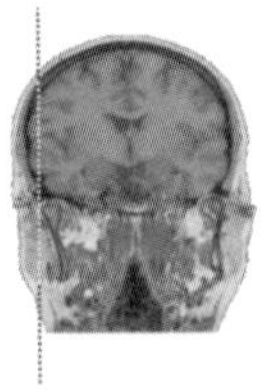

Image 3. (1) Middle temporal gyrus (T2); (2) middle temporal gyrus (T2); (3) parallel sulcus; (4) angular gyrus; (5) superior temporal gyrus (T1); (6) superior temporal gyrus (T1); (7) posterior segment of lateral fissure; (8) supramarginal gyrus; (9) intraparietal sulcus; (10) postcentral gyrus; (11) subcentral gyrus; (12) precentral gyrus; (13) middle segment of lateral fissure; (14) middle temporal gyrus (T2).

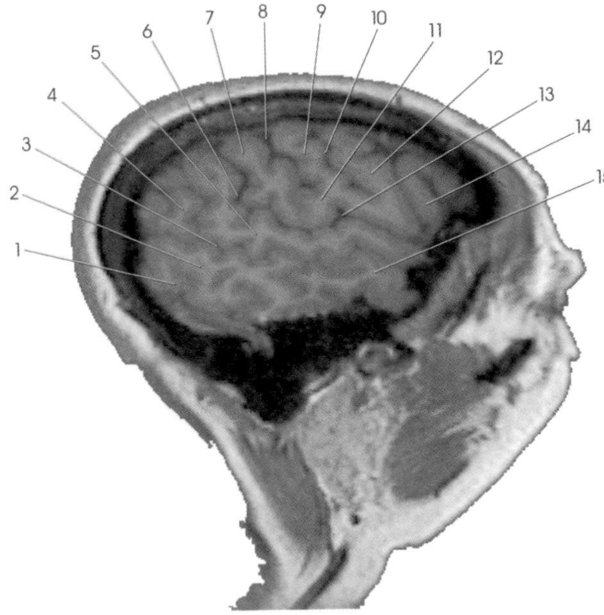


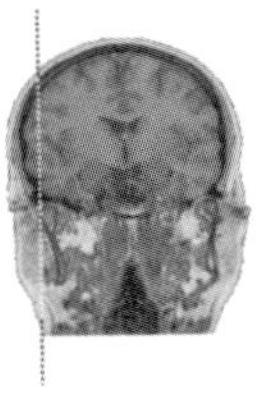

Image 4. (1) Middle temporal gyrus (T2); (2) middle temporal gyrus (T2); (3) parallel sulcus; (4) angular gyrus; (5) Heschl's gyrus; (6) posterior segment of lateral fissure; (7) supramarginal gyrus; (8) intraparietal sulcus; (9) postcentral gyrus; (10) central sulcus; (11) subcentral gyrus; (12) precentral gyrus; (13) middle segment of lateral fissure; (14) inferior frontal gyrus (F3), pars opercularis; (15) middle temporal gyrus (T2).

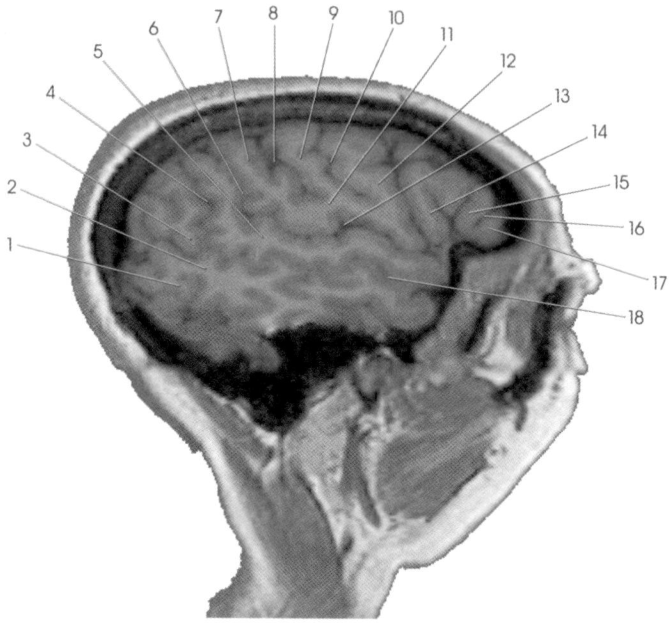


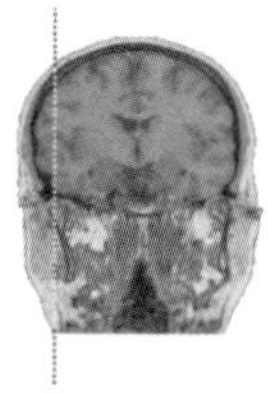

Image 5. (1) Inferior temporal gyrus (T3); (2) middle temporal gyrus (T2); (3) middle occipital gyrus (O2); (4) ascending posterior segment of parallel sulcus; (5) superior temporal gyrus (T1); (6) posterior segment of lateral fissure; (7) supramarginal gyrus; (8) intraparietal sulcus; (9) postcentral gyrus; (10) central sulcus; (11) subcentral gyrus; (12) precentral gyrus; (13) middle segment of lateral fissure; (14) inferior frontal gyrus (F3), pars opercularis; (15) inferior frontal gyrus (F3), pars orbitalis; (16) lateral orbital sulcus; (17) lateral orbital gyrus; (18) parallel sulcus.

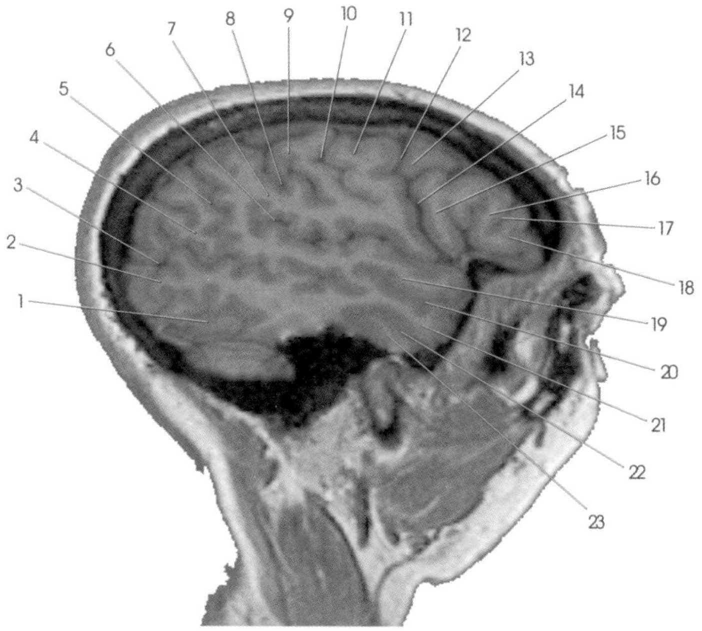


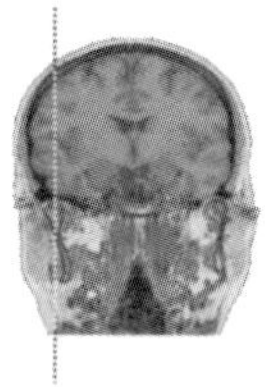

Image 6. (1) Inferior occipital gyrus (O3); (2) middle temporal gyrus (T2); (3) parallel sulcus; (4) angular gyrus; (5) ascending posterior segment of parallel sulcus; (6) lateral fissure; (7) supramarginal gyrus; (8) intraparietal sulcus; (9) postcentral gyrus; (10) central sulcus; (11) precentral gyrus; (12) inferior precentral sulcus; (13) middle frontal gyrus (F2); (14) inferior precentral sulcus; (15) inferior frontal gyrus (F3), pars opercularis; (16) inferior frontal gyrus (F3), pars orbitalis; (17) lateral orbital sulcus; (18) lateral orbital gyrus; (19) superior temporal gyrus (T1); (20) parallel sulcus; (21) middle temporal gyrus (T2); (22) inferior temporal sulcus; (23) inferior temporal gyrus (T3).

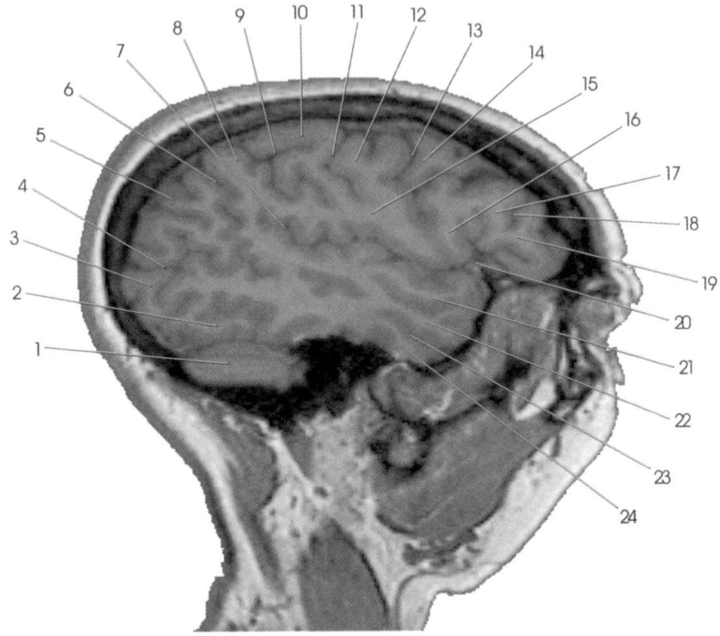


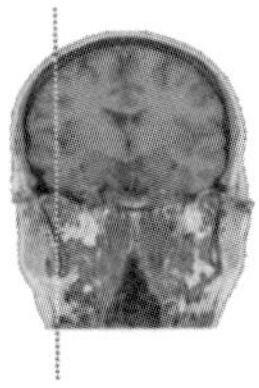

Image 7. (1) Cerebellum; (2) inferior occipital gyrus (O3); (3) middle occipital gyrus (O2); (4) horizontal posterior segment of parallel sulcus; (5) angular gyrus; (6) ascending posterior segment of parallel sulcus; (7) lateral fissure; (8) supramarginal gyrus; (9) postcentral sulcus; (10) postcentral gyrus; (11) central sulcus; (12) precentral gyrus; (13) inferior precentral sulcus; (14) middle frontal gyrus (F2); (15) subcentral gyrus; (16) inferior frontal gyrus (F3), pars opercularis; (17) inferior frontal gyrus (F3), pars orbitalis; (18) lateral orbital sulcus; (19) lateral orbital gyrus; (20) anterior segment of lateral fissure; (21) superior temporal gyrus (T1); (22) middle temporal gyrus (T2); (23) inferior temporal sulcus; (24) inferior temporal gyrus (T3).

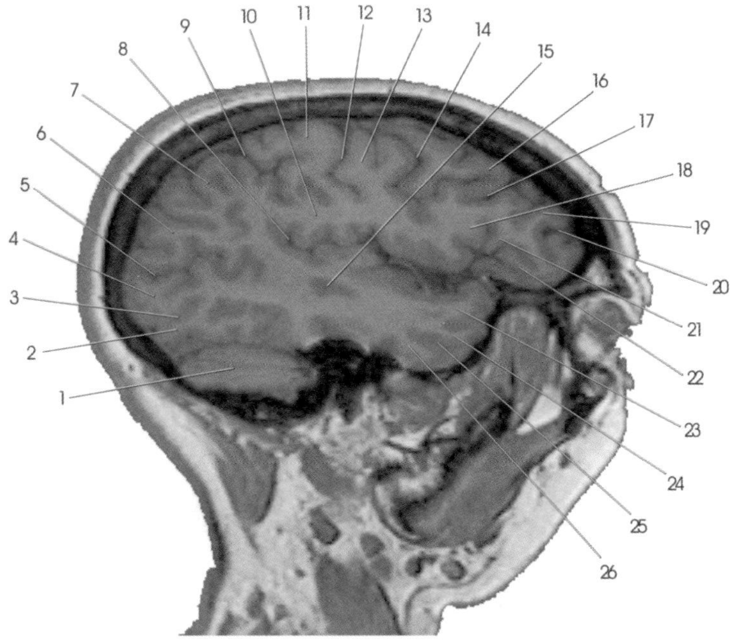


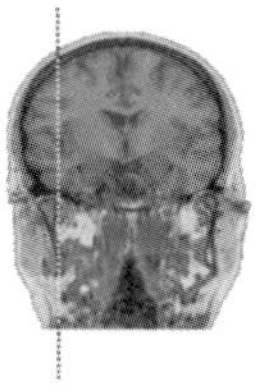

Image 8. (1) Cerebellum; (2) inferior occipital gyrus (O3); (3) lateral occipital sulcus; (4) middle occipital gyrus (O2); (5) transverse occipital sulcus; (6) middle occipital gyrus (O2); (7) angular gyrus; (8) lateral fissure; (9) intraparietal sulcus; (10) subcentral gyrus; (11) postcentral gyrus; (12) central sulcus; (13) precentral gyrus; (14) inferior precentral sulcus; (15) discontinuous parallel sulcus; (16) middle frontal gyrus (F2); (17) inferior frontal sulcus; (18) inferior frontal gyrus (F3), pars opercularis; (19) middle frontal gyrus (F2); (20) lateral orbital sulcus; (21) inferior frontal gyrus (F3), pars orbitalis; (22) lateral orbital gyrus; (23) superior temporal gyrus (T1); (24) middle temporal gyrus (T2); (25) inferior temporal sulcus; (26) inferior temporal gyrus (T3).

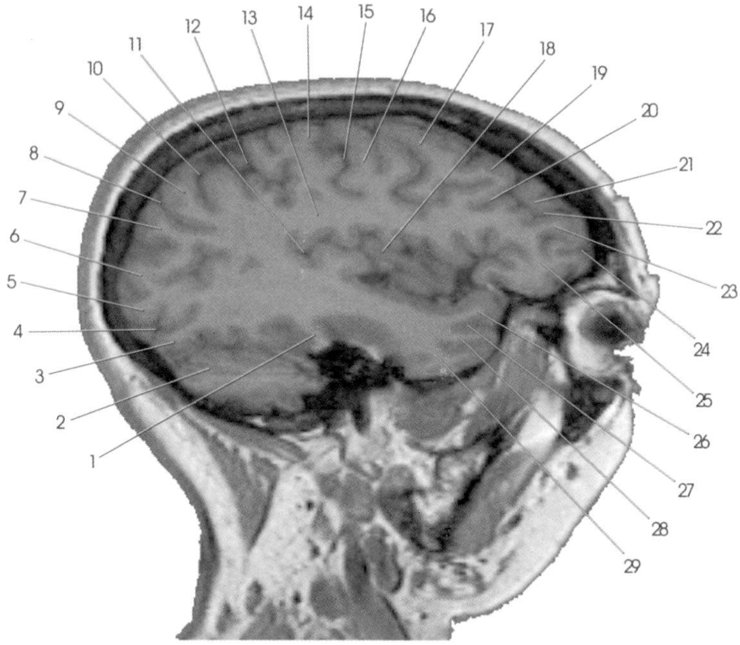


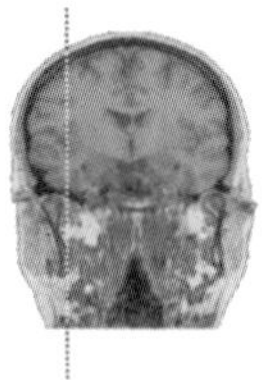

Image 9. (1) Fusiform gyrus; (2) cerebellum; (3) inferior occipital gyrus (O3); (4) lateral occipital sulcus; (5) middle occipital gyrus (O2); (6) middle occipital gyrus (O2); (7) angular gyrus; (8) intraparietal sulcus; (9) angular gyrus; (10) intraparietal sulcus; (11) lateral fissure; (12) intraparietal sulcus; (13) subcentral gyrus; (14) postcentral gyrus; (15) central sulcus; (16) precentral gyrus; (17) middle frontal gyrus (F2); (18) long insular gyri; (19) middle frontal gyrus (F2); (20) inferior frontal sulcus; (21) middle frontal gyrus (F2); (22) middle frontal sulcus; (23) middle frontal gyrus (F2); (24) frontomarginal sulcus; (25) lateral orbital gyrus; (26) superior temporal gyrus (T1); (27) parallel sulcus; (28) middle temporal gyrus (T2); (29) inferior temporal gyrus (T3).

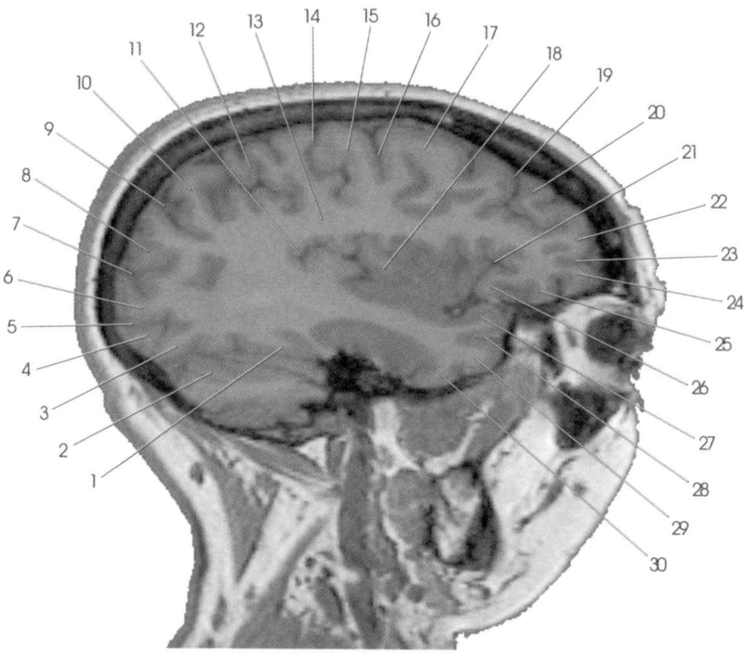


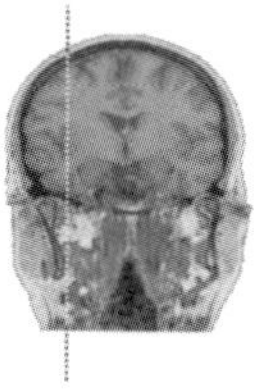

Image 10. (1) Fusiform gyrus; (2) cerebellum; (3) inferior occipital gyrus (O3); (4) inferior occipital gurus (O3); (5) lateral occipital sulcus; (6) middle occipital gyrus (O2); (7) transverse occipital sulcus; (8) middle occipital gyrus (O2); (9) intraparietal sulcus; (10) angular gyrus; (11) lateral fissure; (12) postcentral gyrus; (13) subcentral gyrus; (14) central sulcus; (15) precentral gyrus; (16) superior precentral sulcus; (17) middle frontal gyrus (F2); (18) long insular gyri; (19) inferior frontal sulcus; (20) middle frontal gyrus (F2); (21) anterior segment of lateral fissure; (22) middle frontal gyrus (F2); (23) frontomarginal sulcus; (24) frontomarginal gyrus; (25) lateral orbital gyrus; (26) lateral orbital gyrus; (27) superior temporal gyrus (T1); (28) parallel sulcus; (29) middle temporal gyrus (T2); (30) inferior temporal gyrus (T3).

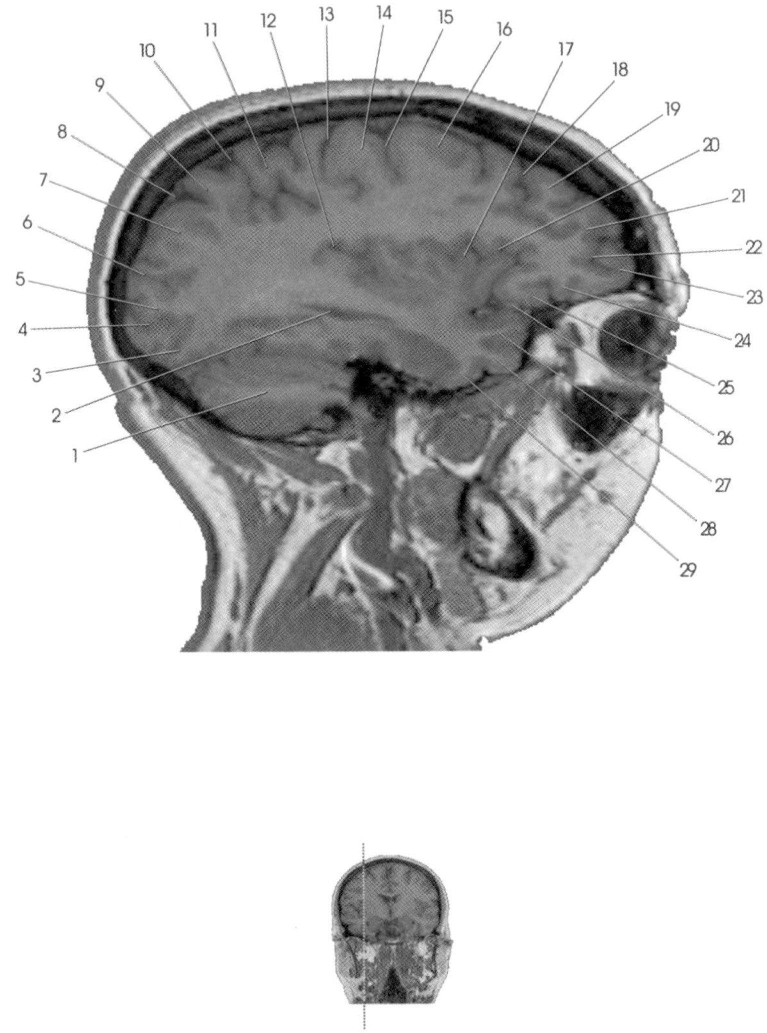


Image 11. (1) Cerebellum; (2) temporal horn of lateral ventricle; (3) inferior occipital gyrus (O3); (4) inferior occipital sulcus; (5) middle occipital gyrus (O2); (6) transverse occipital sulcus; (7) middle occipital gyrus (O2); (8) intraparietal sulcus; (9) superior parietal gyrus (P1); (10) intraparietal sulcus; (11) superior parietal gyrus (P1); (12) lateral fissure; (13) central sulcus; (14) precentral gyrus; (15) superior precentral sulcus; (16) middle frontal gyrus (F2); (17) insular cortex; (18) inferior frontal sulcus; (19) middle frontal gyrus (F2); (20) insular cortex; (21) middle frontal gyrus (F2); (22) frontomarginal sulcus; (23) frontomarginal gyrus; (24) lateral orbital gyrus; (25) lateral orbital gyrus; (26) lateral fissure; (27) superior temporal gyrus (T1); (28) middle temporal gyrus (T2); (29) inferior temporal gyrus (T3).

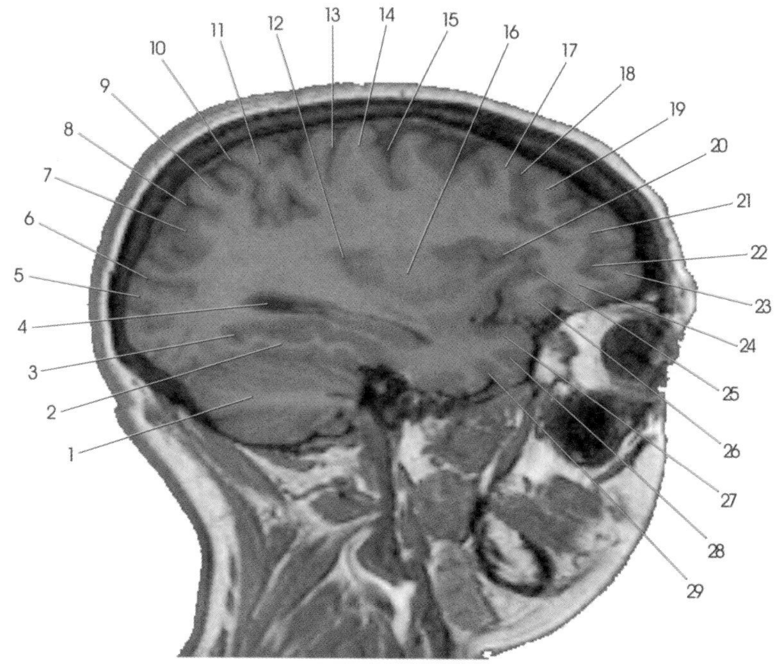


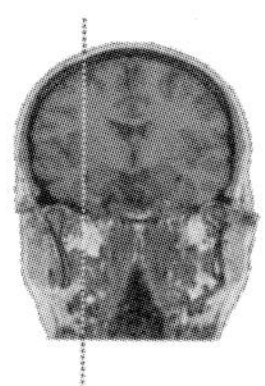

Image 12. (1) Cerebellum; (2) fusiform gyrus; (3) collateral sulcus; (4) temporal horn of lateral ventricle; (5) middle occipital gyrus (O2); (6) transverse occipital sulcus; (7) middle occipital gyrus (O2); (8) intraparietal sulcus; (9) superior parietal gyrus (P1); (10) intraparietal sulcus; (11) superior parietal gyrus (P1); (12) insula; (13) central sulcus; (14) precentral gyrus; (15) superior precentral sulcus; (16) claustrum; (17) middle frontal gyrus (F2); (18) inferior frontal sulcus; (19) middle frontal gyrus (F2); (20) lateral fissure; (21) middle frontal gyrus (F2); (22) frontomarginal sulcus; (23) frontomarginal gyrus; (24) lateral orbital gyrus; (25) ascending ramus of lateral fissure; (26) lateral orbital gyrus; (27) superior temporal gyrus (T1); (28) parallel sulcus; (29) inferior temporal sulcus.

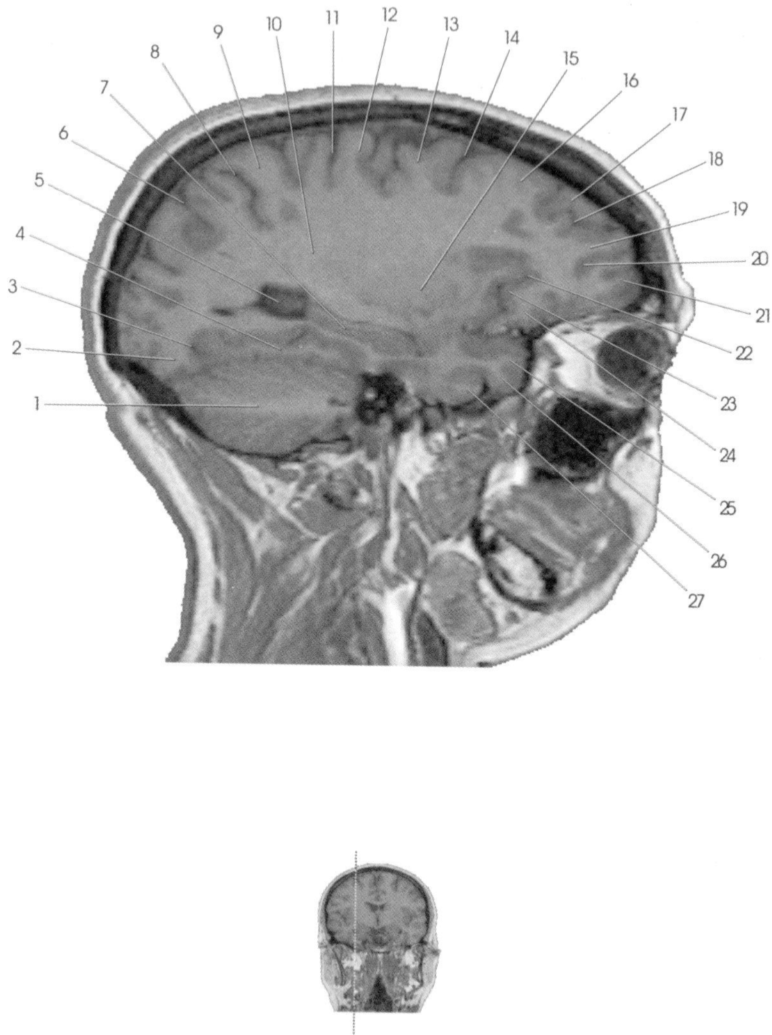


Image 13. (1) Cerebellum; (2) inferior occipital gyrus (O3); (3) collateral sulcus; (4) fusiform gyrus; (5) atria of lateral ventricle; (6) intraparietal sulcus; (7) hippocampus; (8) intraparietal sulcus; (9) superior parietal gyrus (P1); (10) retrolentiform part of internal capsule; (11) central sulcus; (12) precentral gyrus; (13) superior frontal gyrus (F1); (14) superior frontal sulcus; (15) putamen; (16) middle frontal gyrus (F2); (17) superior frontal gyrus (F1); (18) superior frontal sulcus; (19) middle frontal gyrus (F2); (20) frontomarginal sulcus; (21) frontomarginal gyrus; (22) lateral fissure; (23) inferior frontal gyrus (F3); (24) lateral orbital gyrus; (25) superior temporal gyrus (T1); (26) middle temporal gyrus (T2); (27) inferior temporal gyrus (T3).

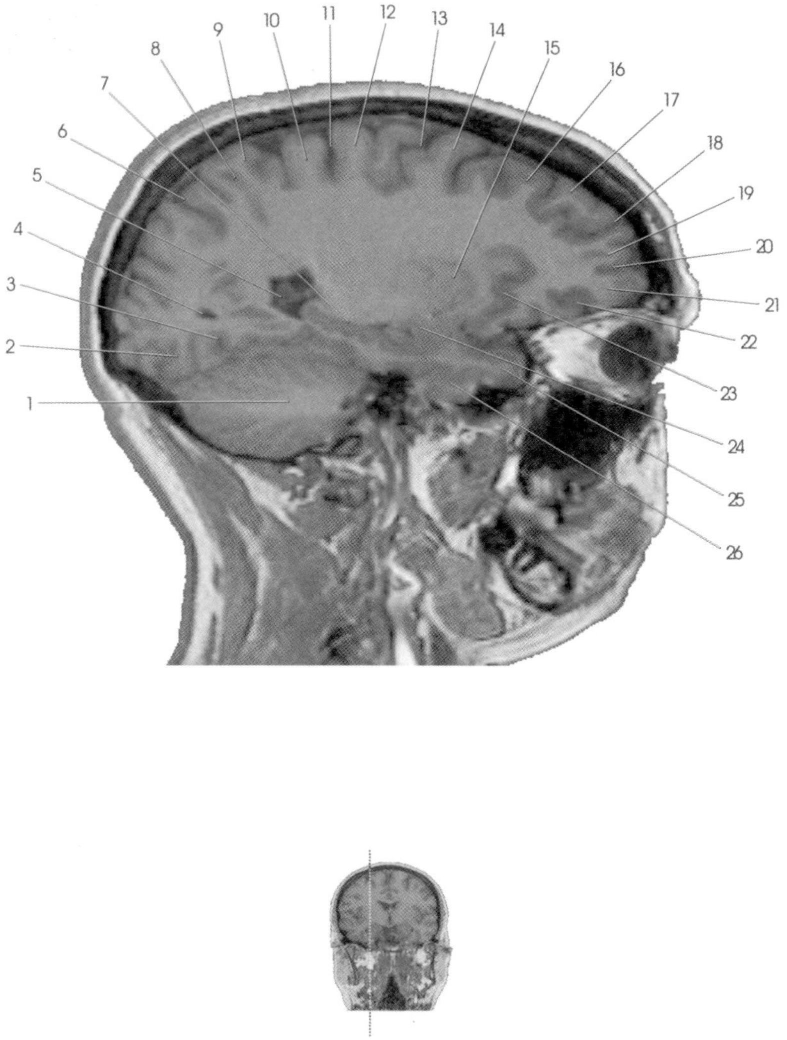


Image 14. (1) Cerebellum; (2) inferior occipital gyrus (O3); (3) collateral sulcus; (4) occipital horn of lateral ventricle; (5) atria of lateral ventricle; (6) intraparietal sulcus; (7) hippocampus; (8) intraparietal sulcus; (9) superior parietal gyrus (P1); (10) postcentral gyrus; (11) central sulcus; (12) precentral gyrus; (13) superior frontal sulcus; (14) superior frontal gyrus (F1); (15) putamen; (16) middle frontal gyrus (F2); (17) superior frontal gyrus (F1); (18) inferior frontal sulcus; (19) middle frontal gyrus (F2); (20) frontomarginal sulcus; (21) frontomarginal gyrus; (22) lateral orbital gyrus; (23) ascending ramus of lateral fissure; (24) amygdala; (25) temporal pole; (26) inferior temporal gyrus (T3).

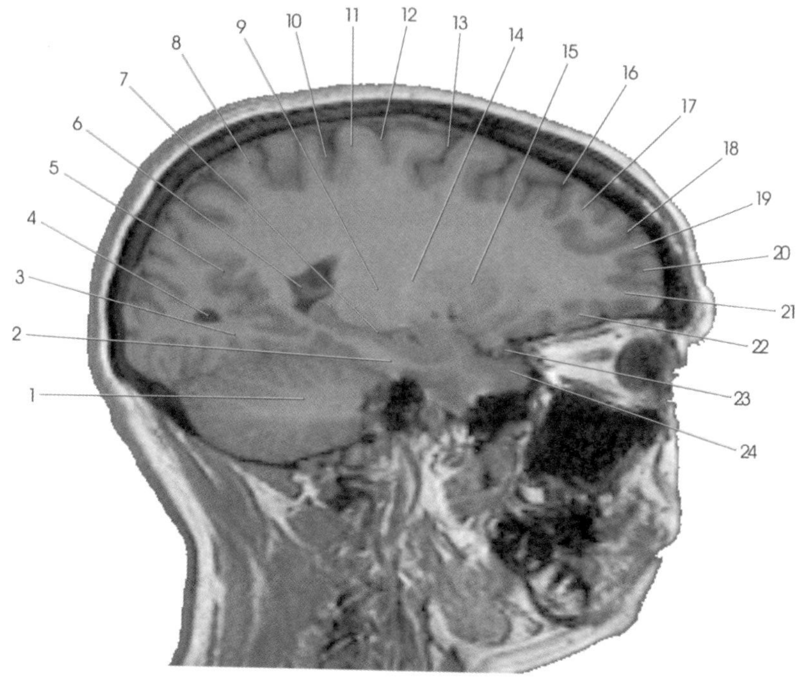


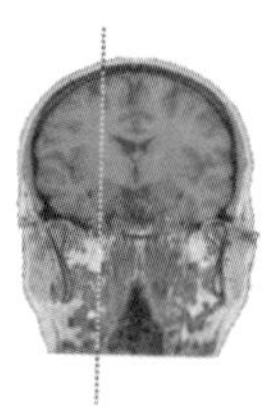

Image 15. (1) Cerebellum; (2) Parahippocampal gyrus; (3) lingual gyrus; (4) occipital horn of lateral ventricle; (5) parieto-occipital sulcus; (6) atria of lateral ventricle; (7) hippocampus; (8) superior parietal gyrus (P1); (9) pulvinar; (10) central sulcus; (11) precentral gyrus; (12) precentral sulcus; (13) superior frontal sulcus; (14) posterior limb of internal capsule; (15) putamen; (16) superior frontal sulcus; (17) superior frontal gyrus (F1); (18) inferior frontal sulcus; (19) middle frontal gyrus (F2); (20) inferior frontopolar gyrus; (21) frontomarginal gyrus; (22) anterior orbital gyrus; (23) lateral fissure; (24) temporal pole.

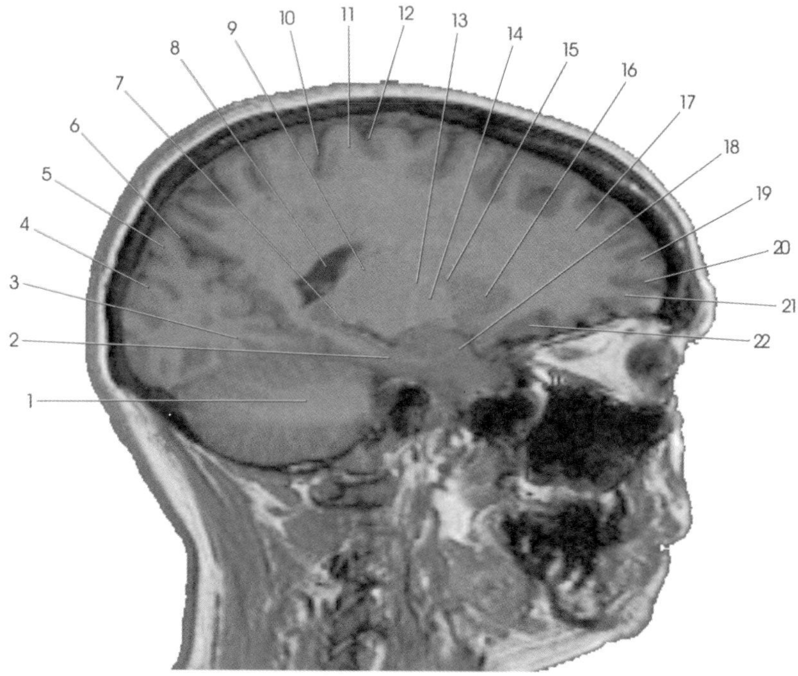


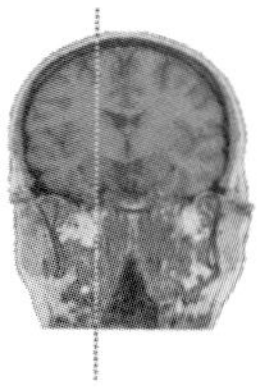

Image 16. (1) Cerebellum; (2) parahippocampal gyrus; (3) lingual gyrus; (4) middle occipital gyrus (O2); (5) superior occipital gyrus (O1); (6) parieto-occipital sulcus; (7) choroidal fissure; (8) atria of lateral ventricle; (9) retrolentiform part of internal capsule; (10) central sulcus; (11) precentral gyrus; (12) precentral sulcus; (13) posterior limb of internal capsule; 14) globus pallidus; (15) transcapsular striatum; (16) putamen; (17) superior frontal gyrus (F1); (18) amygdala; (19) inferior frontopolar gyrus; (20) frontomarginal sulcus; (21) frontomarginal gyrus; (22) medial orbital gyrus.

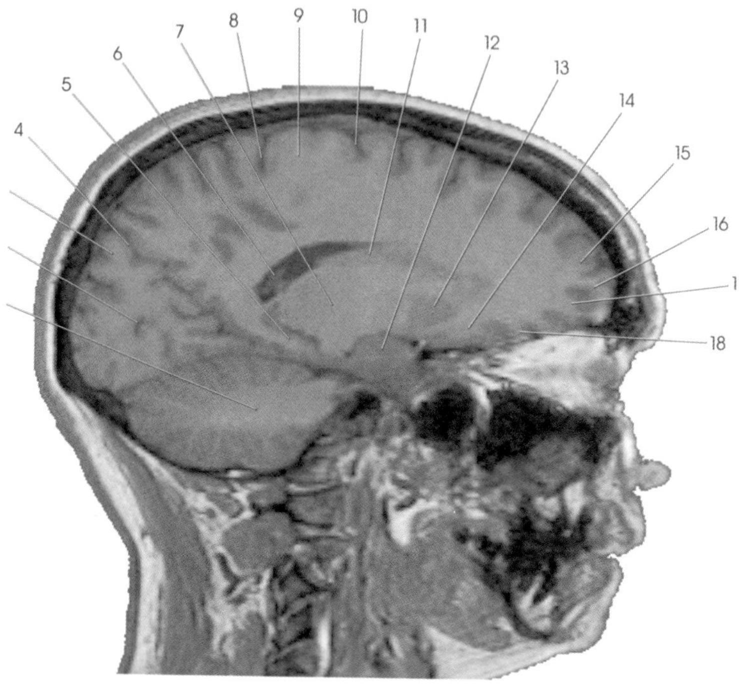


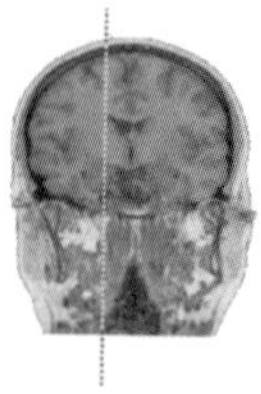

Image 17. (1) Cerebellum; (2) calcarine sulcus; (3) superior occipital gyrus (O1); (4) parieto-occipital sulcus; (5) parahippocampal gyrus; (6) body of lateral ventricle; (7) thalamus; (8) central sulcus; (9) precentral gyrus; (10) precentral sulcus; (11) caudate nucleus; (12) amygdala; (13) putamen; (14) medial orbital gyrus; (15) superior frontopolar gyrus; (16) middle frontopolar gyrus; (17) frontomarginal gyrus; (18) medial orbital gyrus.

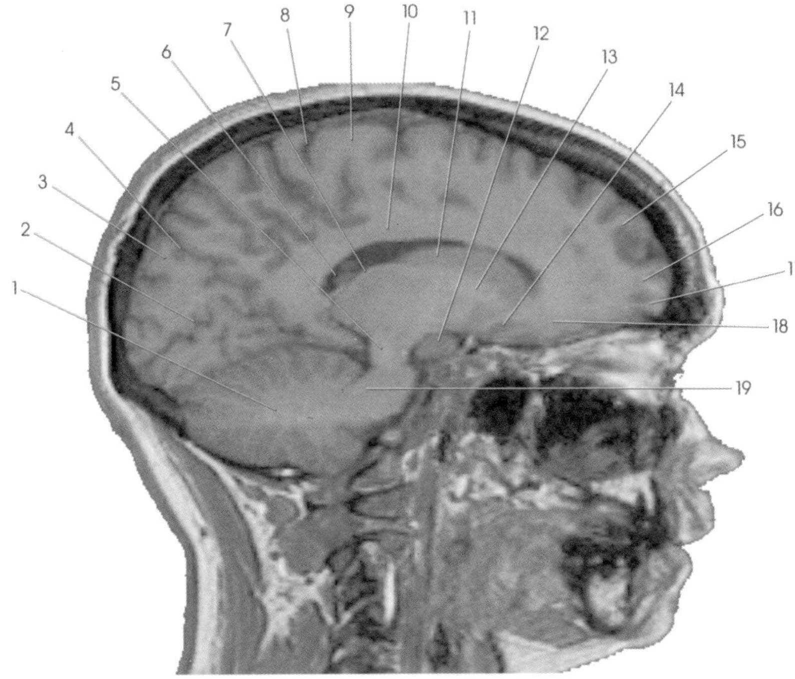


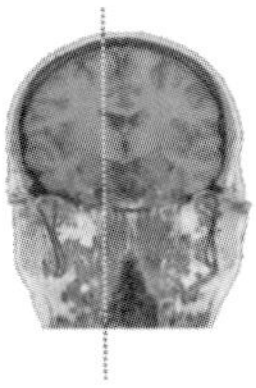

Image 18. (1) Cerebellum; (2) calcarine sulcus; (3) superior occipital gyrus (O1); (4) parieto-occipital sulcus; (5) brachium pontis; (6) body of lateral ventricle; (7) thalamus; (8) central sulcus; (9) paracentral lobule; (10) corpus callosum; (11) caudate nucleus; (12) uncus; (13) genu of internal capsule; (14) olfactory sulcus; (15) superior frontal gyrus (F1); (16) frontopolar gyrus; (17) frontomarginal gyrus; (18) medial orbital gyrus; (19) middle cerebellar peduncle.

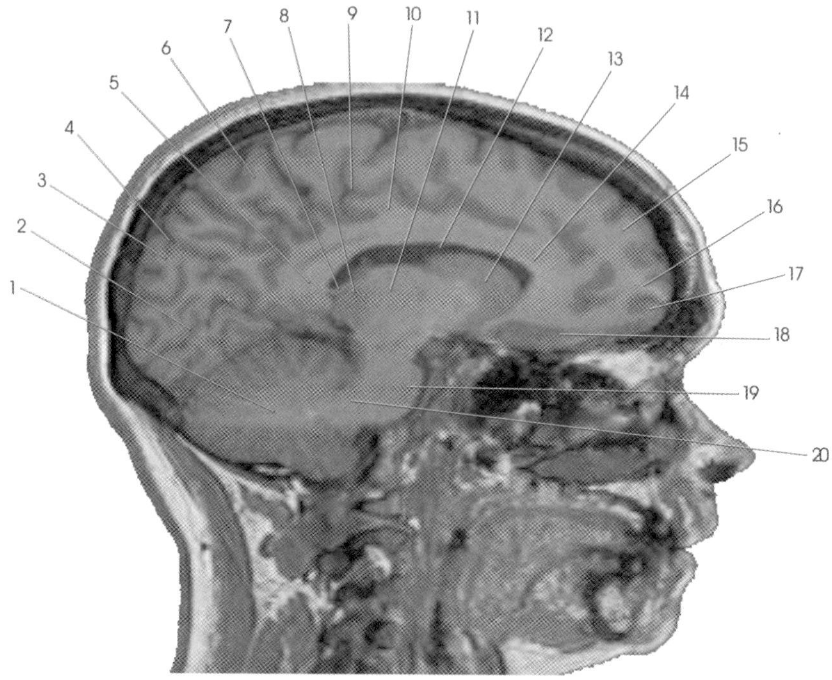


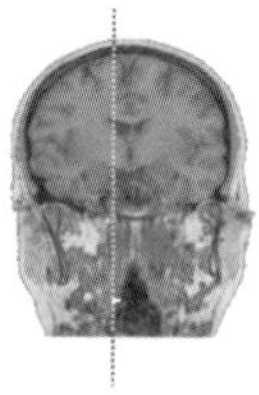

Image 19. (1) Cerebellum; (2) calcarine sulcus; (3) superior occipital gyrus (O1); (4) parieto-occipital sulcus; (5) splenium of corpus callosum; (6) superior parietal gyrus (P1); (7) fornix; (8) pulvinar; (9) cingulate sulcus; (10) cingulate gyrus; (11) thalamus; (12) body of lateral ventricle; (13) body of caudate nucleus; (14) genu of corpus callosum; (15) superior frontal gyrus (F1); (16) frontopolar gyrus; (17) frontomarginal gyrus; (18) medial orbital gyrus; (19) pons; (20) middle cerebellar peduncle.

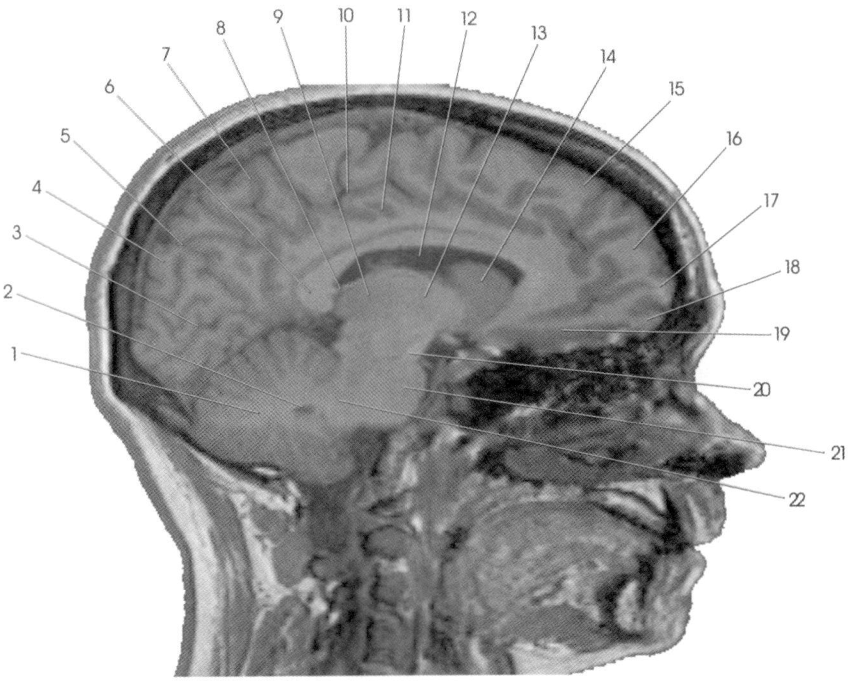


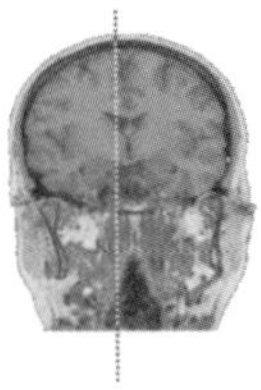

Image 20. (1) Cerebellum; (2) fourth ventricle; (3) calcarine sulcus; (4) cuneus; (5) parieto-occipital sulcus; (6) splenium of corpus callosum; (7) parietal lobule; (8) fornix; (9) thalamus; (10) cingulate sulcus; (11) cingulate gyrus; (12) body of lateral ventricle; (13) thalamus; (14) head of caudate nucleus; (15) superior frontal gyrus (F1); (16) superior frontal gyrus (F1); (17) frontopolar gyrus; (18) frontomarginal gyrus; (19) medial orbital gyrus (O2); (20) cerebral peduncle; (21) pons; (22) middle cerebellar peduncle.

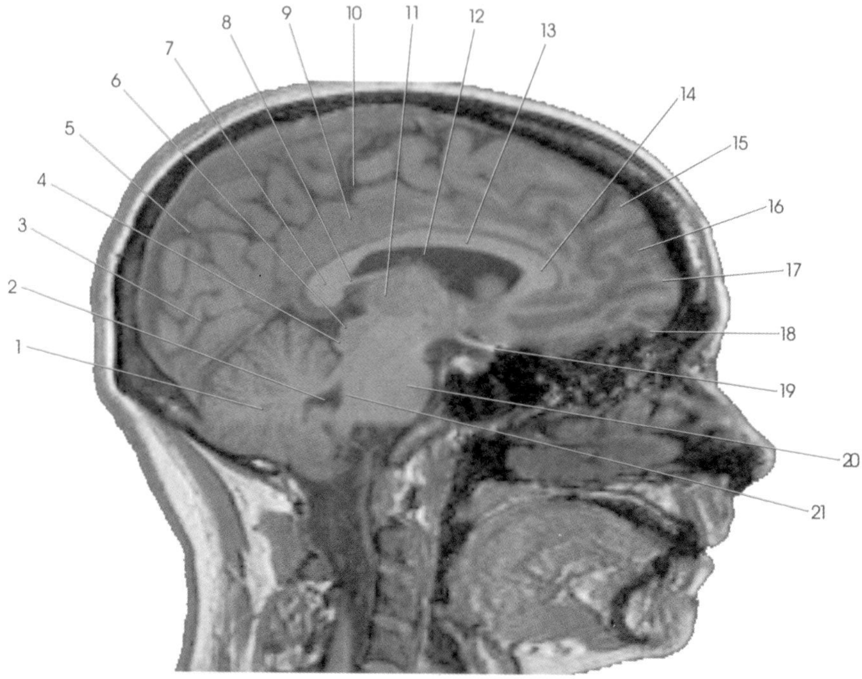


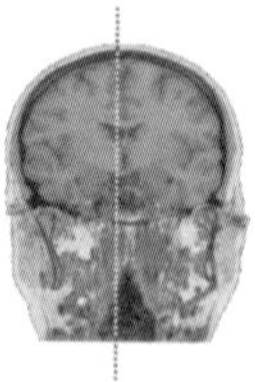

Image 21. (1) Cerebellum; (2) choroid plexus of fourth ventricle; (3) calcarine sulcus; (4) inferior colliculus; (5) parieto-occipital sulcus; (6) superior colliculus; (7) splenium of corpus callosum; (8) fornix; (9) cingulate gyrus; (10) cingulate sulcus; (11) thalamus; (12) body of lateral ventricle; (13) body of corpus callosum; (14) genu of corpus callosum; (15) superior frontal gyrus (F1); (16) superior frontal gyrus (F1); (17) frontopolar gyrus; (18) frontomarginal gyrus; (19) optic tract; (20) pons; (21) fourth ventricle.

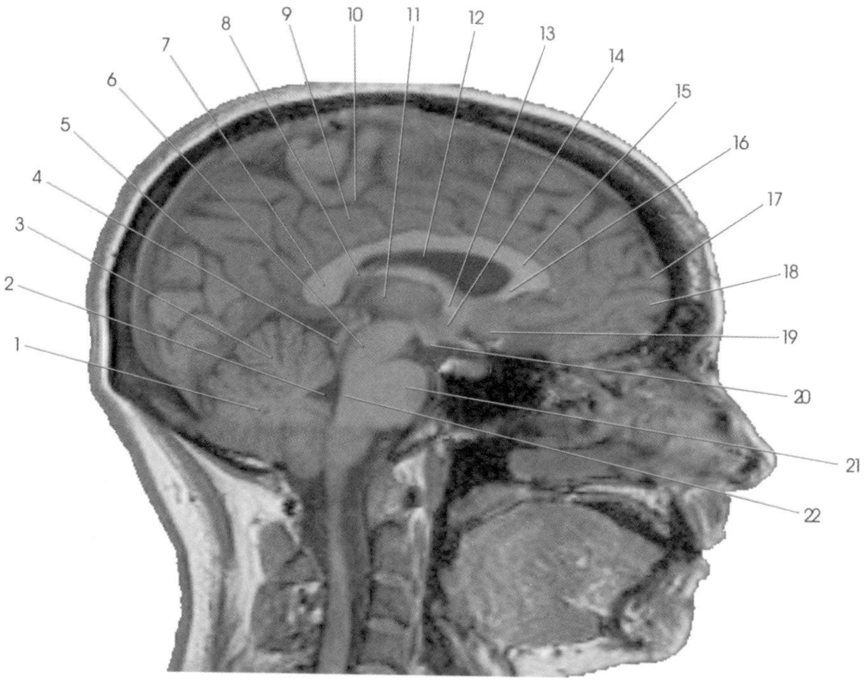


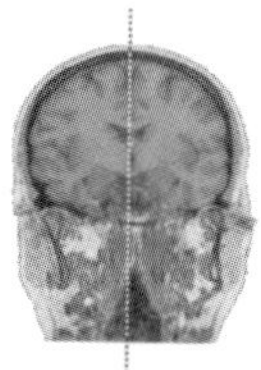

Image 22. (1) Cerebellum; (2) fourth ventricle; (3) arbor vitae of cerebellum; (4) tectum; (5) parieto-occipital sulcus; (6) tegmentum; (7) splenium of corpus callosum; (8) fornix; (9) cingulate gyrus; (10) cingulate sulcus; (11) thalamus; (12) body of lateral ventricle; (13) column of fornix; (14) hypothalamus; (15) genu of corpus callosum; (16) rostrum of corpus callosum; (17) superior frontopolar gyrus; (18) inferior frontopolar gyrus; (19) subcallosal gyrus; (20) mammilary body; (21) pons; (22) middle cerebellar peduncle.

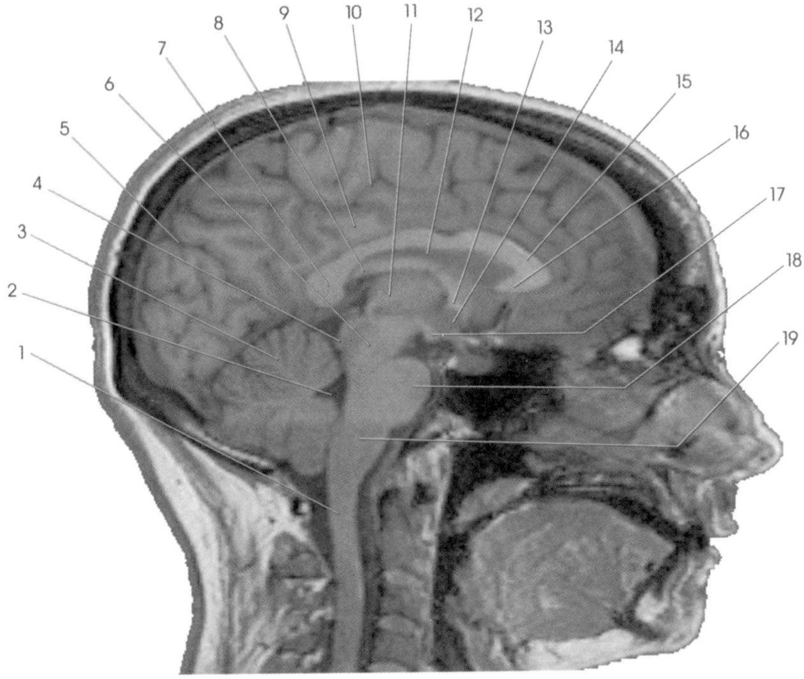


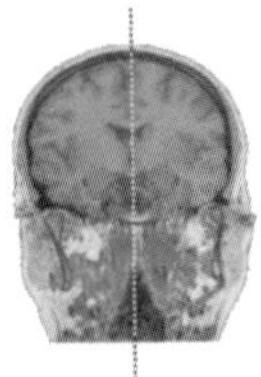

Image 23. (1) Spinal cord; (2) fourth ventricle; (3) arbor vitae of cerebellum; (4) inferior colliculus; (5) parieto-occipital sulcus; (6) superior cerebellar cistern; (7) splenium of corpus callosum; (8) fornix; (9) cingulate gyrus; (10) cingulate sulcus; (11) thalamus; (12) septum pellucidum; (13) column of fornix; (14) hypothalamus; (15) genu of corpus callosum; (16) rostrum of corpus callosum; (17) mamillary body; (18) pons; (19) medulla.

# Coronal Images

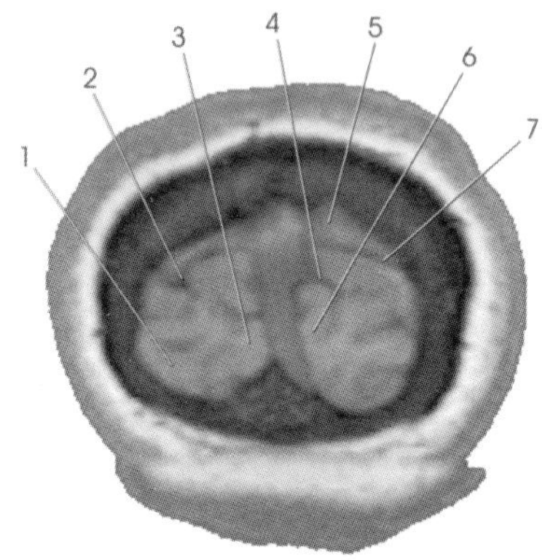


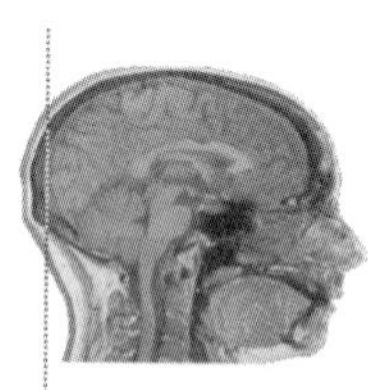

Image 1. (1) Middle occipital gyrus (O2); (2) lateral occipital sulcus; (3) cuneus; (4) calcarine sulcus; (5) superior occipital gyrus (O1); (6) cuneus; (7) lateral occipital sulcus.

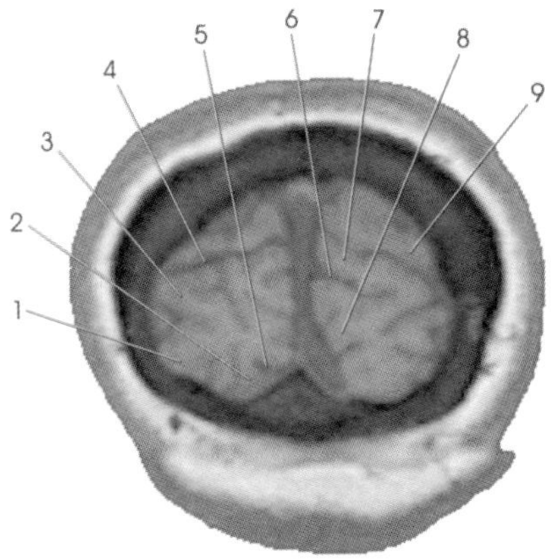


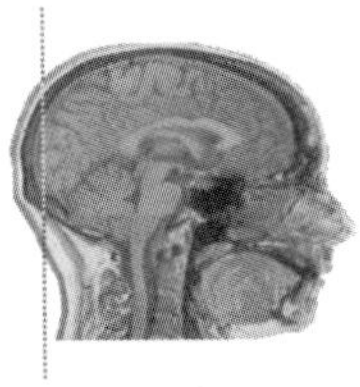

Image 2. (1) Inferior occipital gyrus (O3); (2) fourth occipital gyrus (O4); (3) middle occipital gyrus (O2); (4) lateral occipital sulcus; (5) calcarine sulcus; (6) retrocalcarine sulcus; (7) middle occipital gyrus (O2); (8) cuneus; (9) middle occipital gyrus (O2).

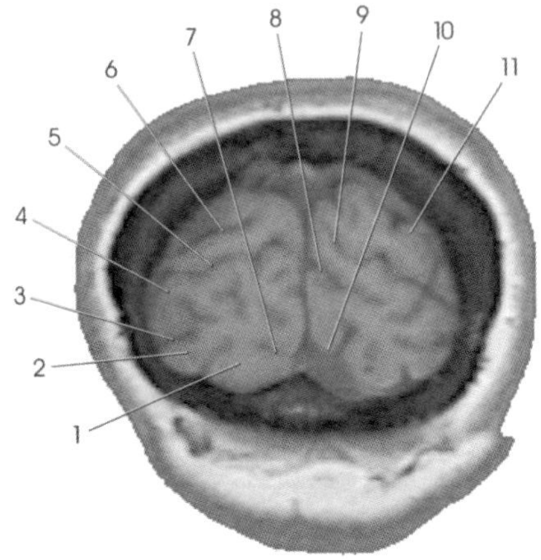


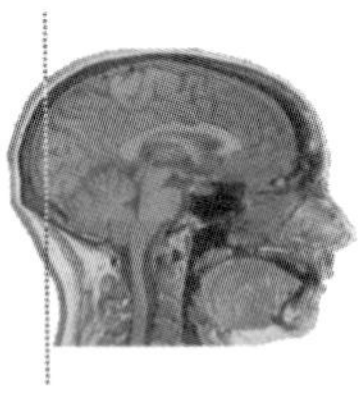

Image 3. (1) Fourth occipital gyrus (O4); (2) inferior occipital gyrus (O3); (3) inferior occipital sulcus; (4) middle occipital gyrus (O2); (5) lateral occipital sulcus; (6) intra-occipital sulcus; (7) calcarine sulcus; (8) retrocalcarine sulcus; (9) middle occipital gyrus (O2); (10) cuneus; (11) middle occipital gyrus (O2).

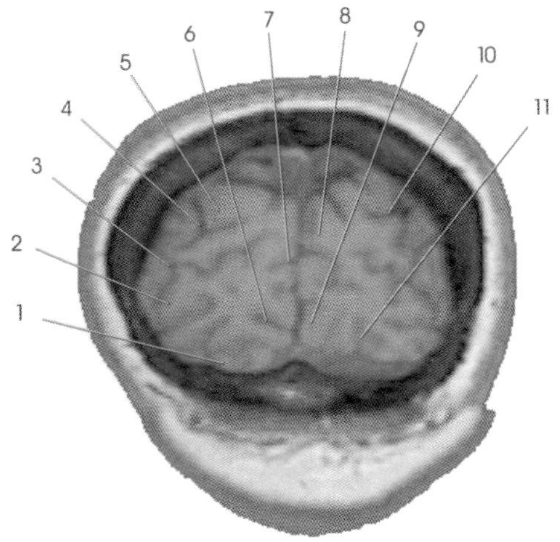


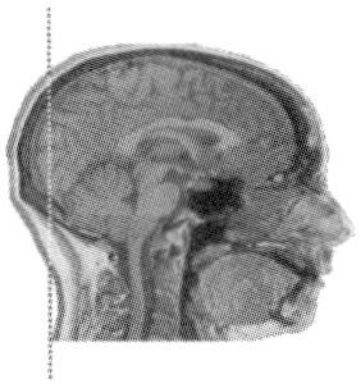

Image 4. (1) Inferior occipital gyrus (O3); (2) middle occipital gyrus (O2), inferior part; (3) lateral occipital sulcus; (4) intra-occipital sulcus; (5) middle occipital gyrus (O2), superior part; (6) calcarine sulcus; (7) cuneus; (8) superior occipital gyrus (O1); (9) cuneus; (10) intra-occipital gyrus; (11) middle occipital gyrus (O2), inferior part.

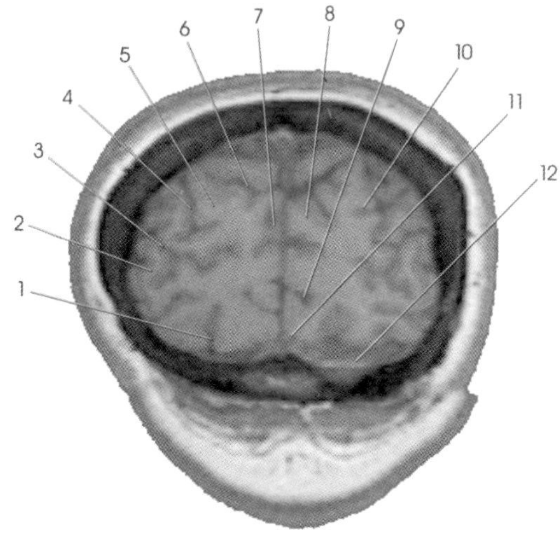


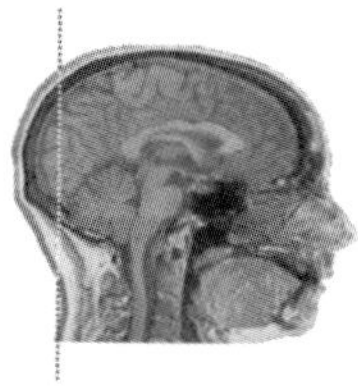

Image 5. (1) posterior segment of collateral sulcus; (2) middle occipital gyrus (O2); (3) lateral occipital sulcus; (4) intraparietal sulcus; (5) superior occipital gyrus (O1); (6) parieto-occipital sulcus; (7) cuneus; (8) cuneus; (9) calcarine sulcus; (10) superior occipital gyrus (O1); (11) lingual gyrus; (12) cerebellum.

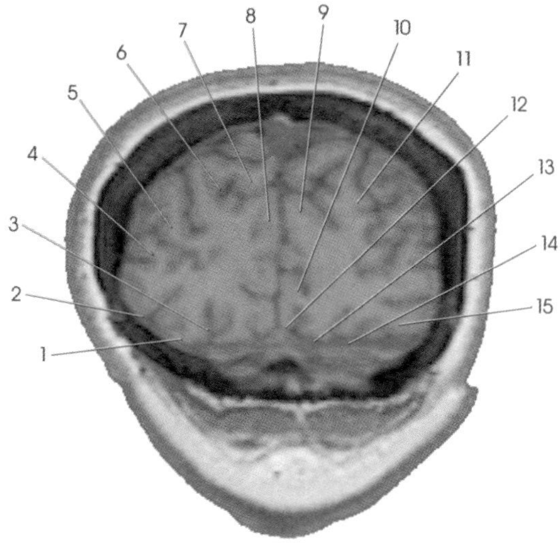


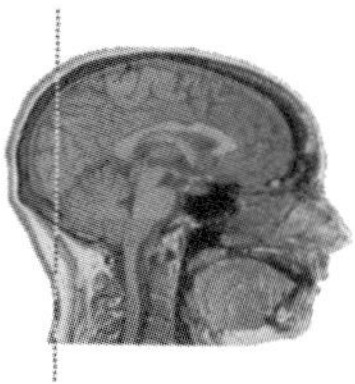

Image 6. (1) Inferior occipital gyrus (O3); (2) lateral occipital sulcus; (3) posterior transverse collateral sulcus; (4) middle occipital gyrus (O2), inferior part; (5) middle occipital gyrus (O2), superior part; (6) parieto-occipital sulcus; (7) superior parietal gyrus (P1); (8) cuneus; (9) cuneus; (10) calcarine sulcus; (11) intraparietal sulcus; (12) lingual gyrus; (13) lingual gyrus; (14) fourth occipital gyrus (O4); (15) inferior occipital gyrus (O3).

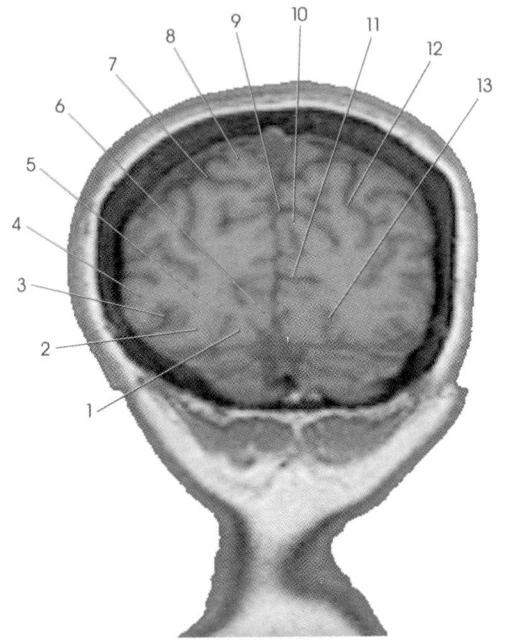


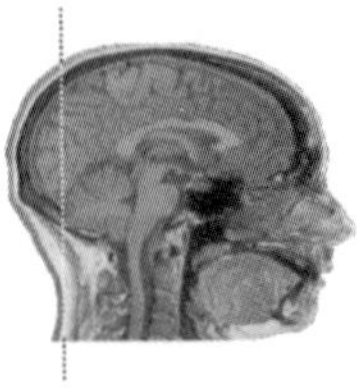

Image 7. (1) Lingual gyrus; (2) inferior occipital gyrus (O3); (3) lateral occipital sulcus; (4) middle occipital gyrus (O2); (5) optic radiations; (6) lingual gyrus; (7) intraparietal sulcus; (8) precuneus; (9) parieto-occipital sulcus; (10) cuneus; (11) calcarine sulcus; (12) intraparietal sulcus; (13) collateral sulcus.

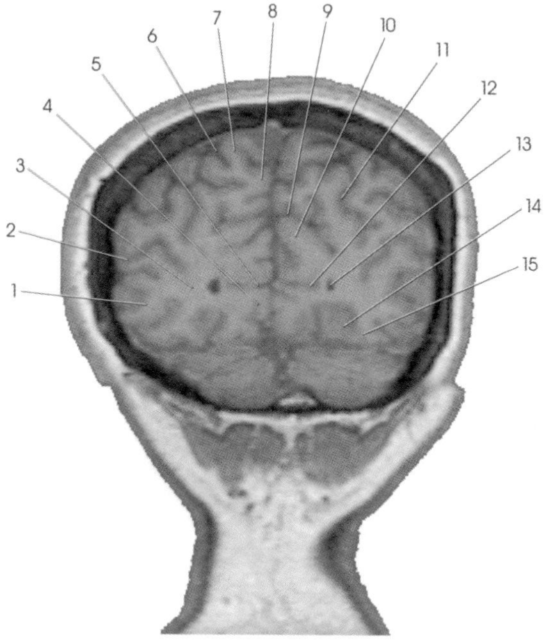


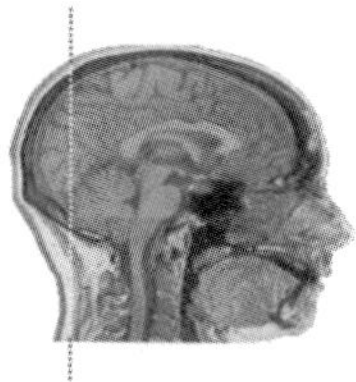

Image 8. (1) Middle occipital gyrus (O2), inferior part; (2) middle occipital gyrus (O2), superior part; (3) optic radiations; (4) lingual gyrus; (5) calcarine sulcus; (6) transverse parietal sulcus; (7) superior parietal gyrus (P1); (8) superior parietal gyrus (P1); (9) parieto-occipital sulcus; (10) precuneus; (11) intraparietal sulcus; (12) calcarine sulcus; (13) occipital horn of lateral ventricle; (14) collateral sulcus; (15) inferior occipital gyrus (O3).

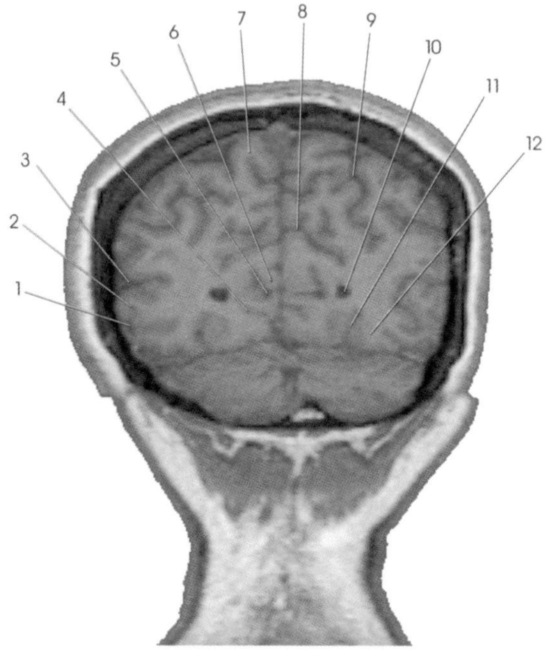


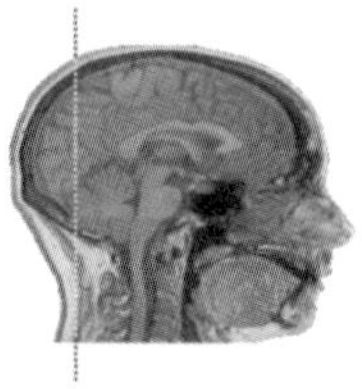

Image 9. (1) Occipitotemporal sulcus; (2) middle occipital gyrus (O2); (3) horizontal posterior segment of parallel sulcus; (4) posterior transverse collateral sulcus; (5) calcarine sulcus; (6) cuneus; (7) superior parietal gyrus (P1); (8) parieto-occipital sulcus; (9) intraparietal sulcus; (10) occipital horn of lateral ventricle; (11) collateral sulcus; (12) inferior occipital gyrus (O3).

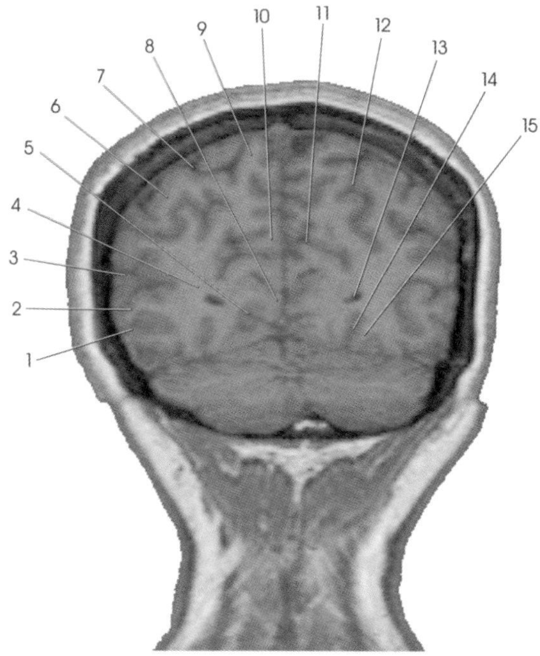


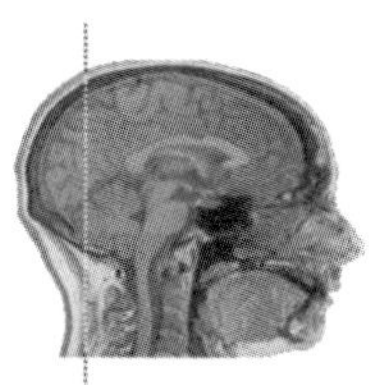

Image 10. (1) Occipitotemporal sulcus; (2) middle occipital gyrus (O2); (3) horizontal posterior segment of parallel sulcus; (4) optic radiations; (5) lingual gyrus; (6) angular gyrus; (7) intraparietal sulcus; (8) lingual gyrus; (9) superior parietal gyrus (P1); (10) precuneus; (11) parieto-occipital sulcus; (12) intraparietal sulcus; (13) occipital horn of lateral ventricle; (14) collateral sulcus; (15) fusiform gyrus.

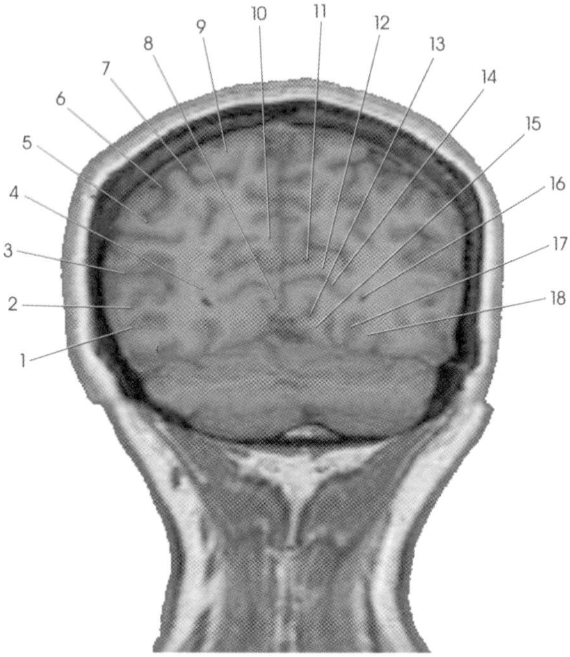


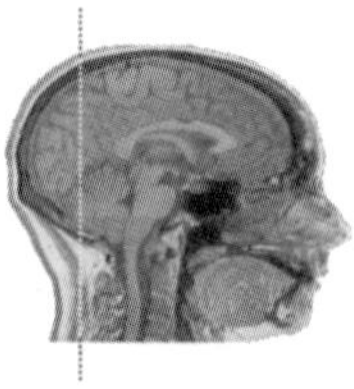

Image 11. (1) Occipitotemporal sulcus; (2) lateral occipital sulcus; (3) horizontal posterior segment of parallel sulcus; (4) optic radiations; (5) angular gyrus; (6) intraparietal sulcus; (7) intraparietal sulcus; (8) lingual gyrus; (9) superior parietal gyrus (P1); (10) precuneus; (11) parieto-occipital sulcus; (12) cuneus; (13) posterior transverse collateral sulcus; (14) calcarine sulcus; (15) fusiform gyrus; (16) occipital horn of lateral ventricle; (17) collateral sulcus; (18) fusiform gyrus.

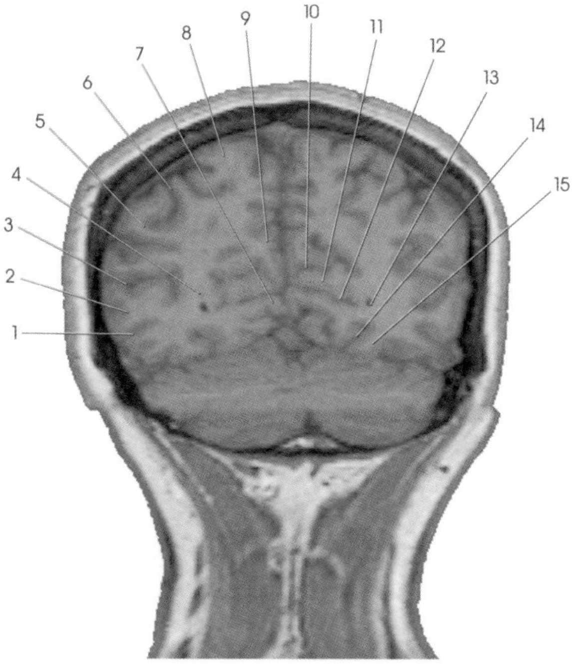


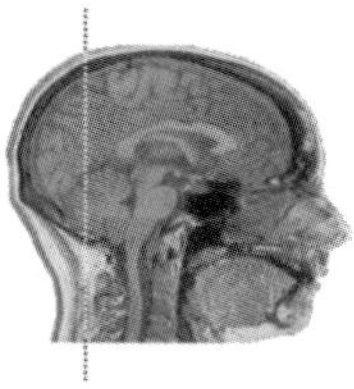

Image 12. (1) Occipitotemporal sulcus; (2) lateral occipital sulcus; (3) horizontal posterior segment of parallel sulcus; (4) optic radiations; (5) angular gyrus; (6) intraparietal sulcus; (7) lingual gyrus; (8) superior parietal gyrus (P1); (9) precuneus; (10) parieto-occipital sulcus; (11) cuneus; (12) calcarine sulcus; (13) occipital horn of lateral ventricle; (14) collateral sulcus; (15) fusiform gyrus.

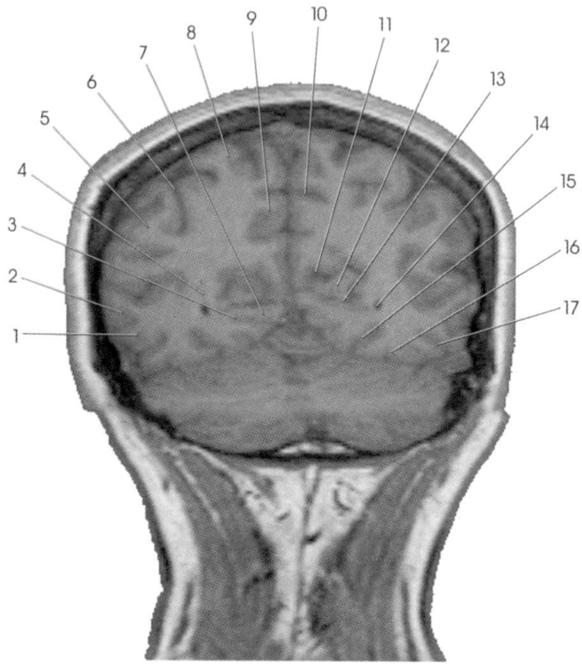


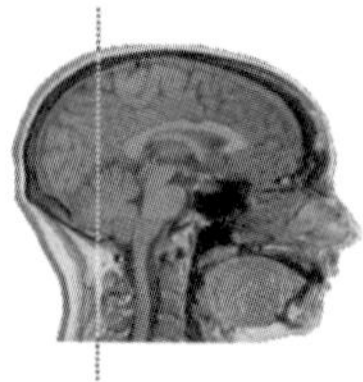

Image 13. (1) Inferior temporal gyrus (T3); (2) middle temporal gyrus (T2); (3) collateral sulcus; (4) optic radiations; (5) angular gyrus; (6) intraparietal sulcus; (7) lingual gyrus; (8) superior parietal gyrus (P1); (9) parietal lobule; (10) intraparietal sulcus; (11) parieto-occipital sulcus; (12) cuneus; (13) calcarine sulcus; (14) occipital horn of lateral ventricle; (15) collateral sulcus; (16) fusiform gyrus; (17) occipitotemporal sulcus.

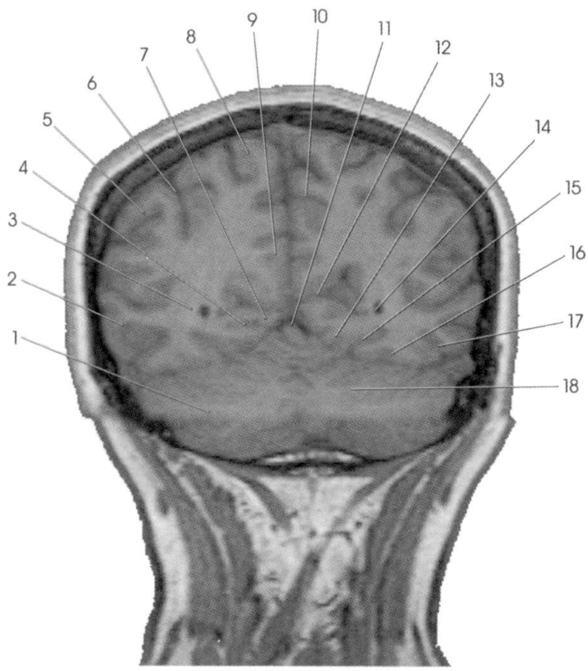


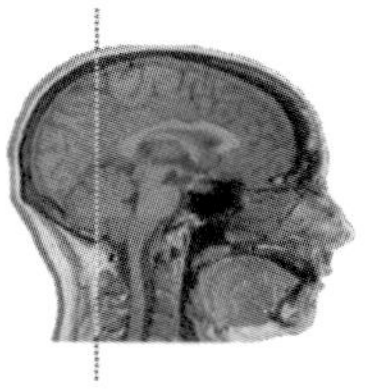

Image 14. (1) Corpus medullare of cerebellum; (2) middle temporal gyrus (T2); (3) optic radiations; (4) lingual sulcus; (5) angular gyrus; (6) intraparietal sulcus; (7) lingual gyrus; (8) paracentral lobule; (9) precuneus; (10) subparietal sulcus; (11) declive; (12) parieto-occipital sulcus; (13) lingual gyrus; (14) occipital horn of lateral ventricle; (15) collateral sulcus; (16) fusiform gyrus; (17) occipitotemporal sulcus; (18) cerebellum.

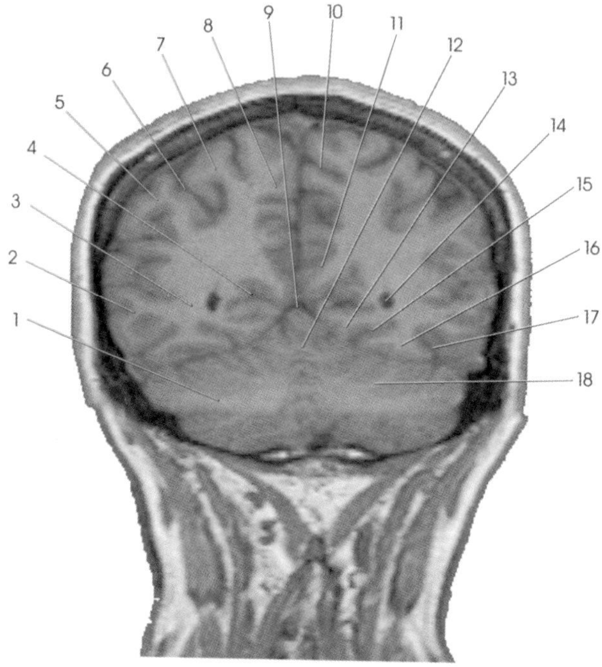


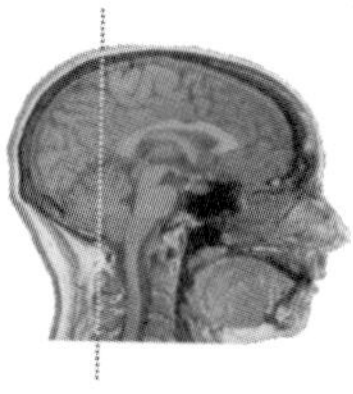

Image 15. (1) Corpus medullare of cerebellum; (2) middle temporal gyrus (T2); (3) optic radiations; (4) calcarine sulcus; (5) superior parietal gyrus (P1); (6) intraparietal sulcus; (7) superior parietal gyrus (P1); (8) paracentral lobule; (9) declive; (10) marginal ramus of cingulate sulcus; (11) isthmus of cingulum; (12) inferior cerebellar decussation; (13) lingual gyrus; (14) occipital horn of lateral ventricle; (15) collateral sulcus; (16) fusiform gyrus; (17) occipitotemporal sulcus; (18) cerebellum.

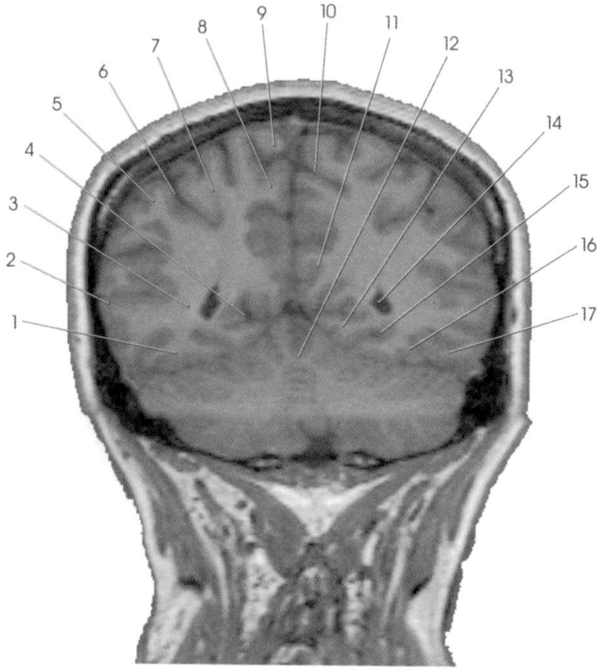


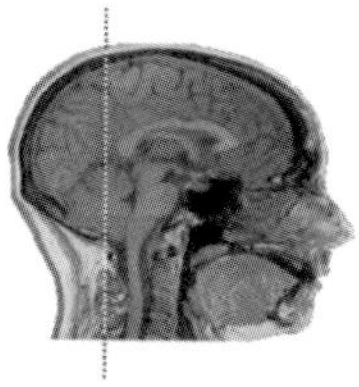

Image 16. (1) Fusiform gyrus; (2) parallel sulcus; (3) optic radiations; (4) junction of calcarine sulcus and parieto-occipital sulcus; (5) supramarginal gyrus; (6) intraparietal sulcus; (7) superior parietal gyrus (P1); (8) parietal lobule; (9) paracentral lobule; (10) ascending branch of cingulate sulcus; (11) precuneus; (12) cerebellar decussation; (13) lingual gyrus; (14) occipital horn of lateral ventricle; (15) collateral sulcus; (16) fusiform gyrus; (17) inferior temporal gyrus (T3).

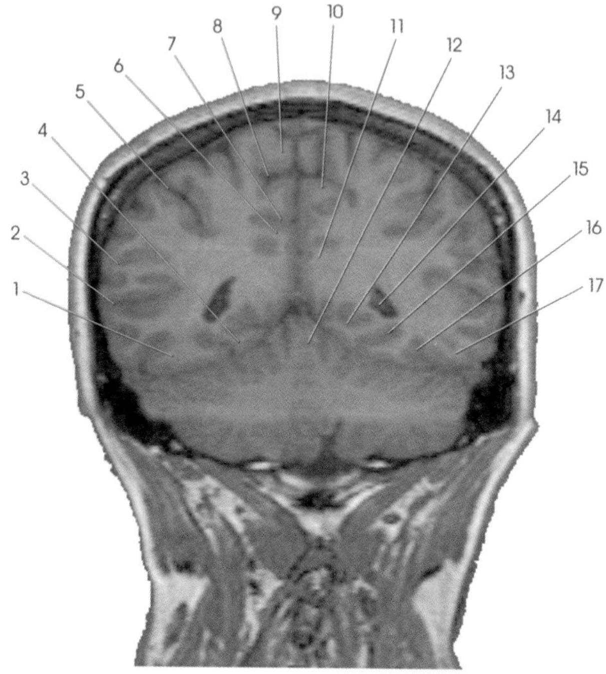


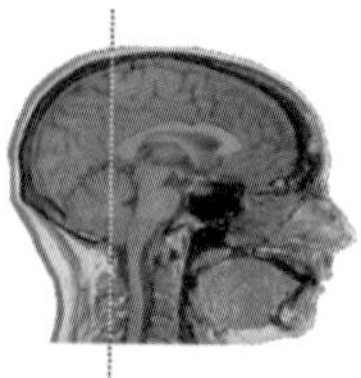

Image 17. (1) Fusiform gyrus; (2) parallel sulcus; (3) supramarginal gyrus; (4) lingual gyrus; (5) intraparietal sulcus; (6) cingulate gyrus; (7) subparietal sulcus; (8) marginal ramus of cingulate sulcus; (9) paracentral lobule; (10) cingulate gyrus; (11) cingulate gyrus; (12) declive; (13) calcarine sulcus; (14) occipital horn of lateral ventricle; (15) collateral sulcus; (16) occipitotemporal sulcus; (17) inferior temporal gyrus (T3).

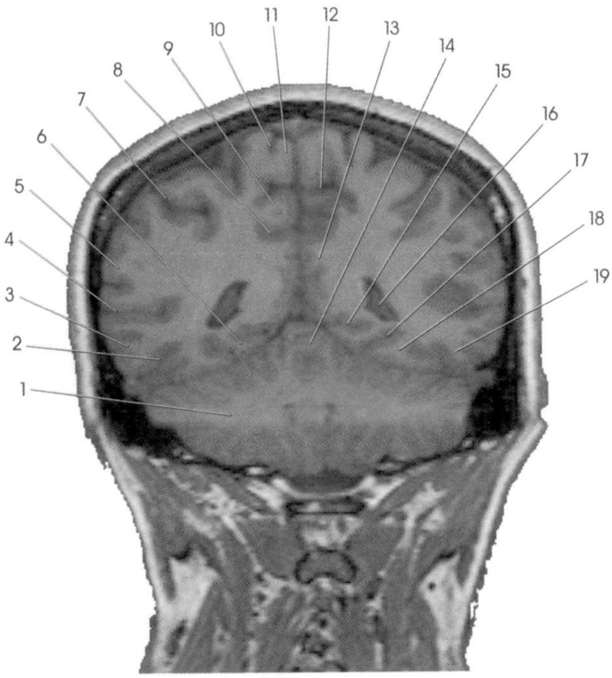


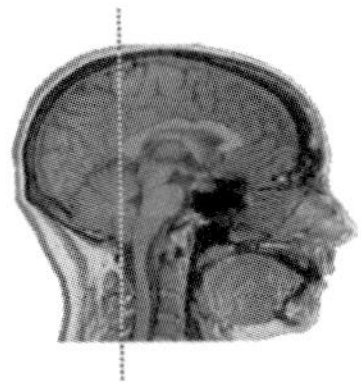

Image 18. (1) Corpus medullare of cerebellum; (2) lateral occipitotemporal sulcus; (3) inferior temporal gyrus (T3); (4) parallel sulcus; (5) supramarginal gyrus; (6) lingual gyrus; (7) intraparietal sulcus; (8) cingulate sulcus; (9) parietal lobule; (10) central sulcus; (11) parietal lobule; (12) marginal ramus of cingulate sulcus; (13) cingulate gyrus; (14) declive; (15) parieto-occipital sulcus; (16) atria of lateral ventricle; (17) collateral sulcus; (18) fusiform gyrus; (19) inferior temporal gyrus (T3).

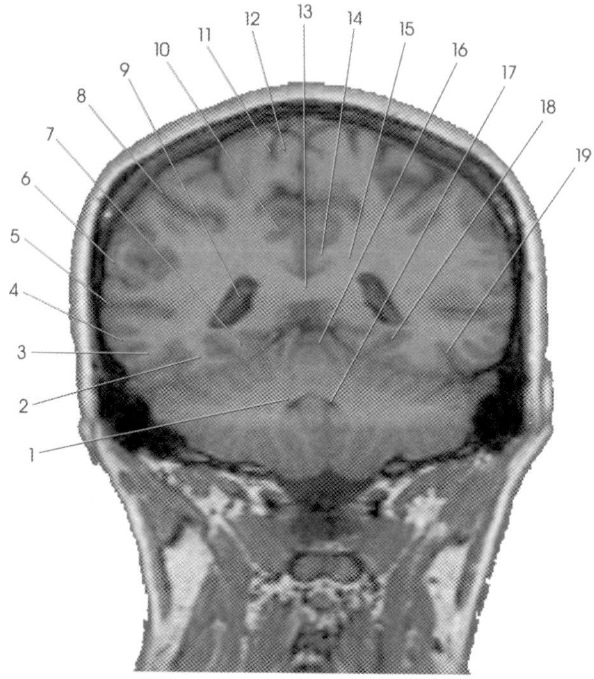


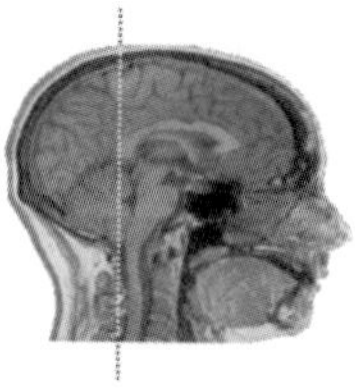

Image 19. (1) Superior cerebellar peduncle; (2) fusiform gyrus; (3) inferior temporal gyrus (T3); (4) inferior temporal gyrus (T3); (5) parallel sulcus; (6) supramarginal gyrus; (7) isthmus of cingulum; (8) intra-parietal sulcus; (9) atria of lateral ventricle; (10) cingulate sulcus; (11) central sulcus; (12) paracentral lobule; (13) splenium of corpus callosum; (14) cingulate gyrus; (15) forceps major of corpus callosum; (16) declive; (17) fourth ventricle; (18) collateral sulcus; (19) occipitotemporal sulcus.

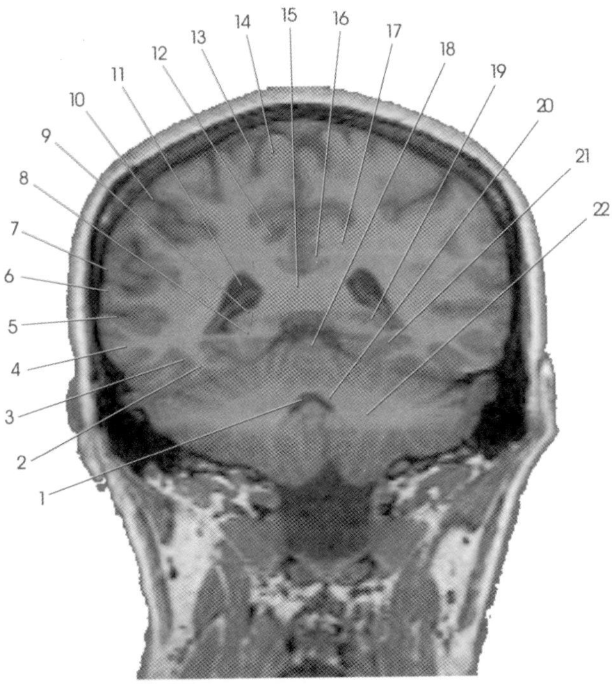


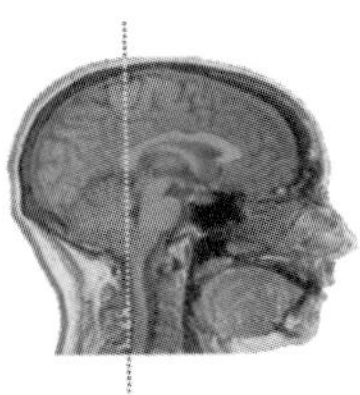

Image 20. (1) Fourth ventricle; (2) fusiform gyrus; (3) occipitotemporal sulcus; (4) middle temporal gyrus (T2); (5) parallel sulcus; (6) superior temporal gyrus (T1); (7) supramarginal gyrus; (8) isthmus; (9) crus of fornix; (10) intraparietal sulcus; (11) atria of lateral ventricle; (12) cingulate sulcus; (13) central sulcus; (14) paracentral lobule; (15) splenium of corpus callosum; (16) cingulate gyrus; (17) forceps major of corpus callosum; (18) anterior vermis of cerebellum; (19) hippocampus; (20) superior cerebellar peduncle; (21) parahippocampal gyrus; (22) middle cerebellar peduncle.

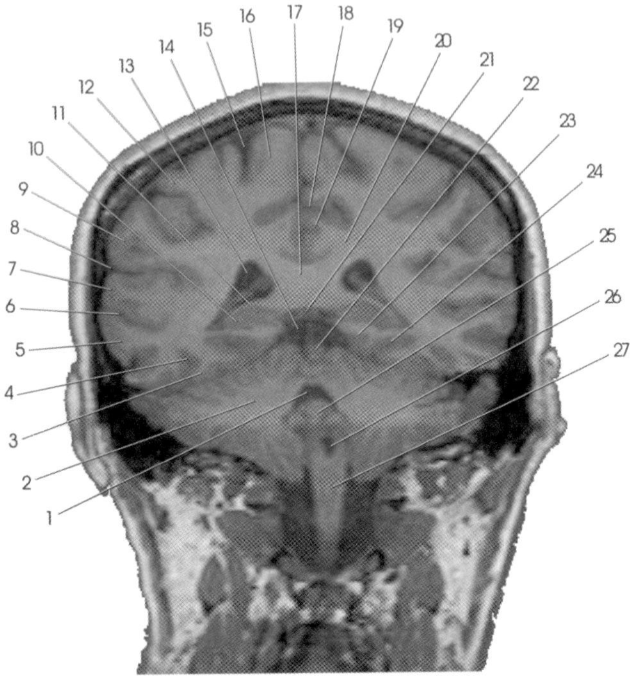


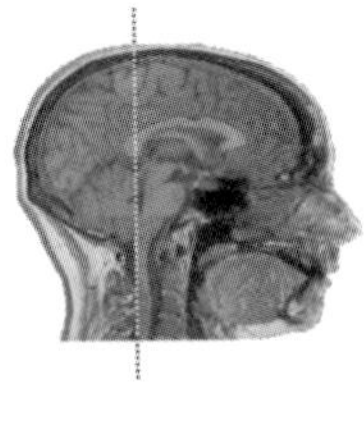

Image 21. (1) Fourth ventricle; (2) middle cerebellar peduncle; (3) fusiform gyrus; (4) occipitotemporal sulcus; (5) middle temporal gyrus (T2); (6) parallel sulcus; (7) superior temporal gyrus (T1); (8) lateral fissure; (9) supramarginal gyrus; (10) hippocampus; (11) crus of fornix; (12) postcentral gyrus; (13) atria of lateral ventricle; (14) superior cerebellar cistern; (15) central sulcus; (16) superior frontal gyrus (F1); (17) splenium of corpus callosum; (18) cingulate sulcus; (19) cingulate gyrus; (20) forceps major of corpus callosum; (21) pineal body; (22) anterior vermis of cerebellum; (23) parahippocampal gyrus; (24) collateral sulcus; (25) nodulus of cerebellum; (26) spinal aqueduct; (27) medulla.

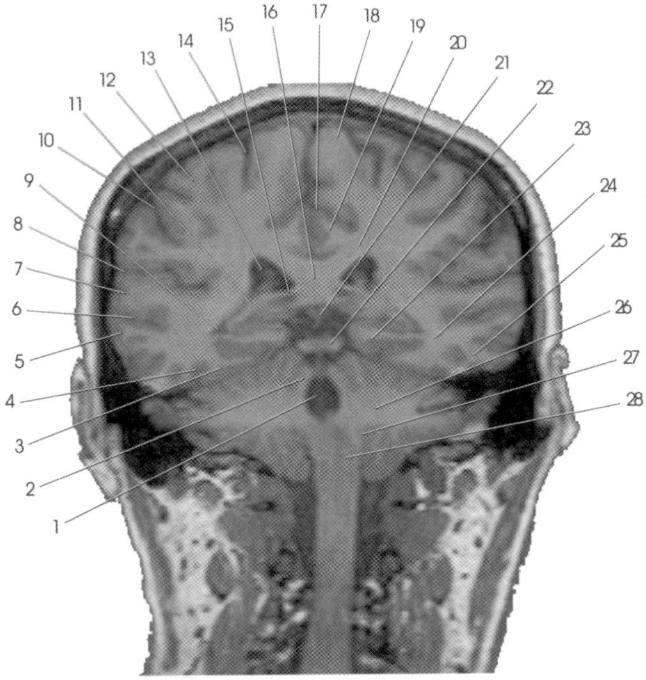


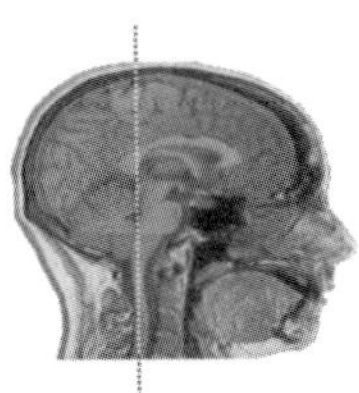

Image 22. (1) Fourth ventricle; (2) superior cerebellar peduncle; (3) fusiform gyrus; (4) occipitotemporal sulcus; (5) middle temporal gyrus (T2); (6) parallel sulcus; (7) superior temporal gyrus (T1); (8) lateral fissure; (9) atria of lateral ventricle; (10) intra-parietal sulcus; (11) pulvinar; (12) postcentral gyrus; (13) body of lateral ventricle; (14) central sulcus; (15) fornix; (16) splenium of corpus callosum; (17) cingulate sulcus; (18) paracentral lobule; (19) cingulate gyrus; (20) forceps major of corpus callosum; (21) pineal body; (22) inferior colliculus; (23) parahippocampal gyrus; (24) inferior longitudinal fasciculous; (25) inferior temporal gyrus (T3); (26) middle cerebellar peduncle; (27) inferior cerebellar peduncle; (28) medulla.

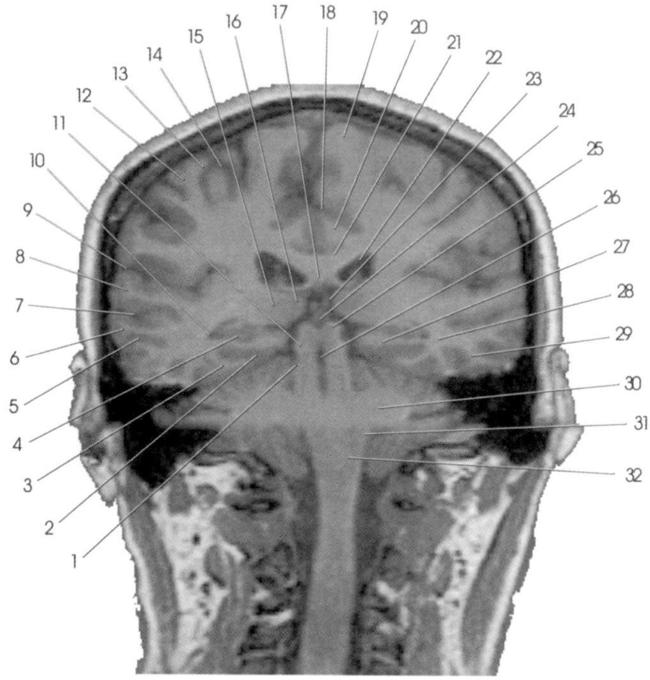


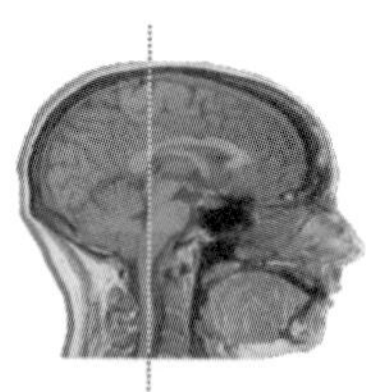

Image 23. (1) Superior cerebellar peduncle; (2) collateral sulcus; (3) fusiform gyrus; (4) hippocampus; (5) inferior temporal sulcus; (6) middle temporal gyrus (T2); (7) parallel sulcus; (8) superior temporal gyrus (T1); (9) lateral fissure; (10) temporal horn of lateral ventricle; (11) inferior colliculus; (12) postcentral gyrus; (13) central sulcus; (14) central sulcus; (15) pulvinar; (16) dorsomedial thalamic nucleus; (17) corpus callosum; (18) cingulate sulcus; (19) superior frontal gyrus (F1); (20) cingulate gyrus; (21) body of corpus callosum; (22) body of lateral ventricle; (23) supra pineal recess; (24) pineal body; (25) superior colliculus; (26) cerebral aqueduct; (27) parahippocampal gyrus; (28) temporal stem; (29) inferior temporal gyrus (T3); (30) middle cerebellar peduncle; (31) inferior cerebellar peduncle; (32) medulla.

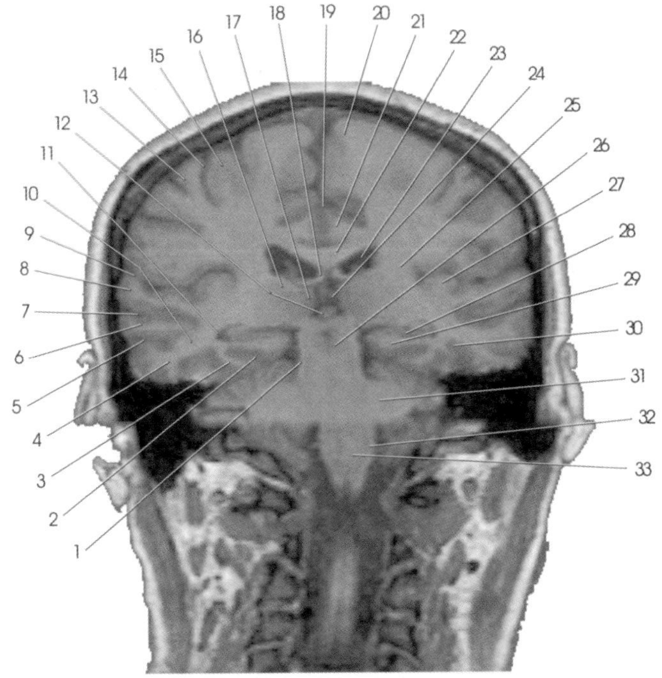


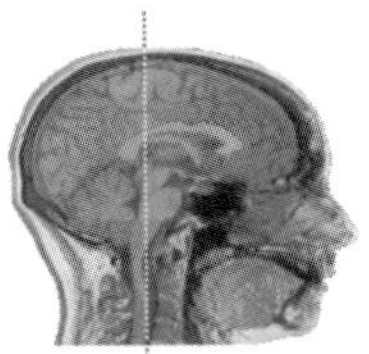

Image 24. (1) Superior cerebellar peduncle; (2) collateral sulcus; (3) fusiform gyrus; (4) inferior temporal gyrus (T3); (5) inferior temporal sulcus; (6) middle temporal gyrus (T2); (7) parallel sulcus; (8) superior temporal gyrus (T1); (9) lateral fissure; (10) temporal stem; (11) temporal horn of lateral ventricle; (12) pineal body; (13) postcentral gyrus; (14) central sulcus; (15) superior frontal gyrus (F1); (16) ventral lateral thalamic nucleus; (17) dorsomedial thalamic nucleus; (18) fornix; (19) cingulate sulcus; (20) superior frontal gyrus (F1); (21) cingulate gyrus; (22) body of corpus callosum; (23) body of lateral ventricle; (24) cistern of velum interpositum; (25) internal capsule; (26) red nucleus; (27) insular cortex; (28) hippocampus; (29) parahippocampal gyrus; (30) occipitotemporal sulcus; (31) middle cerebellar peduncle; (32) inferior cerebellar peduncle; (33) medulla.

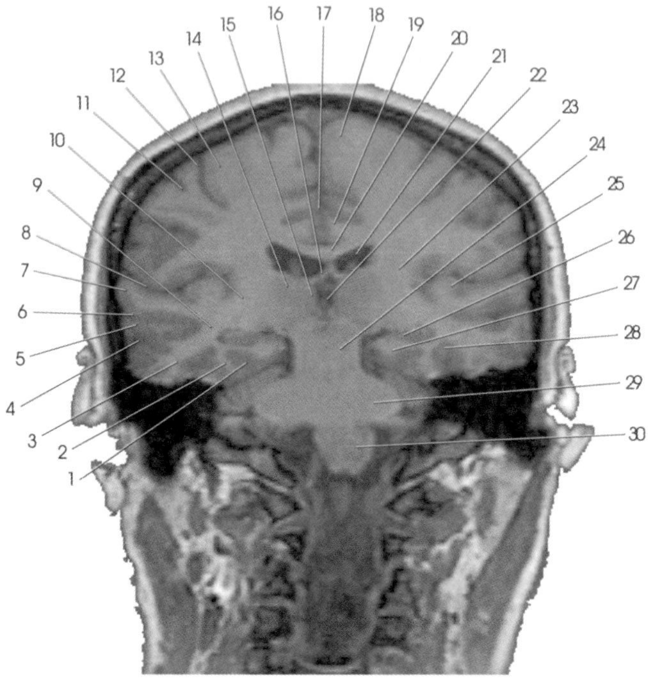


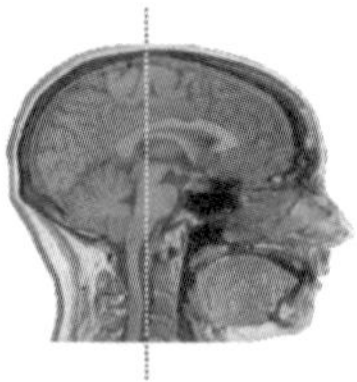

Image 25. (1) Collateral sulcus; (2) fusiform gyrus; (3) inferior temporal gyrus (T3); (4) inferior temporal sulcus; (5) middle temporal gyrus (T2); (6) parallel sulcus; (7) superior temporal gyrus (T1); (8) lateral fissure; (9) temporal stem; (10) internal capsule; (11) postcentral gyrus; (12) central sulcus; (13) precentral gyrus; (14) tail of caudate nucleus; (15) pulvinar; (16) fornix; (17) cingulate sulcus; (18) paracentral lobule; (19) cingulate gyrus; (20) body of corpus callosum; (21) body of lateral ventricle; (22) third ventricle; (23) posterior limb of internal capsule; (24) red nucleus; (25) transverse temporal gyrus; (26) hippocampus; (27) parahippocampal gyrus; (28) occipitotemporal sulcus; (29) cerebellar peduncle; (30) medulla.

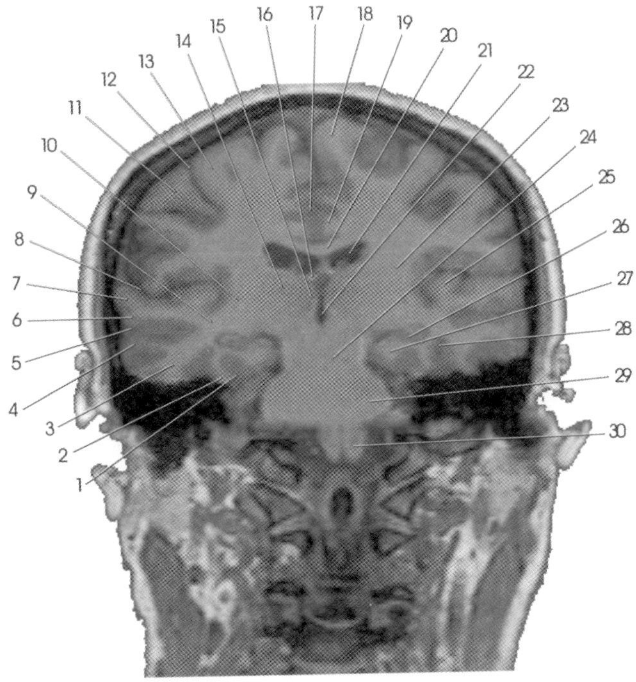


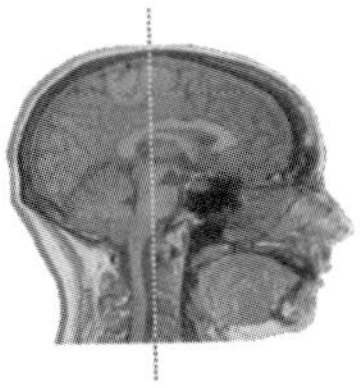

Image 26. (1) Collateral sulcus; (2) fusiform gyrus; (3) inferior temporal gyrus (T3); (4) middle temporal gyrus (T2); (5) parallel sulcus; (6) superior temporal gyrus (T1); (7) planum temporale; (8) lateral fissure; (9) temporal stem; (10) putamen; (11) postcentral gyrus; (12) central sulcus; (13) precentral gyrus; (14) ventral lateral thalamic nucleus; (15) dorsomedial thalamic nucleus; (16) fornix; (17) cingulate sulcus; (18) superior frontal gyrus (F1); (19) cingulate gyrus; (20) body of corpus callosum; (21) body of lateral ventricle; (22) third ventricle; (23) internal capsule; (24) red nucleus; (25) anterior transverse temporal gyrus; (26) hippocampus; (27) parahippocampal gyrus; (28) lateral occipitotemporal sulcus; (29) pons; (30) medulla.

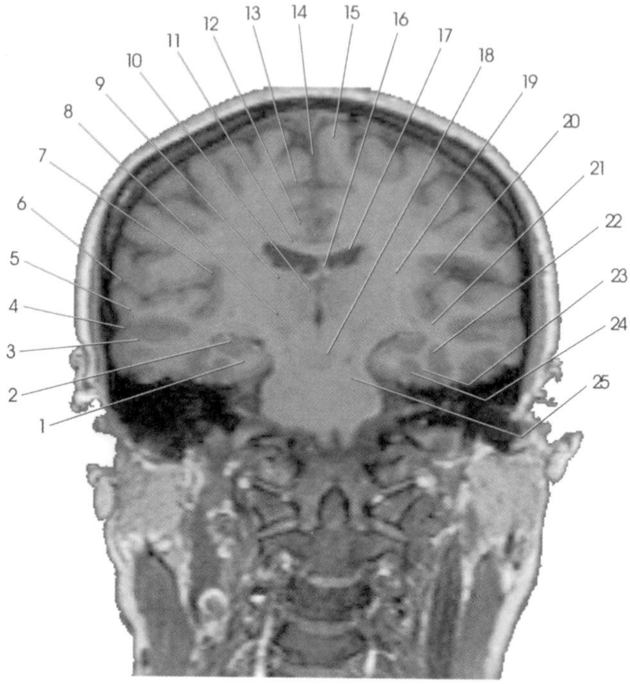


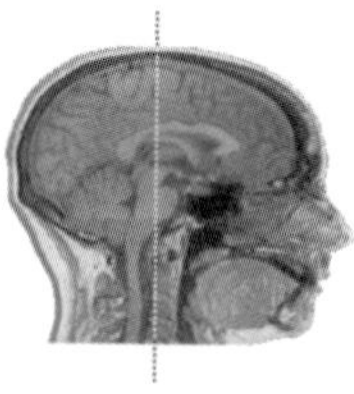

Image 27. (1) Parahippocampal gyrus; (2) hippocampus; (3) middle temporal gyrus (T2); (4) parallel sulcus; (5) superior temporal gyrus (T1); (6) lateral fissure; (7) circular insular sulcus; (8) medial lemniscus; (9) lateral dorsal thalamic nucleus; (10) dorsomedial thalamic nucleus; (11) body of corpus callosum; (12) cingulate gyrus; (13) cingulate sulcus; (14) interhemispheric fissure; (15) superior frontal gyrus (F1); (16) fornix; (17) body of lateral ventricle; (18) red nucleus; (19) internal capsule; (20) insular cortex; (21) temporal stem; (22) lateral occipitotemporal sulcus; (23) fusiform gyrus; (24) collateral sulcus; (25) cerebral peduncle.

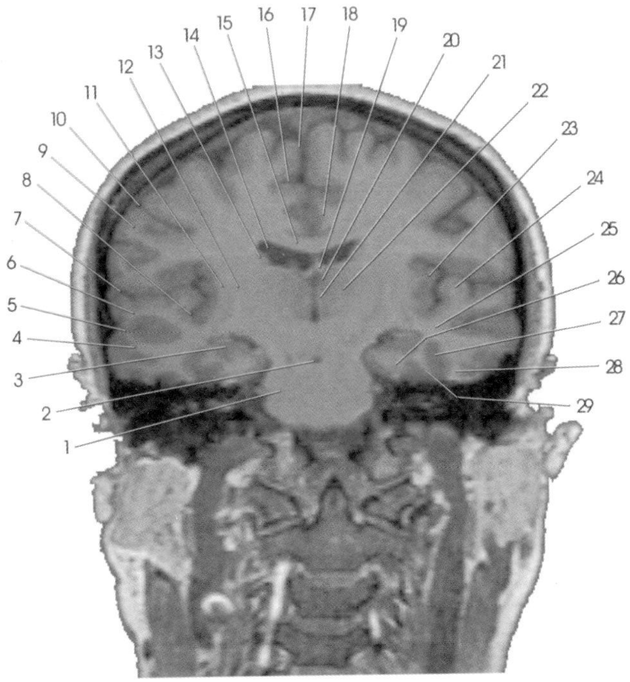


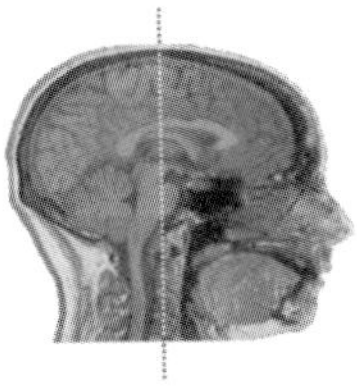

Image 28. (1) Pons; (2) cerebral peduncle; (3) hippocampus; (4) middle temporal gyrus (T2); (5) parallel sulcus; (6) superior temporal gyrus (T1); (7) lateral fissure; (8) circular insular sulcus; (9) precentral gyrus; (10) central sulcus; (11) external capsule; (12) internal capsule; (13) caudate nucleus; (14) body of lateral ventricle; (15) body of corpus callosum; (16) cingulate sulcus; (17) interhemispheric fissure; (18) cingulate gyrus; (19) fornix; (20) third ventricle; (21) dorsomedial thalamic nucleus; (22) ventral lateral thalamic nucleus; (23) insular cortex; (24) transverse temporal gyrus; (25) temporal stem; (26) parahippocampal gyrus; (27) temporo-occipital sulcus; (28) inferior temporal gyrus (T3); (29) fusiform gyrus.

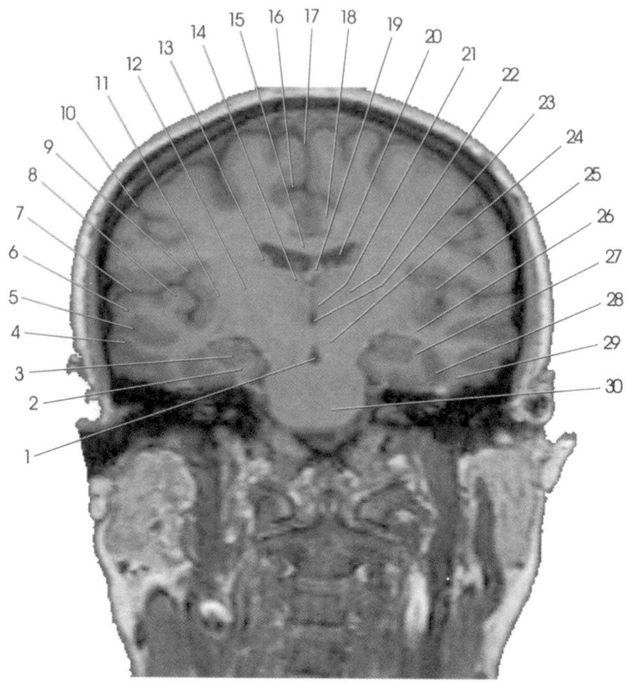


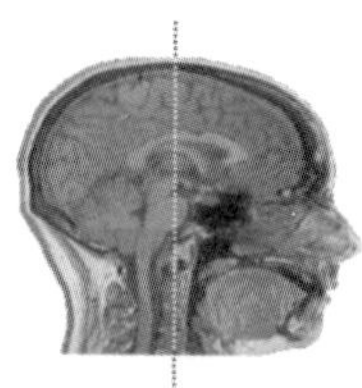

Image 29. (1) Interpeduncular cistern; (2) parahippocampal gyrus; (3) hippocampus; (4) middle temporal gyrus (T2); (5) parallel sulcus; (6) superior temporal gyrus (T1); (7) lateral fissure; (8) anterior transverse temporal gyrus; (9) circular insular sulcus; (10) central sulcus; (11) putamen; (12) internal capsule; (13) head of caudate; (14) lateral dorsal thalamic nucleus; (15) body of corpus callosum; (16) cingulate sulcus; (17) interhemispheric fissure; (18) cingulate gyrus; (19) fornix; (20) body of lateral ventricle; (21) dorsomedial thalamic nucleus; (22) ventral lateral thalamic nucleus; (23) third ventricle; (24) red nucleus; (25) insular cortex; (26) temporal stem; (27) fusiform gyrus; (28) occipitotemporal sulcus; (29) inferior temporal gyrus (T3); (30) pons.

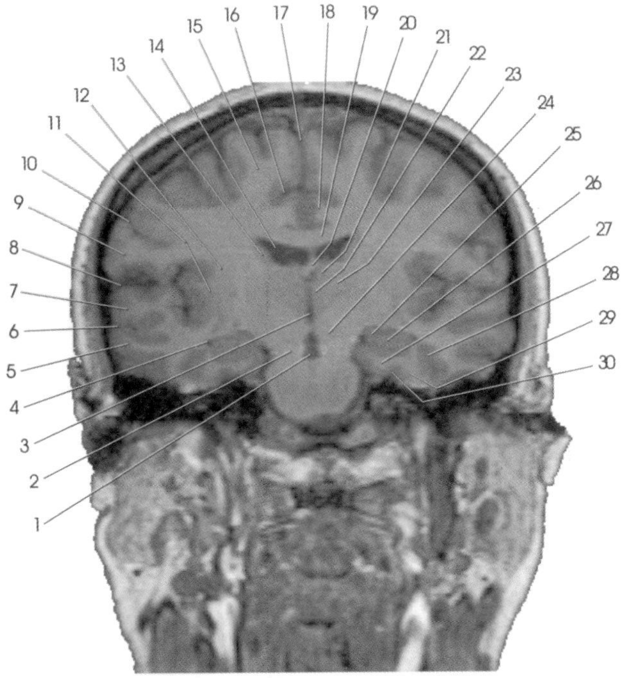


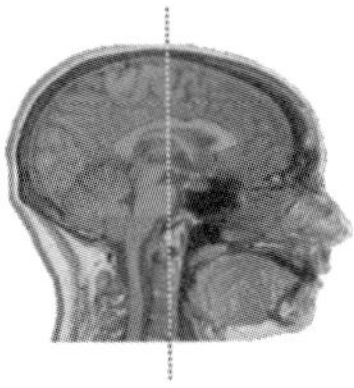

Image 30. (1) Interpeduncular cistern; (2) crus cerebri; (3) third ventricle; (4) temporal horn of lateral ventricle; (5) middle temporal gyrus (T2); (6) parallel sulcus; (7) superior temporal gyrus (T1); (8) lateral fissure; (9) postcentral gyrus; (10) central sulcus; (11) putamen; (12) internal capsule; (13) tail of caudate nucleus; (14) body of lateral ventricle; (15) superior frontal gyrus (F1); (16) cingulate sulcus; (17) interhemispheric fissure; (18) cingulate gyrus; (19) body of corpus callosum; (20) fornix; (21) anterior thalamic nucleus; (22) dorsomedial thalamic nucleus; (23) ventral lateral thalamic nucleus; (24) red nucleus; (25) hippocampus; (26) hippocampus; (27) parahippocampal gyrus; (28) occipitotemporal sulcus; (29) fusiform gyrus; (30) collateral sulcus.

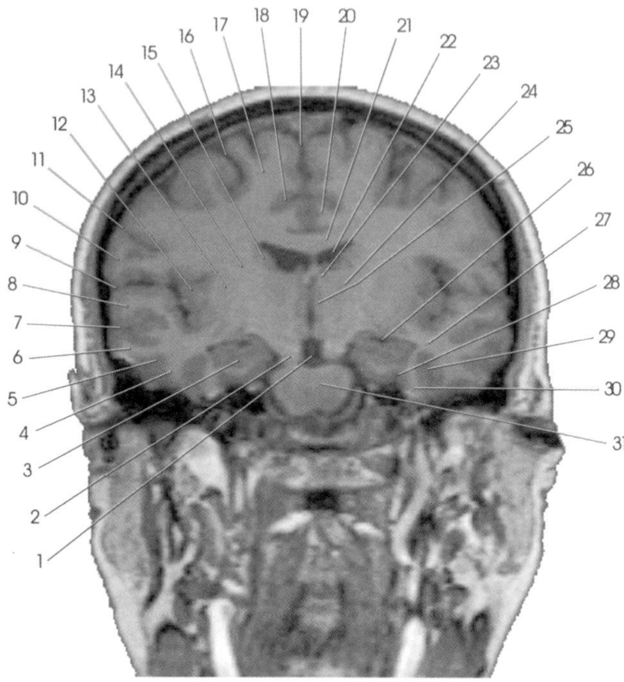


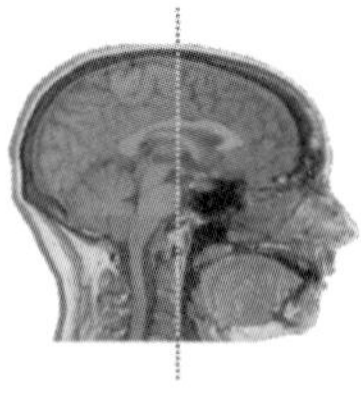

Image 31. (1) Interpeduncular cistern; (2) cerebral peduncle; (3) hippocampus; (4) inferior temporal gyrus (T3); (5) inferior temporal sulcus; (6) middle temporal gyrus (T2); (7) parallel sulcus; (8) superior temporal gyrus (T1); (9) lateral fissure; (10) precentral gyrus; (11) central sulcus; (12) insular cortex; (13) putamen; (14) internal capsule; (15) caudate nucleus; (16) superior precentral sulcus; (17) superior frontal gyrus (F1); (18) cingulate sulcus; (19) interhemispheric fissure; (20) cingulate gyrus; (21) body of corpus callosum; (22) body of lateral ventricle; (28) fornix; (24) ventral lateral thalamic nucleus; (25) dorsomedial thalamic nucleus; (26) temporal horn of lateral ventricle; (27) temporal stem; (28) collateral sulcus; (29) occipitotemporal sulcus; (30) fusiform gyrus; (31) pons.

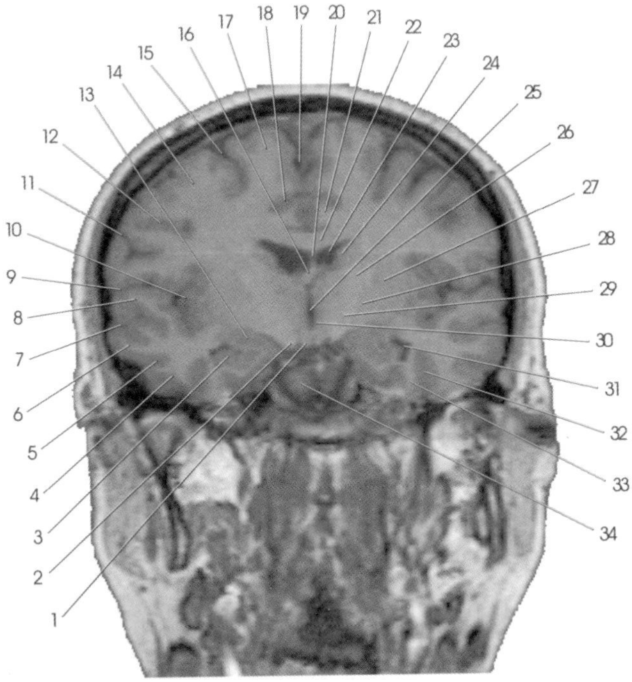


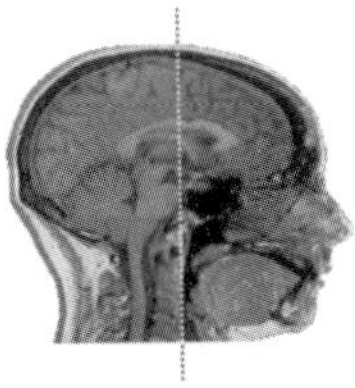

Image 32. (1) Mamillary body; (2) optic tract; (3) hippocampus; (4) inferior temporal gyrus (T3); (5) inferior temporal sulcus; (6) middle temporal gyrus (T2); (7) parallel sulcus; (8) superior temporal gyrus (T1); (9) lateral fissure; (10) insular cortex; (11) central sulcus; (12) inferior precentral sulcus; (13) amygdala; (14) middle frontal gyrus (F2); (15) superior frontal sulcus; (16) fornix; (17) superior frontal gyrus (F1); (18) cingulate sulcus; (19) interhemispheric fissure; (20) septum pellucidum; (21) cingulate gyrus; (22) body of corpus callosum; (23) body of lateral ventricle; (24) caudate nucleus; (25) third ventricle; (26) internal capsule; (27) putamen; (28) pars lateralis of globus pallidus; (29) pars medialis of globus pallidus; (30) hypothalamus; (31) temporal horn of lateral ventricle; (32) lateral occipitotemporal sulcus; (33) fusiform gyrus; (34) pons.

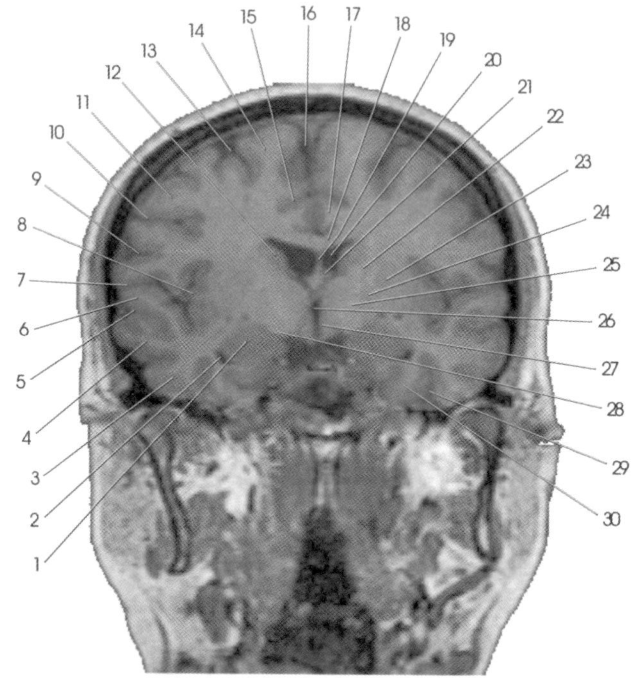


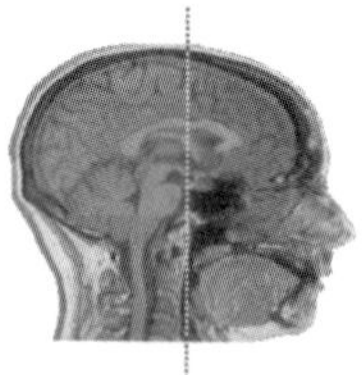

Image 33. (1) Hippocampus; (2) temporal horn of lateral ventricle; (3) inferior temporal gyrus (T3); (4) middle temporal gyrus (T2); (5) parallel sulcus; (6) superior temporal gyrus (T1); (7) lateral fissure; (8) insular cortex; (9) central sulcus; (10) inferior precentral sulcus; (11) middle frontal gyrus (F2); (12) caudate nucleus; (13) superior frontal sulcus; (14) superior frontal gyrus (F1); (15) cingulate sulcus; (16) interhemispheric fissure; (17) cingulate gyrus; (18) body of corpus callosum; (19) septum pellucidum; (20) body of lateral ventricle; (21) fornix; (22) anterior limb of internal capsule; (23) putamen; (24) pars lateralis of globus pallidus; (25) pars medialis of globus pallidus; (26) third ventricle; (27) hypothalamus; (28) optic tract; (29) lateral occipitotemporal sulcus; (30) fusiform gyrus.

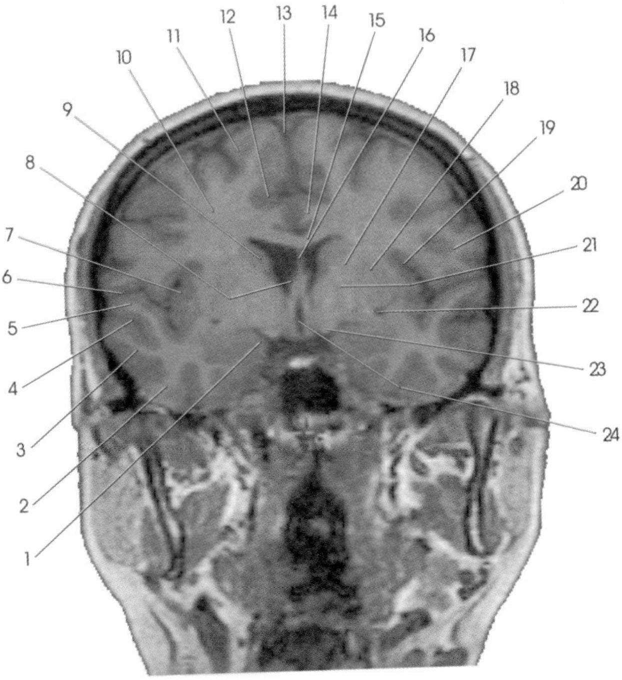


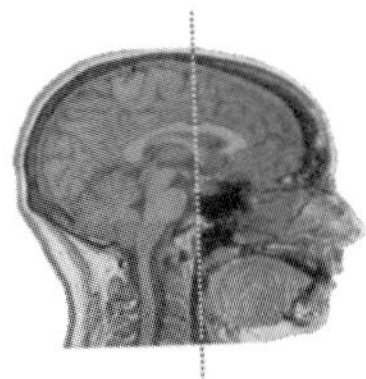

Image 34. (1) Uncus; (2) inferior temporal gyrus (T3); (3) middle temporal gyrus (T2); (4) parallel sulcus; (5) superior temporal gyrus (T1); (6) lateral fissure; (7) insular cortex; (8) fornix; (9) caudate nucleus; (10) corona radiata; (11) superior frontal gyrus (F1); (12) cingulate sulcus; (13) interhemispheric fissure; (14) cingulate gyrus; (15) body of corpus callosum; (16) septum pellucidum; (17) anterior limb of internal capsule; (18) putamen; (19) circular insular sulcus; (20) precentral gyrus; (21) genu of internal capsule; (22) middle cerebral artery; (23) optic tract; (24) chiasmic recess of third ventricle.

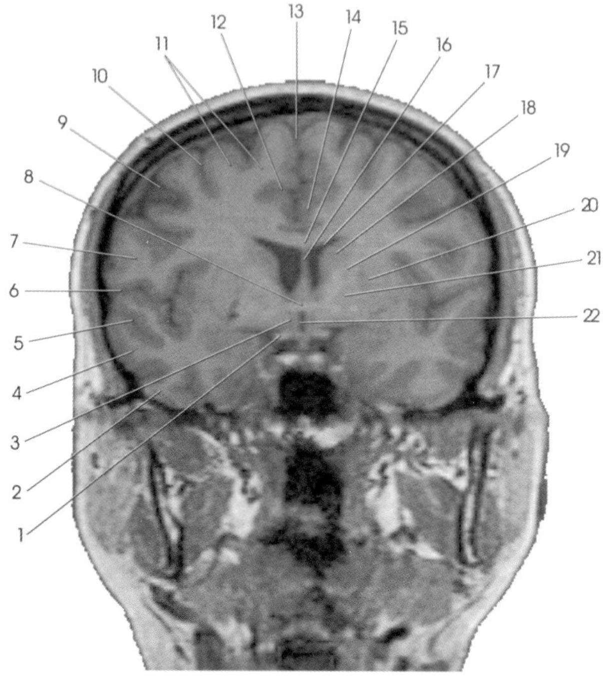


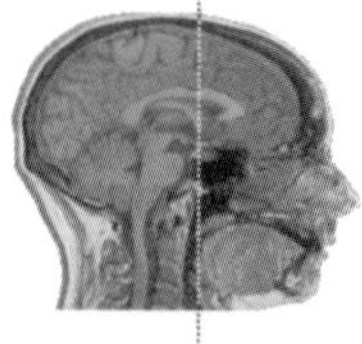

Image 35. (1) Optic chiasm; (2) inferior temporal gyrus (T3); (3) hypothalamus; (4) middle temporal gyrus (T2); (5) parallel sulcus; (6) lateral fissure; (7) precentral gyrus; (8) anterior commisure; (9) inferior precentral sulcus; (10) superior frontal sulcus; (11) superior frontal gyrus (F1); (12) cingulate sulcus; (13) interhemispheric fissure; (14) cingulate gyrus; (15) body of corpus callosum; (16) septum pellucidum; (17) anterior horn of lateral ventricle; (18) caudate nucleus; (19) anterior limb of internal capsule; (20) putamen; (21) globus pallidus; (22) third ventricle.

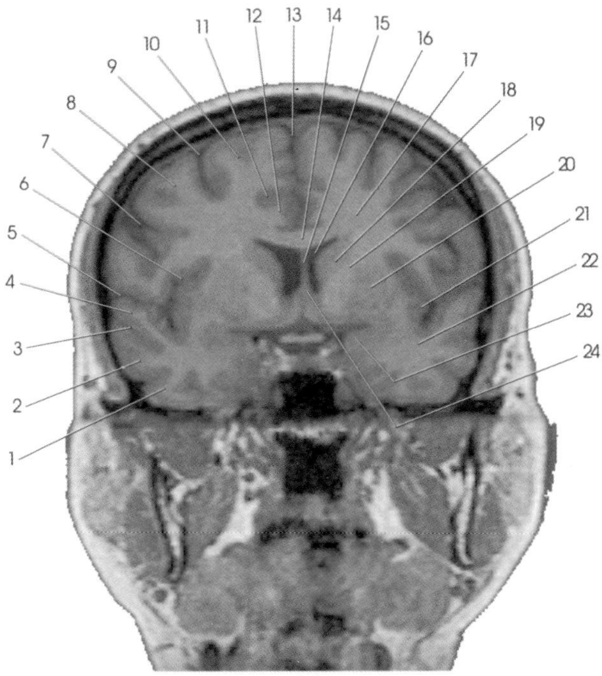


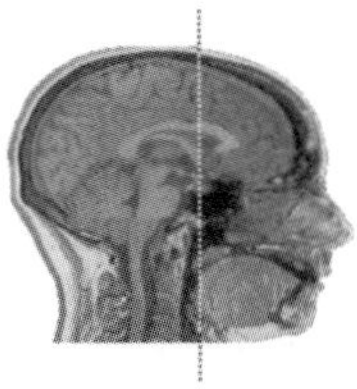

Image 36. (1) Inferior temporal gyrus (T3); (2) middle temporal gyrus (T2); (3) parallel sulcus; (4) superior temporal gyrus (T1); (5) lateral fissure; (6) circular insular sulcus; (7) inferior precentral sulcus; (8) middle frontal gyrus (F2); (9) superior frontal sulcus; (10) superior frontal gyrus (F1); (11) cingulate sulcus; (12) cingulate gyrus; (13) interhemispheric fissure; (14) body of corpus callosum; (15) septum pellucidum; (16) anterior horn of lateral ventricle; (17) corona radiata; (18) caudate nucleus; (19) anterior limb of internal capsule; (20) putamen; (21) insular cortex; (22) temporal stem; (23) amygdala; (24) septal nuclei.

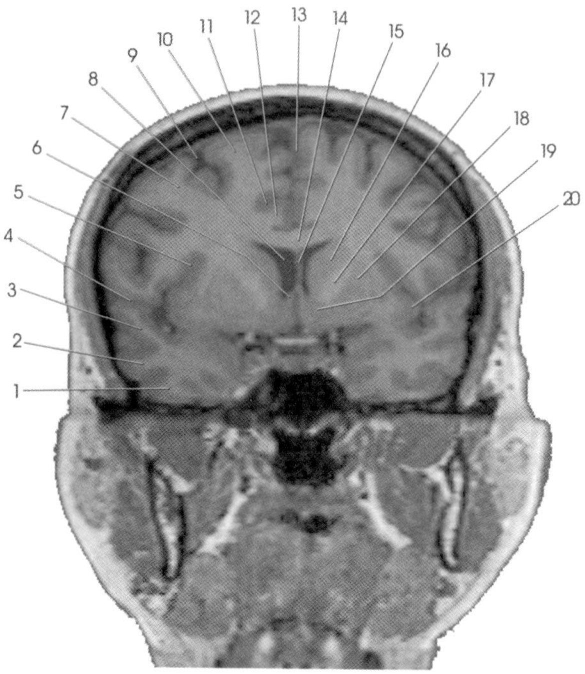


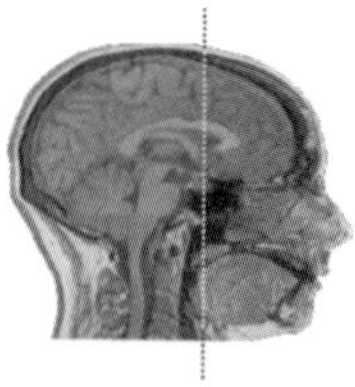

Image 37. (1) Inferior temporal gyrus (T3); (2) middle temporal gyrus (T2); (3) superior temporal gyrus (T1); (4) lateral fissure; (5) circular insular sulcus; (6) septal nucleus; (7) middle frontal gyrus (F2); (8) anterior horn of lateral ventricle; (9) superior frontal sulcus; (10) superior frontal gyrus (F1); (11) cingulate sulcus; (12) cingulate gyrus; (13) interhemispheric fissure; (14) body of corpus callosum; (15) septum pellucidum; (16) caudate nucleus; (17) anterior limb of internal capsule; (18) putamen; (19) nucleus accumbens; (20) insular cortex.

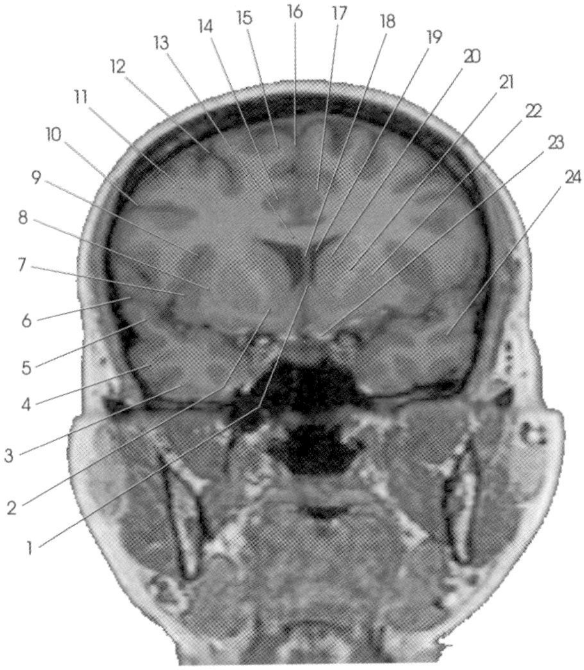


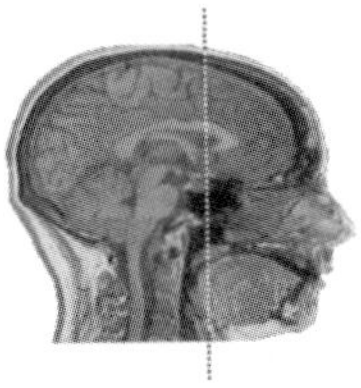

Image 38. (1) Lateral septal nucleus; (2) nucleus accumbens; (3) inferior temporal gyrus (T3); (4) middle temporal gyrus (T2); (5) superior temporal gyrus (T1); (6) lateral fissure; (7) insular cortex; (8) claustrum; (9) circular insular sulcus; (10) inferior frontal sulcus; (11) middle frontal gyrus (F2); (12) superior frontal sulcus; (13) body of corpus callosum; (14) cingulate sulcus; (15) superior frontal gyrus (F1); (16) interhemispheric fissure; (17) cingulate gyrus; (19) septum pellucidum; (19) frontal horn of lateral ventricle; (20) caudate nucleus; (21) internal capsule; (22) putamen; (23) optic nerve; (24) parallel sulcus.

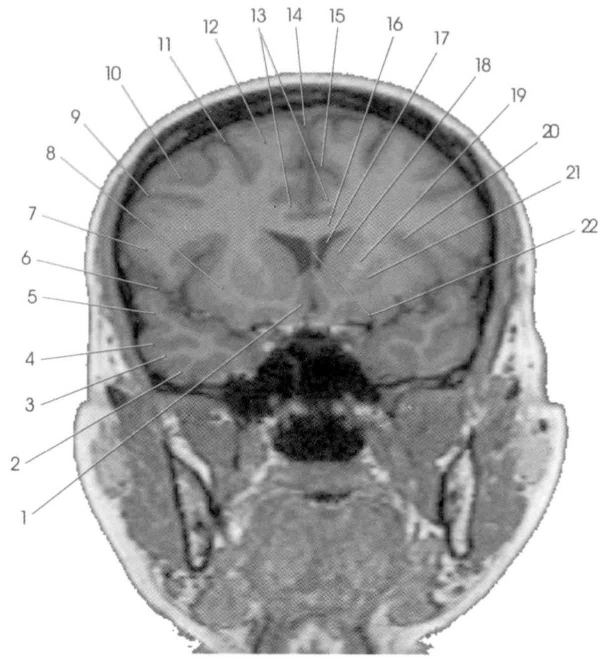


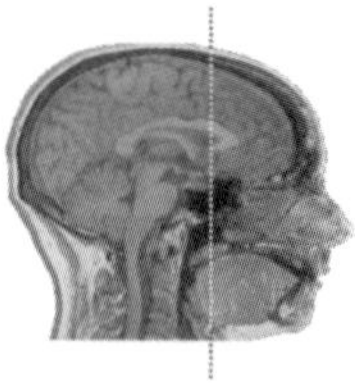

Image 39. (1) Subcallosal gyrus; (2) inferior temporal gyrus (T3); (3) middle temporal gyrus (T2); (4) parallel sulcus; (5) superior temporal gyrus (T1); (6) lateral fissure; (7) inferior frontal gyrus (F3); (8) claustrum; (9) inferior frontal sulcus; (10) middle frontal gyrus (F2); (11) superior frontal sulcus; (12) superior frontal gyrus (F1); (13) cingulate gyrus; (14) interhemispheric fissure; (15) cingulate sulcus; (16) body of corpus callosum; (17) frontal horn of lateral ventricle; (18) caudate nucleus; (19) internal capsule; (20) circular insular sulcus; (21) putamen; (22) septum pellucidum.

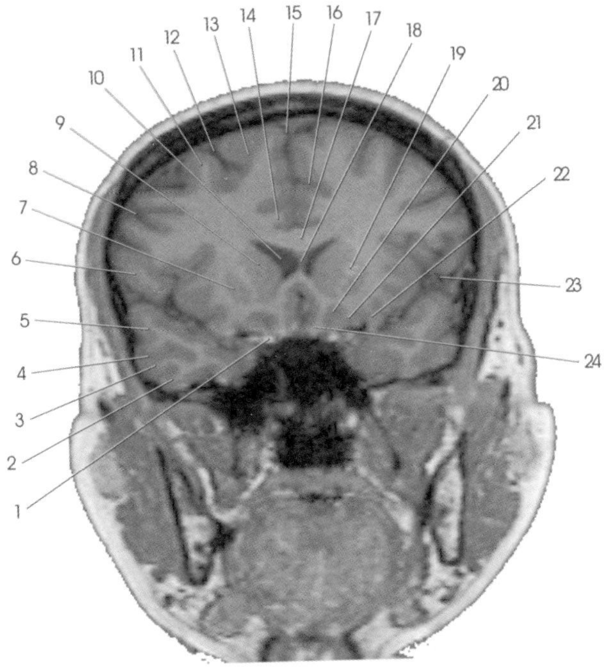


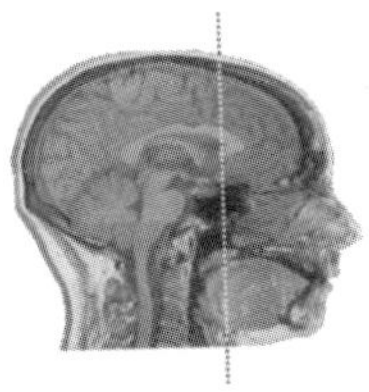

Image 40. (1) Optic nerve; (2) middle temporal gyrus (T2); (3) middle temporal gyrus (T2); (4) parallel sulcus; (5) superior temporal gyrus (T1); (6) inferior frontal gyrus (F3), pars opercularis; (7) putamen; (8) inferior frontal sulcus; (9) caudate nucleus; (10) frontal horn of lateral ventricle; (11) middle frontal gyrus (F2); (12) superior frontal sulcus; (13) superior frontal gyrus (F1); (14) cingulate gyrus; (15) interhemispheric fissure; (16) cingulate sulcus; (17) body of corpus callosum; (18) septum pellucidum; (19) anterior limb of internal capsule; (20) medial orbital sulcus; (21) medial orbital sulcus; (22) posterior orbital gyrus; (23) lateral fissure; (24) gyrus rectus.

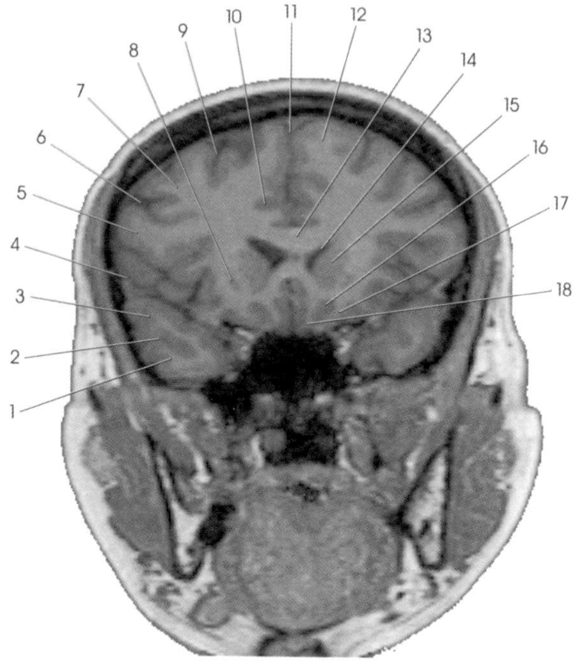


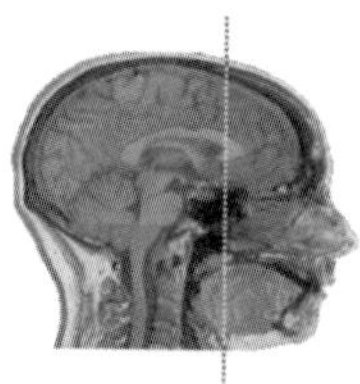

Image 41. (1) Middle temporal gyrus (T2); (2) parallel sulcus; (3) superior temporal gyrus (T1); (4) lateral fissure; (5) inferior frontal gyrus (F3), pars triangularis; (6) inferior frontal sulcus; (7) middle frontal gyrus (F2); (8) putamen; (9) superior frontal sulcus; (10) cingulate sulcus; (11) interhemispheric fissure; (12) superior frontal gyrus (F1); (13) body of corpus callosum; (14) frontal horn of lateral ventricle; (15) caudate nucleus; (16) olfactory sulcus; (17) medial orbital gyrus; (18) gyrus rectus.

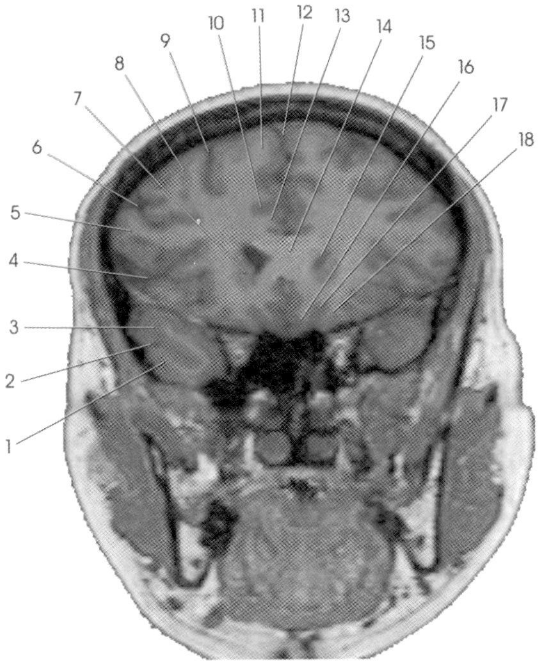


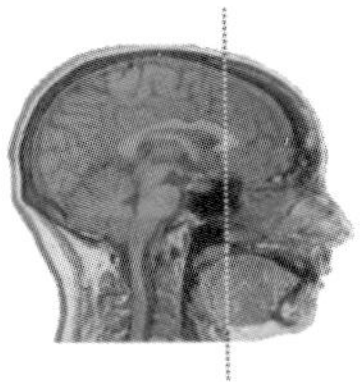

Image 42. (1) Middle temporal gyrus (T2); (2) parallel sulcus; (3) superior temporal gyrus (T1); (4) horizontal ramus of lateral fissure; (5) inferior frontal gyrus (F3), pars triangularis; (6) inferior frontal sulcus; (7) head of caudate nucleus; (8) middle frontal gyrus (F2); (9) superior frontal sulcus; (10) cingulate sulcus; (11) superior frontal gyrus (F1); (12) interhemispheric fissure; (13) cingulate gyrus; (14) genu of corpus callosum; (15) frontal horn of lateral ventricle; (16) gyrus rectus; (17) olfactory sulcus; (18) posterior orbital gyrus.

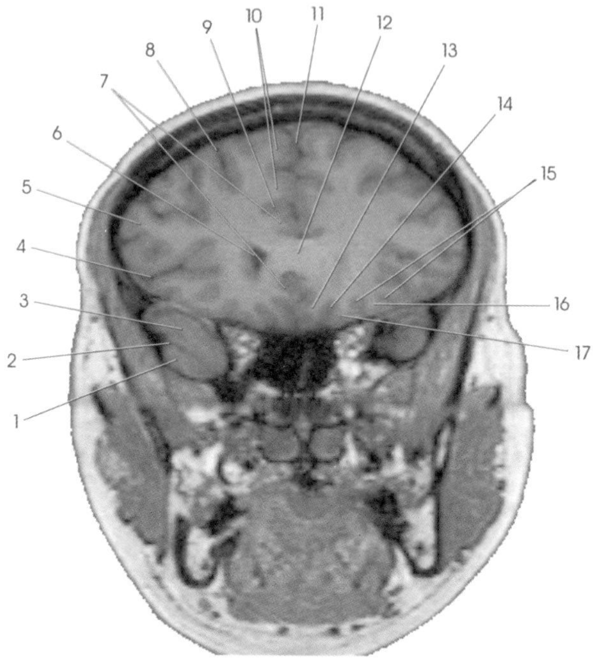


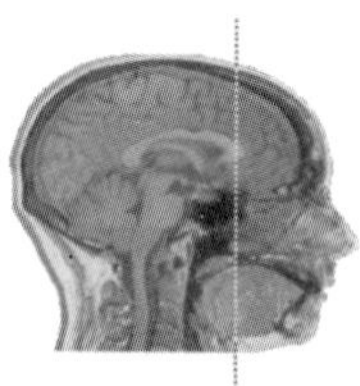

Image 43. (1) Middle temporal gyrus (T2); (2) parallel sulcus; (3) superior temporal gyrus (T1); (4) lateral fissure; (5) inferior frontal gyrus (F3), pars triangularis; (6) frontal horn of lateral ventricle; (7) cingulate gyrus; (8) superior frontal sulcus; (9) cingulate sulcus; (10) superior frontal gyrus (F1); (11) interhemispheric fissure; (12) genu of corpus callosum; (13) gyrus rectus; (14) medial orbital sulcus; (15) H-shaped orbital sulcus; (16) anterior orbital gyrus; (17) medial orbital gyrus.

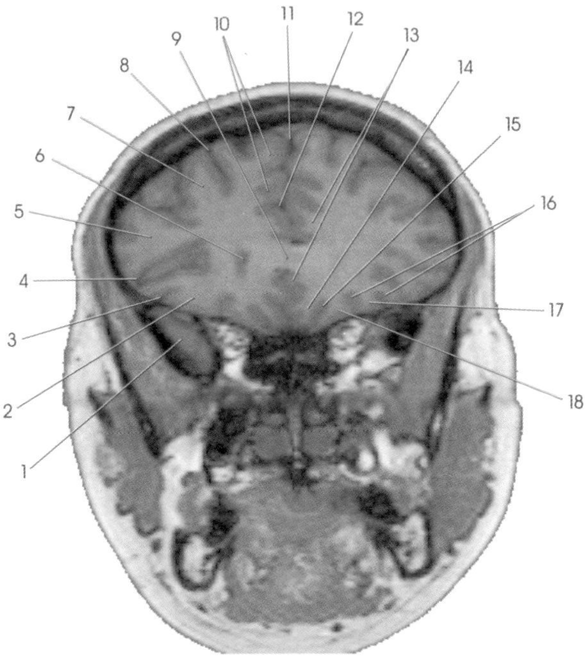


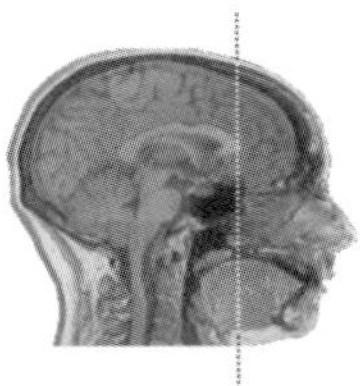

Image 44. (1) Temporal pole; (2) posterior orbital gyrus; (3) orbital sulcus; (4) horizontal ramus of lateral fissure; (5) inferior frontal gyrus (F3), pars triangularis; (6) frontal horn of lateral ventricle; (7) middle frontal gyrus (F2); (8) superior frontal sulcus; (9) genu of corpus callosum; (10) superior frontal gyrus (F1); (11) interhemispheric fissure; (12) cingulate sulcus; (13) cingulate gyrus; (14) gyrus rectus; (15) olfactory sulcus; (16) H-shaped orbital sulcus; (17) posterior orbital gyrus; (18) medial orbital gyrus.

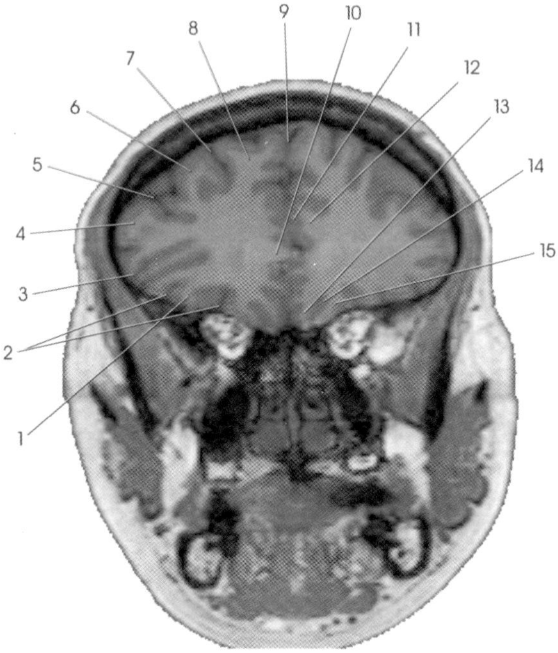


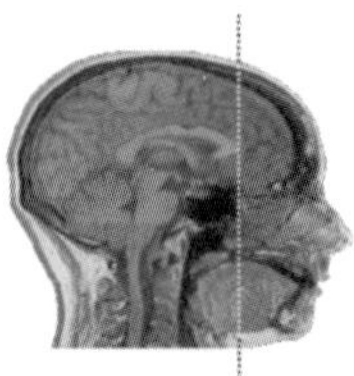

Image 45. (1) Medial orbital gyrus; (2) orbital sulci; (3) horizontal ramus of lateral fissure; (4) inferior frontal gyrus (F3), pars triangularis; (5) inferior frontal sulcus; (6) middle frontal gyrus (F2); (7) superior frontal sulcus; (8) superior frontal gyrus (F1); (9) interhemispheric fissure; (10) genu of corpus callosum; (11) cingulate sulcus; (12) cingulate gyrus; (13) gyrus rectus; (14) olfactory sulcus; (15) medial orbital gyrus.

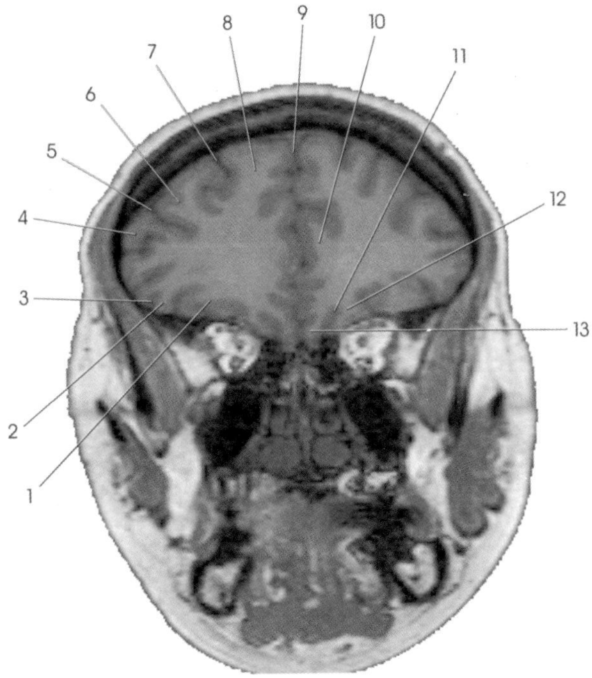


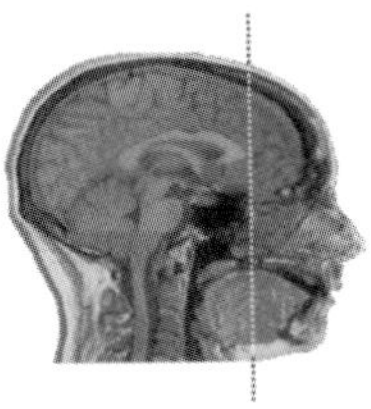

Image 46. (1) Posterior orbital gyrus; (2) lateral orbital gyrus; (3) lateral orbital sulcus; (4) inferior frontal gyrus (F3), pars orbitalis; (5) inferior frontal sulcus; (6) middle frontal gyrus (F2); (7) superior frontal sulcus; (8) superior frontal gyrus (F1); (9) interhemispheric fissure; (10) cingulate gyrus; (11) olfactory sulcus; (12) medial orbital gyrus; (13) gyrus rectus.

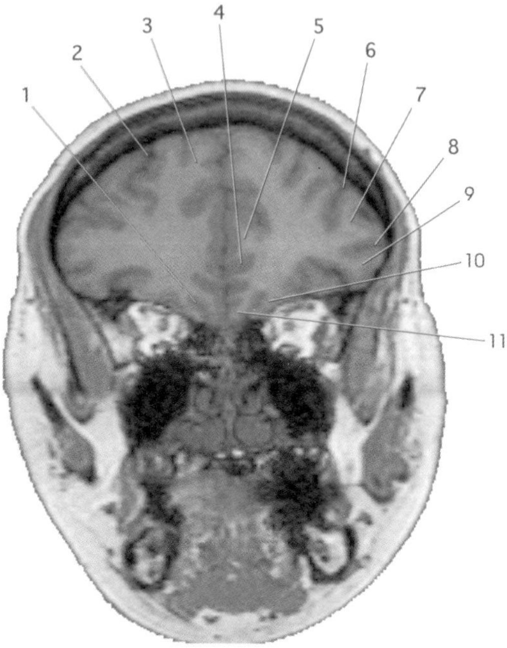


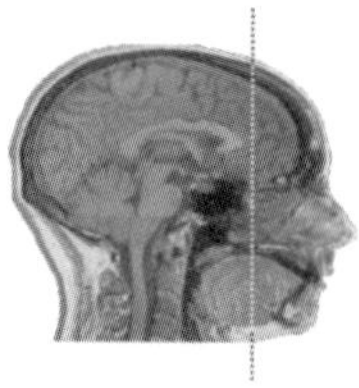

Image 47. (1) Olfactory sulcus; (2) superior frontal sulcus; (3) superior frontal gyrus (F1); (4) cingulate sulcus; (5) cingulate gyrus; (6) inferior frontal sulcus; (7) inferior frontal gyrus (F3); (8) lateral orbital sulcus; (9) lateral orbital gyrus; (10) medial orbital gyrus; (11) gyrus rectus.

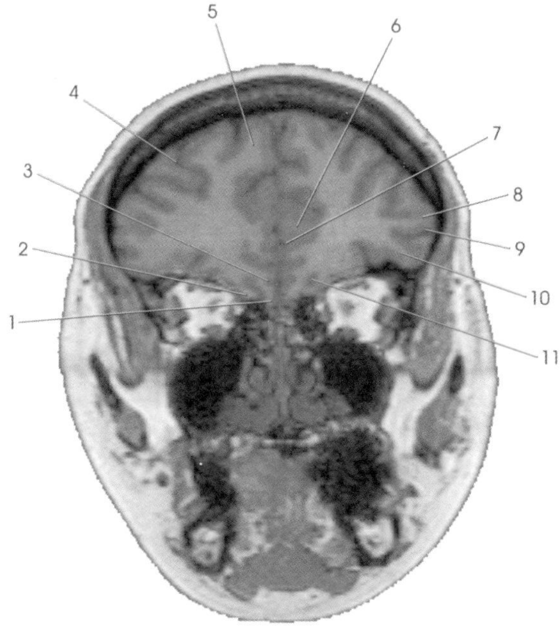


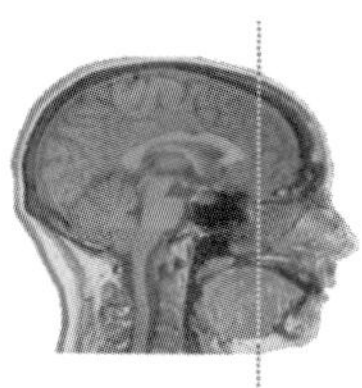

Image 48. (1) Gyrus rectus; (2) medial orbital gyrus; (3) suborbital sulcus; (4) superior frontal sulcus; (5) superior frontal gyrus (F1); (6) cingulate gyrus; (7) cingulate sulcus; (8) inferior frontal gyrus (F3); (9) lateral orbital sulcus; (10) lateral orbital gyrus; (11) olfactory sulcus.

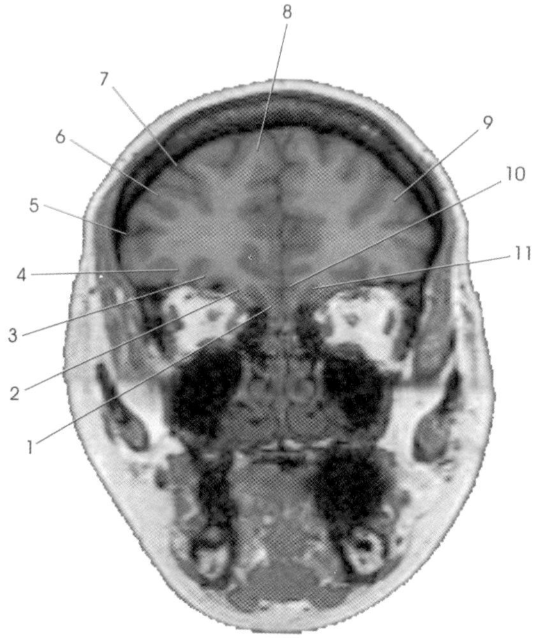


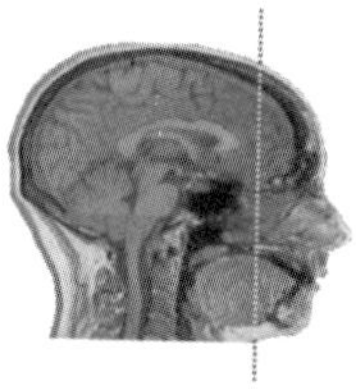

Image 49. (1) Gyrus rectus; (2) medial orbital gyrus; (3) anterior orbital gyrus; (4) lateral orbital gyrus; (5) inferior frontal sulcus; (6) middle frontal gyrus (F2); (7) superior frontal sulcus; (8) superior frontal gyrus (F1); (9) middle frontal gyrus (F2); (10) susorbital sulcus; (11) olfactory sulcus.

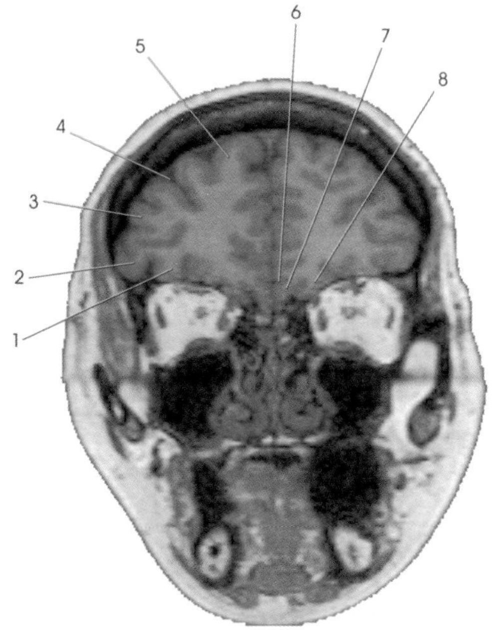


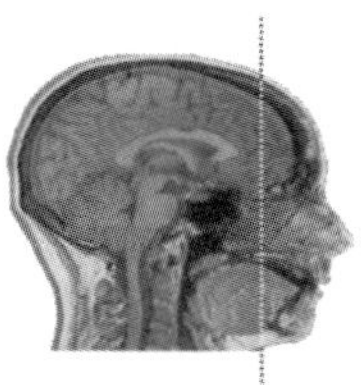

Image 50. (1) Anterior orbital gyrus; (2) lateral orbital gyrus; (3) middle frontal gyrus (F2); (4) superior frontal sulcus; (5) superior frontal gyrus (F1); (6) susorbital sulcus; (7) gyrus rectus; (8) medial orbital gyrus.

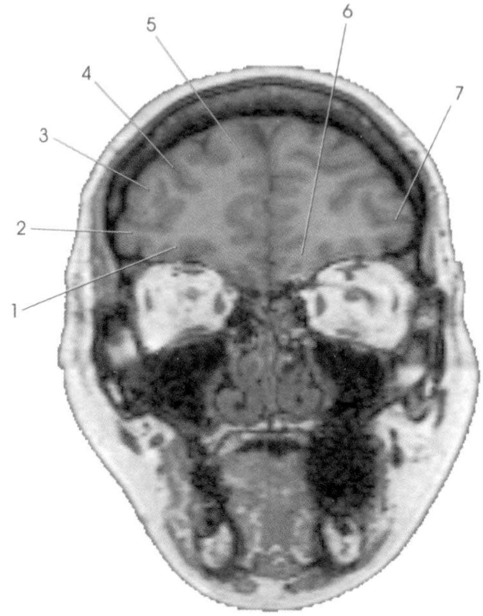


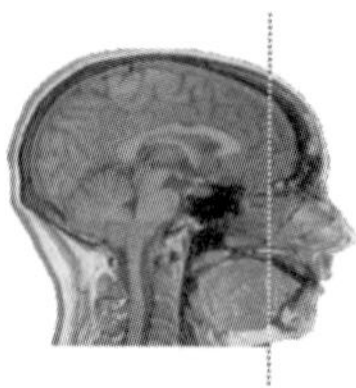

Image 51. (1) Anterior orbital gyrus; (2) lateral orbital gyrus; (3) middle frontal gyrus (F2); (4) superior frontal sulcus; (5) superior frontal gyrus (F1); (6) medial orbital gyrus; (7) lateral orbital sulcus.

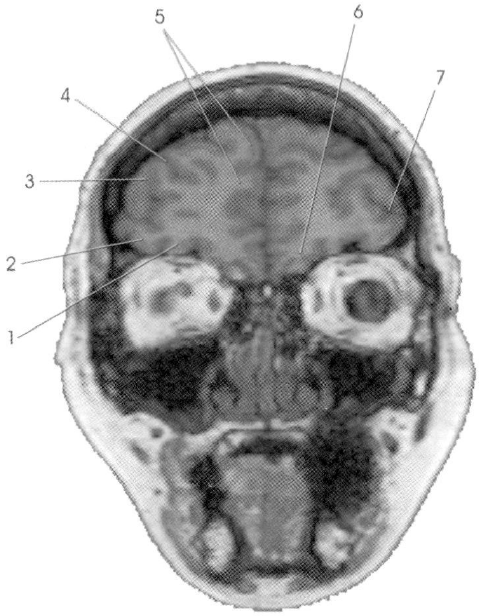


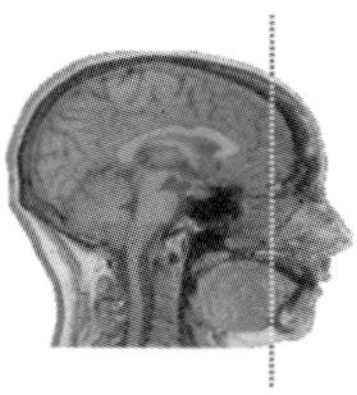

Image 52. (1) Anterior orbital gyrus; (2) lateral orbital gyrus; (3) middle frontal gyrus (F2); (4) superior frontal sulcus; (5) superior frontal gyrus (F1); (6) medial orbital gyrus; (7) lateral orbital sulcus.

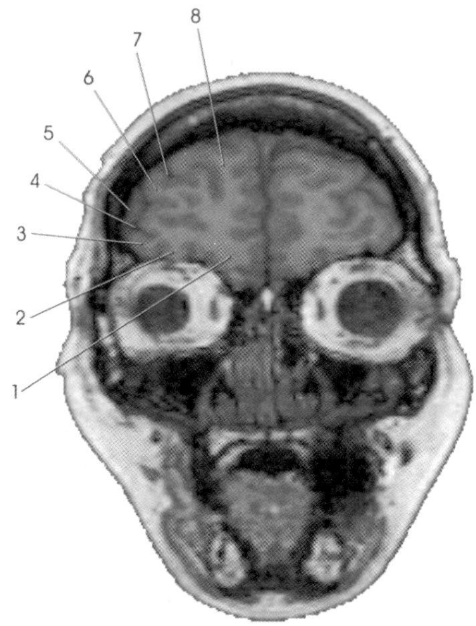


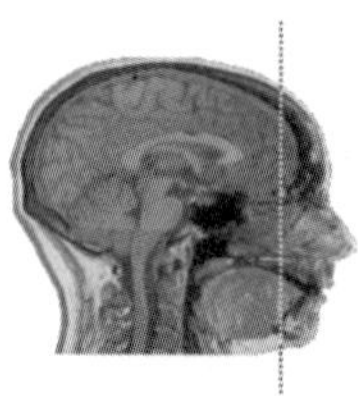

Image 53. (1) Medial orbital gyrus; (2) anterior orbital gyrus; (3) lateral orbital gyrus; (4) lateral orbital sulcus; (5) lateral orbital gyrus; (6) middle frontal gyrus (F2); (7) superior frontal sulcus; (8) superior frontal gyrus (F1).

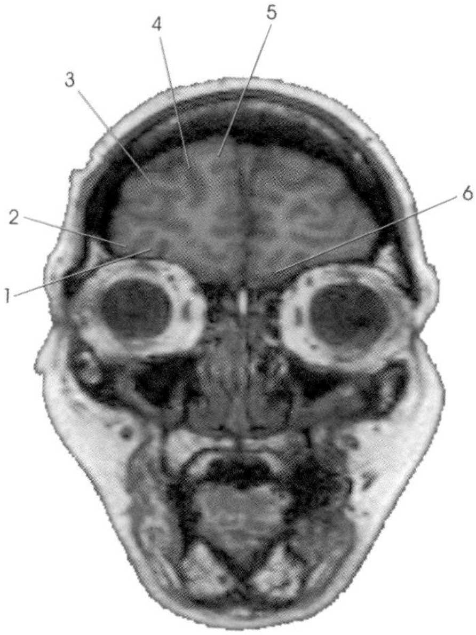


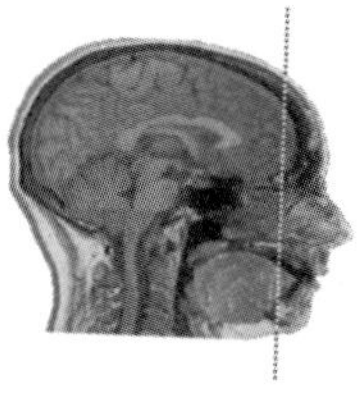

Image 54. (1) Anterior orbital gyrus; (2) lateral orbital gyrus; (3) middle frontal gyrus (F2); (4) superior frontal sulcus; (5) superior frontal gyrus (F1); (6) medial orbital gyrus.

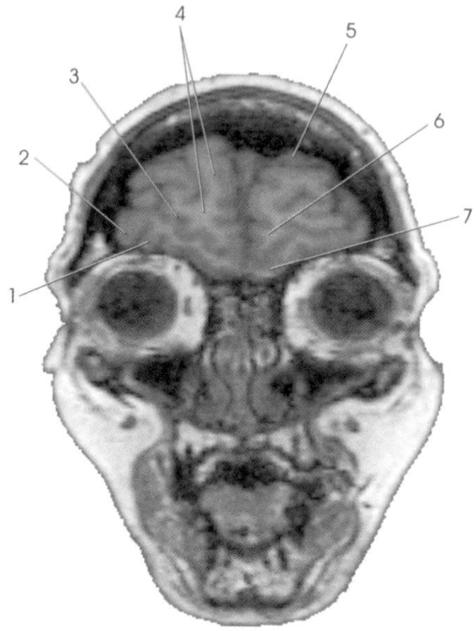


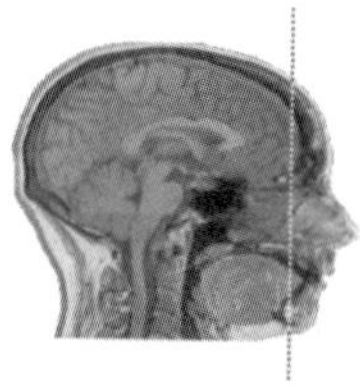

Image 55. (1) Frontomarginal gyrus; (2) frontomarginal gyrus; (3) frontomarginal sulcus; (4) transverse frontopolar gyri; (5) superior frontal gyrus (F1); (6) inferior transverse frontopolar gyrus; (7) gyrus rectus.

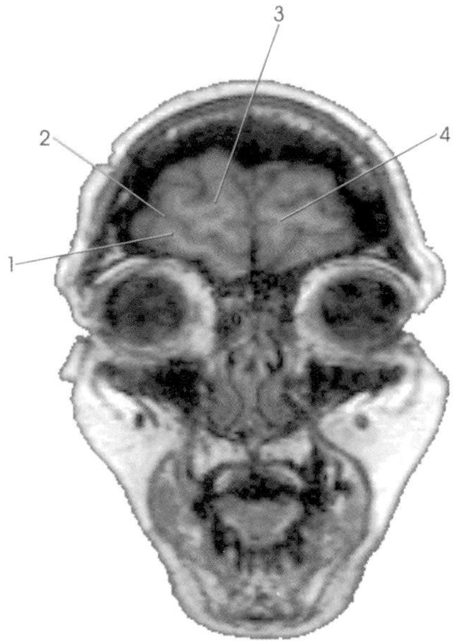


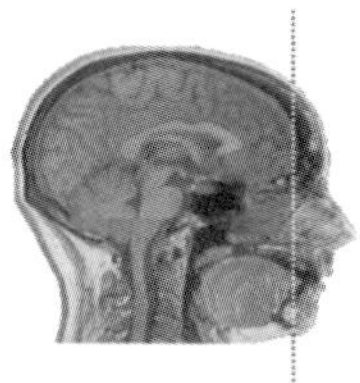

Image 56. (1) Frontomarginal gyrus; (2) frontomarginal sulcus; (3) transverse frontopolar gyrus; (4) transverse frontopolar gyrus.

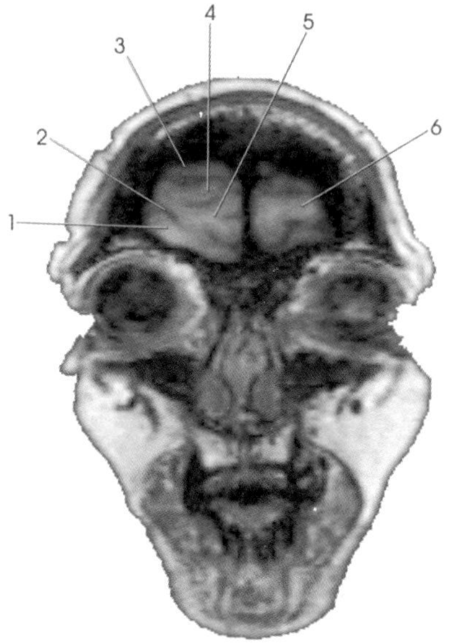


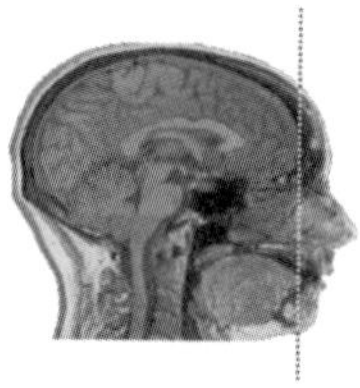

Image 57. (1) Frontomarginal gyrus; (2) frontomarginal sulcus; (3) transverse frontopolar gyrus; (4) transverse frontopolar gyrus; (5) transverse frontopolar gyrus; (6) transverse frontopolar gyrus.

# Index

ISBN 0-306-45228-6
90000
9 780306 452284